T0310392

FUNCTIONAL DIFFERENTIAL EQUATIONS

PURE AND APPLIED MATHEMATICS

A Wiley Series of Texts, Monographs, and Tracts

Founded by RICHARD COURANT

A complete list of the titles in this series appears at the end of this volume.

FUNCTIONAL DIFFERENTIAL EQUATIONS

Advances and Applications

CONSTANTIN CORDUNEANU
YIZENG LI
MEHRAN MAHDAVI

Published by John Wiley & Sons, Inc., Hoboken, New Jersey
Published simultaneously in Canada

Library of Congress Cataloging-in-Publication Data:

Names: Corduneanu, C., author. | Li, Yizeng, 1949– author. | Mahdavi, Mehran, 1959– author.
Title: Functional differential equations : advances and applications / Constantin Corduneanu, Yizeng Li, Mehran Mahdavi.
Description: Hoboken, New Jersey : John Wiley & Sons, Inc., [2016] | Series: Pure and applied mathematics | Includes bibliographical references and index.
Identifiers: LCCN 2015040683 | ISBN 9781119189473 (cloth)
Subjects: LCSH: Functional differential equations.
Classification: LCC QA372 .C667 2016 | DDC 515/.35–dc23 LC record available at http://lccn.loc.gov/2015040683

Set in 10.5/13pts Times by SPi Global, Pondicherry, India

Printed in the United States of America

10 9 8 7 6 5 4 3 2 1

This book is dedicated to the memory of three pioneers of the Theory of Functional Differential Equations:

A.D. MYSHKIS, the author of the first book on non-classical differential equations

N.N. KRASOVSKII, for substantial contributions, particularly for extending Liapunov's method in Stability Theory, to the case of delay equations with finite delay, thus opening the way to the methods of Functional Analysis

J.K. HALE, for creating a vast theory in the study of functional differential equations, by using the modern tools of Functional Analysis

Each created a research school, with many distinguished contributors

CONTENTS

PREFACE

The origin of this book is in the research seminar organized by the first co-author (alphabetically), during the period September 1990 to May 1994, in the Department of Mathematics at The University of Texas at Arlington. The second and the third co-authors were, at that time, Ph.D. students working under the guidance of the first co-author. The seminar was also attended by another Ph.D. student, Zephyrinus Okonkwo, who had been interested in stochastic problems, and sporadically by other members of the Department of Mathematics (among them, Dr. Richard Newcomb II). Visitors also occasionally attended and presented their research results. We mention V. Barbu, Y. Hamaya, M. Kwapisz, I. Gyori, and Cz. Olech. Cooperation between the co-authors continued after the graduation of the co-authors Li and Mahdavi, who attained their academic positions at schools in Texas and Maryland, respectively. All three co-authors continued to pursue the topics that were presented at the seminar and that are included in this book through their Ph.D. theses as well as a number of published papers (single author or jointly). The list of references in this book contains almost all of the contributions that were made during the past two decades (1994–2014). Separation among the co-authors certainly contributed to the extended period that was needed to carry out the required work.

The book is a monograph, presenting only part of the results available in the literature, mainly mathematical ones, without any claim related to the coverage of the whole field of functional differential equations (FDEs). Paraphrasing the ancient dictum "mundum regunt numeri," one can instead say, "mundum regunt aequationes." That is why, likely, one finds a large number of reviews dedicated to this field of research, in all the publications concerned with the review of current literature (*Mathematical Reviews*, *Zentralblatt fur Mathematik*, *Referativnyi Zhurnal*, a. o.).

Over 550 items are listed in the Bibliography, all of which have been selected from a much larger number of publications available to the co-authors (i.e., university libraries, preprints, or reprints obtained from authors, papers received for review in view of insertion into various journals, or on the Internet).

The principles of selection were the *connections* with the topics considered in the book, but also the *inclusion* of information for themes similar to those treated by the co-authors, but not covered in the book because of the limitations imposed by multiple factors occurring during the preparation of the final text.

Mehran Mahdavi was in charge of the technical realization of the manuscript, a task he carried out with patience and skill. The other co-authors thank him for voluntarily assuming this demanding task, which he accomplished with devotion and competence.

Some developing fields of research related to the study of FDEs, but not included in this book's presentations, are the *stochastic* equations, the *fuzzy* equations, and the *fractional-order* functional equations. Some sporadic references are made to *discrete argument* functional equations, also known as *difference equations*. While their presence is spotted in many publications, Springer recently dedicated an entire journal to the subject, *Advances in Difference Equations* (R. P. Agarwal, editor). In addition, Clarendon Press is publishing papers on functional differential equations of the *time scale* type, which is encountered in some applications.

The plan that was followed in presenting the discussed topics is imposed by the different types of applications the theory of FDEs is dealing with, including the following:

1) Existence, uniqueness, estimates of solutions, and some behavior, when they are globally defined (e.g., domain invariance).
2) Stability of solutions is also of great interest for those applying the results in various fields of science, engineering, economics, and others.
3) Oscillatory motion/solutions, a feature occurring in many real phenomena and in man-made machines.

Chapter 1 is introductory and is aimed at providing the readers with the tools necessary to conduct the study of various classes of FDEs. Chapter 2 is primarily concerned with the existence of solutions, the uniqueness (not always present), and estimate for solutions, in view of their application to specific problems. Chapter 3, the most extended, deals with problems of stability, particularly for ordinary DEs (in which case the theory is the most advanced and can provide models for other classes of FDEs). In this chapter, the interest is of concern not only to mathematicians, but also to other scientists, engineers, economists, and others deeply engaged in related applied fields. Chapter 4 deals with oscillatory properties, especially of almost periodic type (which includes periodic). The choice of spaces of almost periodic functions are not, as usual, the classical Bohr type, but a class of spaces forming a scale, starting with the simplest space (Poincaré) of those almost periodic functions whose Fourier series are absolutely convergent and finishes with the space of Besicovitch B_2, the richest one known, for which we have enough meaningful tools. Chapter 5 contains results of any nature, available for the so-called *neutral equations*. There are several types of FDEs belonging to this class, which can be roughly defined as the class of those equations that are not solved with respect to the highest-order derivative involved.

For those types of equations, this book proves existence results, some kinds of behavior (e.g., boundedness), and stability of the solutions (especially asymptotic stability).

Appendix A, written by C. Corduneanu, introduces the reader to what is known about generalized Fourier series of the form $\sum_{k=1}^{\infty} a_k e^{i\lambda_k(t)}$ with $a_k \in \mathcal{C}$ and $\lambda_k(t)$ some real-valued function on R. Such series intervene in studying various applied problems and appears naturally to classify them as belonging to the third stage of development of the Fourier analysis (after periodicity and almost periodicity). The presentation is descriptive, less formal, and somewhat a survey of problems occurring in the construction of new spaces of oscillatory functions.

Since the topics discussed in this book are rather specialized with respect to the general theory of FDEs, the book can serve as source material for graduate students in mathematics, science, and engineering. In many applications, one encounters FDEs that are not of a classical type and, therefore, are only rarely taken into consideration for teaching. The list of references in this book contains many examples of this situation. For instance, the case of equations in population dynamics is a good illustration (see Gopalsamy [225]). Also, the book by Kolmanovskii and Myshkis [292] provides a large number of applications for FDEs, which may interest many categories of readers.

ACKNOWLEDGMENTS

We would like to express our appreciation to John Wiley & Sons, Inc., for publishing our book and to Ms. Susanne Steitz-Filler, senior editor, and the editorial and production staff at Wiley.

Constantin Corduneanu
Cordun@exchange.uta.edu

Yizeng Li
Yizeng.Li@tccd.edu

Mehran Mahdavi
mmahdavi@bowiestate.edu

1

INTRODUCTION, CLASSIFICATION, SHORT HISTORY, AUXILIARY RESULTS, AND METHODS

Generally speaking, a functional equation is a relationship containing an unknown element, usually a function, which has to be determined, or at least partially identifiable by some of its properties. Solving a functional equation (FE) means finding a *solution*, that is, the unknown element in the relationship. Sometimes one finds several solutions (solutions set), while in other cases the equation may be deprived of a solution, particularly when one provides the class/space to which it should belong.

Since a relationship could mean the equality, or an inequality, or even the familiar "belongs to," designated by $=$, $\in$, $\subset$ or $\subseteq$, the description given earlier could also include the functional inequalities or the functional inclusions, rather often encountered in the literature. Actually, in many cases, their theory is based on the theory of corresponding equations with which they interact. For instance, the selection of a single solution from a solution set, especially in case of inclusions.

In this book we are mainly interested in FEs, in the proper/usual sense. We send the readers to adequate sources for cases of related categories, like inequalities or inclusions.

Functional Differential Equations: Advances and Applications, First Edition.
Constantin Corduneanu, Yizeng Li and Mehran Mahdavi

1.1 CLASSICAL AND NEW TYPES OF FEs

The classical types of FEs include the *ordinary differential equations* (ODEs), the *integral equations (IEs) of Volterra* or Fredholm and the *integro-differential equations* (IDEs). These types, which have been thoroughly investigated since Newton's time, constitute the classical part of the vast field of FEs, or *functional differential equations* (FDEs).

The names Bernoulli, Newton, Riccati, Euler, Lagrange, Cauchy (analytic solutions), Dini, and Poincaré as well as many more well-known mathematicians, are usually related to the classical theory of ODE. This theory leads to a large number of applications in the fields of science, engineering, economics, in cases of the modeling of specific problems leading to ODE.

A large number of books/monographs are available in the classical field of ODE: our list of references containing at least those authored by Halanay [237], Hale [240], Hartman [248], Lefschetz [323], Petrovskii [449], Sansone and Conti [489], Rouche and Mawhin [475], Nemytskii and Stepanov [416], and Coddington and Levinson [106].

Another classical type of FEs, closely related to the ODEs, is the class of IEs, whose birth is related to Abel in the early nineteenth century. They reached an independent status by the end of nineteenth century and the early twentieth century, with Volterra and Fredholm. Hilbert is constituting his theory of linear IEs of Fredholm's type, with symmetric kernel, providing a successful start to the spectral theory of completely continuous operators and orthogonal function series.

Classical sources in regard to the basic theory of integral equations include books/monographs by Volterra [528], Lalesco [319], Hilbert [261], Lovitt [340], Tricomi [520], Vath [527]. More recent sources are Corduneanu [135], Gripenberg et al. [228], Burton [80, 84], and O'Regan and Precup [430].

A third category of FEs, somewhat encompassing the differential and the IEs, is the class of IDEs, for which Volterra [528] appears to be the originator. It is also true that E. Picard used the integral equivalent of the ODE $\dot{x}(t) = f(t,x(t))$, under initial condition $x(t_0) = x_0$, Cauchy's problem, namely

$$x(t) = x_0 + \int_{t_0}^{t} f(s,x(s))\,ds,$$

obtaining classical existence and uniqueness results by the method of successive approximations.

A recent reference, mostly based on classical analysis and theories of DEs and IEs, is Lakshmikantham and R. M. Rao [316], representing a rather comprehensive picture of this field, including some significant applications and indicating further sources.

The extended class of FDEs contains all preceding classes, as well equations involving operators instead of functions (usually from R into R). The classical categories are related to the use of the so-called Niemytskii operator, defined by the formula $(Fu)(t) = f(t, u(t))$, with $t \in R$ or in an interval of R, while in the case of FDE, the right-hand side of the equation

$$\dot{x}(t) = (Fx)(t),$$

implies a more general type of operator F. For instance, using Hale's notation, one can take $(Fx)(t) = f(t, x(t), x_t)$, where $x_t(s) = x(t+s)$, $-h \leq s \leq 0$ represents a restriction of the function $x(t)$, to the interval $[t-h, t]$. This is the finite delay case. Another choice is

$$(Fx)(t) = (Vx)(t),\ t \in [t_0, T],$$

where V represents an *abstract Volterra operator* (see definition in Chapter 2), also known as *causal operator*.

Many other choices are possible for the operator F, leading to various classes of FDE. Bibliography is very rich in this case, and exact references will be given in the forthcoming chapters, where we investigate various properties of equations with operators.

The first book entirely dedicated to FDE, in the category of delay type (finite or infinite) is the book by A. Myshkis [411], based on his thesis at Moscow State University (under I. G. Petrovskii). This book was preceded by a survey article in the Uspekhi Mat. Nauk, and one could also mention the joint paper by Myshkis and Elsgoltz [412], reviewing the progress achieved in this field, due to both authors and their followers. The book Myshkis [411] is the first dedicated entirely to the DEs with delay, marking the beginnings of the literature dealing with non-traditional FEs.

The next important step in this direction has been made by N. N. Krasovskii [299], English translation of 1959 Russian edition. In his doctoral thesis (under N. G. Chetayev), Krasovskii introduced the method of *Liapunov functionals* (not just functions!), which permitted a true advancement in the theory of FDEs, especially in the nonlinear case and stability problems. The research school in Ekaterinburg has substantially contributed to the progress of the theory of FDEs (including Control Theory), and names like Malkin, Barbashin, and Krasovskii are closely related to this progress.

The third remarkable step in the development of the theory of FDE has been made by Jack Hale, whose contribution should be emphasized, in respect to the constant use of the arsenal of Functional Analysis, both linear and nonlinear. A first contribution was published in 1963 (see Hale [239]), utilizing the theory of semigroups of linear operators on a Banach function space.

This approach allowed Hale to develop a theory of linear systems with finite delay, in the time-invariant framework, dealing with adequate concepts that naturally generalize those of ODE with constant coefficients (e.g., characteristic values of the system/equation). Furthermore, many problems of the theory of nonlinear ODE have been formulated and investigated for FDE (stability, bifurcation, and others (a.o.)). The classical book of Hale [240] appears to be the first in this field, with strong support of basic results, some of them of recent date, from functional analysis.

In the field of applications of FDE, the book by Kolmanovskii and Myshkis [292] illustrates a great number of applications to science (including biology), engineering, business/economics, environmental sciences, and medicine, including the stochastic factors. Also, the book displays a list of references with over 500 entries.

In concluding this introductory section, we shall mention the fact that the study of FDE, having in mind the nontraditional types, is the focus for a large number of researchers around the world: Japan, China, India, Russia, Ukraine, Finland, Poland, Romania, Greece, Bulgaria, Hungary, Austria, Germany, Great Britain, Italy, France, Morocco, Algeria, Israel, Australia and the Americas, and elsewhere.

The *Journal of Functional Differential Equations* is published at the College of Judea and Samaria, but its origin was at Perm Technical University (Russia), where N. V. Azbelev created a school in the field of FDE, whose former members are currently active in Russia, Ukraine, Israel, Norway, and Mozambique.

Many other journals are dedicated to the papers on FDE and their applications. We can enumerate titles like *Nonlinear Analysis (Theory, Methods & Applications)*, published by Elsevier; *Journal of Differential Equations; Journal of Mathematical Analysis and Applications*, published by Academic Press; *Differentsialuye Uravnenja* (Russian: English translation available); and *Funkcialaj Ekvacioj* (Japan). Also, there are some electronic journals publishing papers on FDE: *Electronic Journal of Qualitative Theory of Differential Equations*, published by Szeged University; EJQTDE, published by Texas State University, San Marcos.

1.2 MAIN DIRECTIONS IN THE STUDY OF FDE

This section is dedicated to the description of various types of problems arising in the investigation of FDE, at the mathematical side of the problem as well as the application of FDE in various fields, particularly in science and engineering.

A first problem occurring in relationship with an FDE is the *existence* or *absence* of a solution. The solution is usually sought in a certain class of functions (scalar, vector, or even Banach space valued) and "a priori" limitations/restrictions may be imposed on it.

In most cases, besides the "pure" existence, we need *estimates* for the solutions. Also, it may be necessary to use the numerical approach, usually approximating the real values of the solution. Such approximations may have a "local" character (i.e., valid in a neighborhood of the initial/starting value of the solution, assumed also unique), or they may be of "global" type, keeping their validity on the whole domain of definition of the solution.

Let us examine an example of a linear FDE, of the form

$$\dot{x}(t) = (Lx)(t) + f(t),\ t \in [0,T], \tag{1.1}$$

with $L : C([0,T],R^n) \to C([0,T],R^n)$ a linear, casual continuous map, while $f \in C([0,T],R^n)$. As shown in Corduneanu [149; p. 85], the unique solution of equation (1.1), such that $x(0) = x^0 \in R^n$, is representable by the formula

$$x(t) = X(t,0)x^0 + \int_0^t X(t,s)f(s)\,ds,\ t \in [0,T]. \tag{1.2}$$

In (1.2), the Cauchy matrix is given, on $0 \le s \le t \le T$, by the formula

$$X(t,s) = I + \int_s^t \tilde{k}(t,u)\,du, \tag{1.3}$$

where $\tilde{k}(t,s)$ stands for the conjugate kernel associated to the kernel $k(t,s)$, the latter being determined by the relationship

$$\int_0^t (Lx)(s)\,ds = \int_0^t k(t,s)x(s)\,ds,\ t \in [0,T]. \tag{1.4}$$

For details, see the reference indicated earlier in the text.

Formula (1.2) is helpful in finding various *estimates* for the solution $x(t)$ of the initial value problem considered previously.

Assume, for instance, that the Cauchy matrix $X(t,s)$ is bounded on $0 \le s \le t \le T$ by M, that is, $|X(t,s)| \le M$; hence $|X(t,0)| \le m \le M$, then (1.2) yields the following estimate for the solution $x(t)$:

$$|x(t)| \le m|x^0| + M \int_0^T |f(s)|\,ds, \tag{1.5}$$

with $T < \infty$ and f continuous on $[0,T]$. We derive from (1.5) the estimate

$$\sup_{0 \leq t \leq T} |x(t)| \leq m|x^0| + M|f|_{L^1}, \tag{1.6}$$

which means an upper bound of the norm of the solutions, in terms of data.

We shall also notice that (1.6) keeps its validity in case $T = \infty$, that is, we consider the problem on the semiaxis R_+. This example shows how, assuming also $f \in L^1(R_+, R^n)$, all solutions of (1.1) remain bounded on the positive semiaxis.

Boundedness of all solutions of (1.1), on the positive semiaxis, is also assured by the conditions $|X(t,0)| \leq m$, $t \in R_+$, and

$$\int_0^t |X(t,s)|\, ds \leq M, \ |f(t)| \leq A < \infty, \ t \in R_+.$$

The readers are invited to check the validity of the following estimate:

$$\sup_{t \geq 0} |x(t)| \leq m|x^0| + AM, \ t \in R_+. \tag{1.7}$$

Estimates like (1.6) or (1.7), related to the concept of boundedness of solutions, are often encountered in the literature. Their significance stems from the fact that the motion/evolution of a man-made system takes place in a bounded region of the space. Without having estimates for the solutions of FDE, it is practically impossible to establish properties of these solutions.

One of the best examples in this regard is constituted by the property of *stability* of an equilibrium state of a system, described by the FDE under investigation. At least, theoretically, the problem of stability of a given motion of a system can be reduced to that of an equilibrium state. Historically, Lagrange has stated a result of stability for the equilibrium for a mechanical system, in terms of a variational property of its energy. This idea has been developed by A. M. Liapunov [332] (1857–1918), who introduced the method of an auxiliary function, later called *Liapunov function* method. Liapunov's approach to *stability* theory is known as one of the most spectacular developments in the theory of DE and then for larger classes of FDE, starting with N. N. Krasovskii [299].

The *comparison method*, on which we shall rely (in Chapter 3), has brought new impetus to the investigation of stability problems. The schools created by V. V. Rumiantsev in Moscow (including L. Hatvani and V. I. Vorotnikov), V. M. Matrosov in Kazan, then moved to Siberia and finally to Moscow, have developed a great deal of this method, concentrating mainly on the ODE case. Also, V. Lakshmikantham and S. Leela have included many contributions in their treaty [309]. They had many followers in the United States and India, publishing a conspicuous number of results and developments of this method. One of the last contributions to this topic [311], authored by Lakshmikantham, Leela, Drici, and McRae, contains the general theory of equations with causal operators, including stability problems.

The *comparison method* consists of the simultaneous use of Liapunov functions (functionals), and differential inequalities. Started in its general setting by R. Conti [110], it has been used to prove global existence criteria for ODE. In short time, the use has extended to deal with uniqueness problems for ODE by F. Brauer [75] and Corduneanu [114, 115] for stability problems. The method is still present in the literature, with contributions continuing those already included in classical references due to Sansone and Conti [489], Hahn [235], Rouche and Mawhin [475], Matrosov [376–378], Matrosov and Voronov [387], Lakshimikantham and Leela [309], and Vorotnikov [531].

A historical account on the development of the stability concept has been accurately given by Leine [325], covering the period from Lagrange to Liapunov. The mechanical/physical aspects are emphasized, showing the significance of the stability concept in modern science. The original work of Liapunov [332] marks a crossroad in the development of this concept, with so many connections in the theory of evolutionary systems occurring in the mathematical description in contemporary science.

In Chapter 3, we shall present stability theory for ODE and FDE, particularly for the equations with finite delay. The existing literature contains results related to the infinite delay equations, a theory that has been originated by Hale and Kato [241]. An account on the status of the theory, including stability, is to be found in Corduneanu and Lakshmikantham [167]. We notice the fact that a theory of stability, for general classes of FDE, has not yet been elaborated. As far as special classes of FDE are concerned, the book [84] by T. Burton presents the method of Liapunov functionals for integral equations, by using modern functional analytic methods. The book [43] by Barbashin, one of the first in this field, contains several examples of constructing Liapunov functions/functionals.

The converse theorems in stability theory, in the case of ODE, have been obtained, in a rather general framework, by Massera [373], Kurzweil [303], and Vrkoč [532]. Early contributions to stability theory of ODE were brought by followers of Liapunov, (see Chetayev [103] and Malkin [356]). In Chapter 3, the readers will find, besides some basic results on stability, more bibliographical indications pertaining to this rich category of problems.

As an example, often encountered in some books containing stability theory, we shall mention here the classical result (Poincaré and Liapunov) concerning the differential system $\dot{x}(t) = A(t)x(t)$, $t \in R_+$, $x : R_+ \to R^n$, and $A : R_+ \to \mathcal{L}(R^n, R^n)$ a continuous map. If we admit the commutativity condition

$$A(t)\int_0^t A(s)\,ds = \left(\int_0^t A(s)\,ds\right)A(t),\ t\in R_+, \tag{1.8}$$

then the solution, under initial condition $x(0)=x^0$, can be represented by

$$x(t,x^0) = x^0\, e^{\int_0^t A(s)\,ds},\ t\geq 0. \tag{1.9}$$

From this representation formula one derives, without difficulty, the following results:

Stability of the solution $x=\theta=$ the zero vector in R^n is equivalent to boundedness, on R_+, of the matrix function $\int_0^t A(s)\,ds$.

Asymptotic stability of the solution $x=\theta$ is equivalent to the condition

$$\lim_{t\to\infty} e^{\int_0^t A(s)\,ds} = O = \text{the zero matrix.} \tag{1.10}$$

Both statements are elementary consequences of formula (1.8). The definitions of various types of stability will be done in Chapter 3. We notice here that the already used terms, *stability* and *asymptotic stability*, suggest that the first stands for the property of the motion to remain in the neighborhood of the equilibrium point when small perturbations of the initial data are occurring, while the second term tells us that besides the property of stability (as intuitively described earlier), the motion is actually "tending" or approaching indefinitely the equilibrium state, when $t\to\infty$.

Remark 1.1 *The aforementioned considerations help us derive the celebrated stability result, known as Poincaré–Liapunov stability theorem for linear differential systems with constant coefficients.*

Indeed, if $A(t)\equiv A=$ constant is an $n\times n$ matrix, with real or complex coefficients, with characteristic equation $\det(\lambda I-A)=0$, $I=$ the unit matrix of type $n\times n$, then we denote by $\lambda_1,\lambda_2,\dots,\lambda_k$ its distinct roots ($k\leq n$). From the elementary theory of DEs with constant coefficients, we know that the entries of the matrix e^{At} are quasi-polynomials of the form

$$\sum_{j=1}^{k} e^{\lambda_j t} p_j(t), \tag{1.11}$$

with $p_j(t), j=1,2,\dots,k$, some algebraic polynomials.

Since the commutativity condition (1.8) is valid when $A(t)\equiv A=$ constant, there results that (1.10) can hold if and only if the condition

$$Re\lambda_j<0,\ j=1,2,\dots,k, \tag{1.12}$$

is satisfied. Condition (1.12) is frequently used in stability theory, particularly in the case of linear systems encountered in applications, but also in the case of nonlinear systems of the form

$$\dot{x}(t) = Ax(t) + f(t, x(t)), \tag{1.13}$$

when f—using an established odd term—is of "higher order" with respect to x (say, for instance, $f(t,x) = x^{\frac{3}{2}} \sin t$).

We will conclude this section with the discussion of another important property of motion, encountered in nature and man-made systems. This property is known as *oscillation* or *oscillatory motion*. Historically, the *periodic oscillations* (of a pendulum, for instance) have been investigated by mathematicians and physicists.

Gradually, more complicated oscillatory motions have been observed, leading to the apparition of *almost periodic* oscillations/vibrations. In the third decade of the twentieth century, Harald Bohr (1887–1951), from Copenhagen, constructed a wider class than the periodic one, called *almost periodic*.

In the last decade of the twentieth century, motivated by the needs of researchers in applied fields, even more complex *oscillatory motions* have emerged. In the books by Osipov [432] and Zhang [553, 554], new spaces of oscillatory functions/motions have been constructed and their applications illustrated.

In case of the *Bohr–Fresnel* almost periodic functions, a new space has been constructed, its functions being representable by generalized Fourier series of the form

$$\sum_{k=1}^{\infty} a_k e^{i(\alpha t^2 + \beta_k t)}, \ t \in R, \tag{1.14}$$

with $a_k \in C$ and $\alpha, \beta_k \in R$, $k \geq 1$.

In the construction of Zhang, the attached generalized Fourier series has the form

$$\sum_{k=1}^{\infty} a_k e^{i q_k(t)}, \ t \in R, \tag{1.15}$$

with $a_k \in C$, $k \geq 1$, and $q_k \in Q(R)$, $Q(R)$ denoting the algebra of polynomial functions of the form

$$q(t) = \sum_{j=1}^{m} \lambda_j t^{\alpha_j} \text{ for } t \geq 0, \tag{1.16}$$

and $q(t) = -q(-t)$ for $t < 0$; $\lambda_j \in R$, while $\alpha_1 > \alpha_2 > \cdots > \alpha_m > 0$ denote arbitrary reals.

The functions (on R) obtained by uniform approximation with generalized trigonometric polynomials of the form

$$P_n(t) = \sum_{k=1}^{n} a_k e^{iq_k(t)} \tag{1.17}$$

are called *strong limit power functions* and their space is denoted by $\mathcal{SLP}(R,\mathcal{C})$.

A discussion of these generalizations of the classical trigonometric series and attached "sum" are presented in Appendix. The research work is getting more and more adepts, contributing to the development of this *third stage* in the history of oscillatory motions/functions.

In order to illustrate, including some applications to FDEs, the role of almost periodic oscillations/motions, we have chosen to present in Chapter 4 only the case of AP_r-almost periodic functions, $r \in [1,2]$, constituting a relatively new class of almost periodic functions, related to the theory of oscillatory motions. Their construction is given, in detail, in Chapter 4, as well as several examples from the theory of FDEs.

Concerning the first two stages in the development of the theory of oscillatory functions, the existing literature includes the treatises of Bary [47] and Zygmund [562]. These present the main achievements of the *first stage* of development (from Euler and Fourier, to contemporary researchers). With regard to the *second stage* in the theory of almost periodic motions/functions, there are many books/monographs dedicated to the development, following the fundamental contributions brought by Harald Bohr. We shall mention here the first books presenting the basic facts, Bohr [72] and Besicovitch [61], Favard [208], Fink [213], Corduneanu [129,156], Amerio and Prouse [21], and Levitan [326], Levitan and Zhikov [327]. These references contain many more indications to the work of authors dealing with the theory of almost periodic motions/functions. They will be mentioned in Section 4.9.

As an example of an almost periodic function, likely the first in the literature but without naming it by its name, seems to be due to Poincaré [454], who dealt with the representations of the form

$$f(t) = \sum_{k=1}^{\infty} a_k \sin \lambda_k t,\ t \in R. \tag{1.18}$$

Supposing that the series converges uniformly to f on R (which situation can occur, for instance, when $\sum_{k=1}^{\infty} |a_k| < \infty$), Poincaré found the formula for the coefficients a_k, introducing simultaneously the concept of mean value of a function on R:

$$M\{f\} = \lim_{\ell\to\infty} (2\ell)^{-1} \int_{-\ell}^{\ell} f(t)\,dt. \tag{1.19}$$

This concept was used 30 years later by H. Bohr, to build up the theory of almost periodic functions (complex-valued). The coefficients were given by the formula

$$a_k = \lim_{\ell\to\infty} (2\ell)^{-1} \int_{-\ell}^{\ell} f(t)\sin\lambda_k t\,dt. \tag{1.20}$$

1.3 METRIC SPACES AND RELATED CONCEPTS

One of the most frequent tools encountered in modern mathematical analysis is a *metric space*, introduced at the beginning of the twentieth century by Maurice Fréchet (in his Ph.D. thesis at Sorbonne). This concept came into being after G. Cantor laid the bases of the set theory, opening a new era in mathematics. The simple idea, exploited by Fréchet, was to consider a “distance” between the elements of an abstract set.

Definition 1.1 *A set S, associated with a map $d : S\times S \to R_+$, is called a metric space, if the following axioms are adopted:*

1) $d(x,y)\geq 0$, *with* $=$ *only when* $x=y$;
2) $d(x,y)=d(y,x)$, $x,y\in S$;
3) $d(x,y)\leq d(x,z)+d(z,y)$, $x,y,z\in S$.

Several consequences can be drawn from Definition 1.1. Perhaps, the most important is contained in the following definition:

Definition 1.2 *Consider a sequence of elements/points $\{x_n; n\geq 1\}\subset S$. If*

$$\lim_{n\to\infty} d(x_n,x)=0, \tag{1.21}$$

then one says that the sequence $\{x_n; n\geq 1\}$ converges to x in S.

Then x is called the limit of the sequence.

It is common knowledge that the limit of a convergent sequence in S is *unique*.

Since the concept of a metric space has gained wide acceptance in *Mathematics, Science and Engineering*, we will send the readers to the book of

Friedman [214] for further elementary properties of metric spaces and the concept of convergence.

It is important to mention the fact that the concept of convergence/limit helps to define other concepts, such as *compactness* of a subset $M \subset S$. Particularly, the concept of a *complete* metric space plays a significant role.

Definition 1.3 *The metric space* (S,d) *is called complete, if any sequence* $\{x_n; n \geq 1\}$ *satisfying the Cauchy condition, "for each* $\epsilon > 0$*, there exists an integer* $N = N(\epsilon)$*, such that* $d(x_n,x_m) < \epsilon$ *for* $n,m \geq N(\epsilon)$*, is convergent in* (S,d)*."*

Definition 1.4 *The metric space* (S,d) *is called compact, according to Fréchet, iff any sequence* $\{x_n; n \geq 1\} \subset S$ *contains a convergent subsequence* $\{x_{n_k}; k \geq 1\}$*, that is, such that* $\lim_{k\to\infty} d(x_{n_k},x) = 0$*, for some* $x \in S$*.*

Definition 1.4 leads easily to other properties of a compact metric space. For instance, the diameter of a compact metric space S is finite: $\sup\{d(x,y); x,y \in S\} < \infty$. Also, every compact metric space is complete.

We rely on other properties of the metric spaces, sending the readers to the aforementioned book of Friedman [214], which contains, in a concise form, many useful results we shall use in subsequent sections of this book. Other references are available in the literature: see, for instance, Corduneanu [135], Zeidler [551], Kolmogorov and Fomin [295], Lusternik and Sobolev [343], and Deimling [190].

Almost all books mentioned already contain applications to the theory of FEs, particularly to differential equations and to integral equations. Other sources can be found in the titles referenced earlier in the text.

The metric spaces are a particular case of *topological spaces*. The latter represent a category of mathematical objects, allowing the use of the concept of *limit*, as well as many other concepts derived from that of limit (of a sequence of a function, limit point of a set, closure of a set, closed set, open set, a.o.)

If we take the definition of a topological space by means of the axioms for the family of *open* sets, then in case of metric spaces the open sets are those subsets A of the space S, defined by the property that any point x of A belongs to A, together with the "ball" of arbitrary small radius r, $\{y : d(x,y) < r\}$.

It is easy to check that the family of all open sets, of a metric space S, verifies the following axioms (for a topological space):

1. The *union* of a family of open sets is also an open set.
2. The *intersection* of a finite family of open sets is also an open set.
3. The space S and the empty set $\emptyset$ belong to the family of open sets.

Such a family, satisfying axioms 1, 2, and 3, induces a topology τ on S. Returning to the class of metric spaces, we shall notice that the couple (S,d) is inducing a topology on S and, therefore, any property of topological nature of this space is the product of the metric structure (S,d). The converse problem, to find conditions on a topological space to be the product of a metric structure, known as *metrizability*, has kept the attention of mathematicians for several decades of the past century, being finally solved. The result is known as the theorem of Nagata–Smirnov.

Substantial progress has been made, with regard to the enrichment of a metric structure, when Banach [39] introduced the new concept of *linear metric space*, known currently as *Banach space*.

Besides the metric structure/space (S,d), one assumes that S is a linear space (algebraically) over the field of reals R, or the field of complex numbers $\mathcal{C}$. Moreover, there must be some compatibility between the metric structure and the algebraic one. Accordingly, the following system of axioms is defining a *Banach space*, denoted $(S, \|\cdot\|)$, with $\|\cdot\|$ a map from S into R_+, $x \to \|x\|$, called a *norm*.

I. S is a linear space over R, in additive notation.

II. S is a *normed* space, that is, there is a map, from S into R_+, $x \to \|x\|$, satisfying the following conditions:

1) $\|x\| \geq 0$ for $x \in S$, $\|x\| = 0$ iff $x = \theta$;
2) $\|ax\| = |a|\ \|x\|$, for $a \in R$, $x \in S$;
3) $\|x+y\| \leq \|x\| + \|y\|$ for $x, y \in S$.

It is obvious that $d(x,y) = \|x-y\|$, $x, y \in S$, is a distance/metric on S.

III. (S,d), with d defined earlier, is a complete metric space.

Also, traditional notations for a Banach space, frequently encountered in literature, are $(B, \|\cdot\|)$ or $(X, \|\cdot\|)$, in the latter case, the generic element of X being denoted by x.

The most commonly encountered Banach space is the vector space R^n (or C^n), the norm being usually defined by

$$\|x\| = (|x_1|^2 + |x_2|^2 + \cdots + |x_n|^2)^{\frac{1}{2}}$$

and called the Euclidean norm. Another norm is defined by $\|x\|_1 = \max(|x_1|, |x_2|, \ldots, |x_n|)$.

Both norms mentioned previously lead to the same kind of convergence in R^n, because $\|x\|_1 \leq \|x\| \leq \sqrt{n}\ \|x\|_1$, $x \in B$. This is the usual convergence on coordinates, that is, $\lim(x_1, x_2, \ldots, x_n) = (\lim x_1, \lim x_2, \ldots, \lim x_n)$.

A special type of Banach space is the *Hilbert space*. The prototype has been constructed by Hilbert, and it is known as $\ell^2(R)$, or $\ell^2(\mathcal{C})$, space. This fact occurred long before Banach introduced his concept of space in the 1920s. The ℓ^2-space appeared in connection with the theory of orthogonal function series, generated by the Fredholm–Hilbert theory of integral equations with symmetric kernel (in the complex case, the condition is $k(t,s) = \overline{k(s,t)}$). It is also worth mentioning that the first book on Hilbert spaces, authored by M. Stone [508], shortly preceded the first book on Banach space theory, Banach [39], 1932. The Banach spaces reduce to Hilbert spaces, in the real case, if and only if the rule of the parallelogram is valid:

$$\|x+y\|^2 + \|x-y\|^2 = 2(\|x\|^2 + \|y\|^2),\ x,y \in B. \tag{1.22}$$

Of course, the parallelogram involved is the one constructed on the vectors x and y as sides.

What is really specific for Hilbert spaces is the fact that the concept of *inner product* is defined for $x,y \in H$, as follows: it is a map from $H \times H$ into R (or $\mathcal{C}$), such that

1) $<x,y>=<y,x>$;
2) $<x,x>\geq 0$, the value 0 leading to $x=\theta$;
3) $<ax+by,z>=a<x,z>+b<y,z>$, with $a,b \in R$, $x,y,z \in H$.

In the complex case, one should change 1) to $<x,y>=\overline{<y,x>}$, 2) remains the same, and 3) must be changed accordingly.

If one starts with a Banach space satisfying condition (1.22), then the inner product is given by

$$<x,y>=\frac{1}{4}(\|x+y\|^2 - \|x-y\|^2),\ x,y \in B. \tag{1.23}$$

Conditions 1), 2), and 3), stated in the text, can be easily verified by the product $<x,y>$ given by formula (1.23).

A condition verified by the inner product is known as Cauchy inequality, and it looks

$$|<x,y>| \leq \|x\|\ \|y\|,\ x,y \in H. \tag{1.24}$$

It is easily obtained starting from the obvious inequality $\|x+ay\|^2 \geq 0$, which is equivalent to $\|x\|^2 + 2a<x,y> + a^2\|y\|^2 \geq 0$, which, regarded as a quadratic polynomial in a, must take only nonnegative values. This would be possible only in case the discriminant is nonpositive, that is, $|<x,y>|^2 \leq \|x\|^2\ \|y\|^2$, which implies (1.24). Using (1.24), prove that $\|x+y\| \leq \|x\| + \|y\|$, for any $x,y \in H$.

In concluding this section, we will define another special case of linear metric spaces, whose metric is invariant to translations. These spaces are known as linear Fréchet spaces. We will use them in Chapter 2.

If instead of a norm, satisfying conditions *II* and *III* in the definition of a Banach space, we shall limit the imposed properties to $\|x\| \geq 0$, accepting the possibility that there may be elements $x \neq \theta$, we obtain what is called a semi-norm. One can operate with a semi-norm in the same way we do with a norm, the difference appearing in the part that the limit of a convergent sequence is not necessarily unique.

Here precisely, the semi-norm is defined by the means of the following axioms related to a linear space E:

1) $\|x\| \geq 0$ for $x \in E$;
2) $\|ax\| = |a| \, \|x\|$, $a \in R$, $x \in E$;
3) $\|x+y\| \leq \|x\| + \|y\|$, $x, y \in E$.

In order to define a metric/distance on E, we need this concept: a family of semi-norms on E is called *sufficient*, if and only if from $|x|_k = 0$, $k \geq 1$, there results $x = \theta \in E$.

By means of a countable family/sequence of semi-norms, one can define on E the metric by

$$d(x,y) = \sum_{j=1}^{\infty} 2^{-j} \frac{|x-y|_j}{1+|x-y|_j}, \; x,y \in E. \tag{1.25}$$

Indeed, $d(x,y) \geq 0$ for $x,y \in E$ and $d(x,y) = 0$ imply $|x-y|_j = 0, j \geq 1$. Hence, $x - y = \theta$, which means that the distance between two elements is zero, if and only if the elements coincide. The symmetry is obvious while the triangle inequality for $d(x,y)$ follows from the elementary inequality for nonnegative reals, $|a+b|(1+|a+b|)^{-1} \leq |a|(1+|a|) + |b|(1+|b|)$.

It is interesting to mention the fact that $d(x,y)$ is *bounded* on $E \times E$ by 1, regardless of the (possible) situation when each $|x|_j, j \geq 1$ is unbounded on E. This is related to the fact that a metric $d(x,y)$, on E, generates another *bounded* metric $d_1(x,y) = d(x,y)[1+d(x,y)]^{-1}$, with the same kind of convergence in E.

1.4 FUNCTIONS SPACES

Since our main preoccupation in this book is the study of solutions of various classes of FEs (existence, uniqueness, and local or global behavior), it is useful to give an account on the type of functions spaces we will encounter in subsequent chapters. As proceeded in the preceding sections, we will not provide all the details, but we will indicate adequate sources available in the existing literature.

We shall dwell on the spaces of *continuous functions* on R, or intervals in R, using the notations that are established in literature. Generally speaking, by $C(A,B)$ we mean the space of continuous maps from A into B, when continuity has a meaning. An index may be used for C, in case we have an extra property to be imposed. This is a list of spaces, consisting of continuous maps, we shall encounter in the book.

$C([a,b],R^n)$ will denote the Banach space of continuous maps from $[a,b]$ into R^n, with the norm

$$|x|_C = \sup\{|x(t)|;\, t \in [a,b]\}, \tag{1.26}$$

where $|\cdot|$ is the Euclidean norm in R^n. This space is frequently encountered in problems related to FEs, especially when we look for continuous solutions. But even in case of ODEs, the Cauchy problem $\dot{x} = f(t,x)$, $x(t_0) = x_0$, the proof is conducted by showing the existence of a continuous solution to the integral equation $x(t) = x_0 + \int_{t_0}^{t} f(s,x(s))\,ds$. Of course, the differentiability of $x(t)$ follows from the special form of the integral equation, equivalent to Cauchy problem, within the class of continuous functions. A basic property of the space $C([a,b],R^n)$, necessary in the sequel, is the famous criterion of *compactness*, for subsets $M \subset C$, known under the names of Ascoli-Arzelà criterion of compactness in $C([a,b],R^n)$: necessary and sufficient conditions, for the compactness of a set $M \subset C([a,b],R^n)$ are the boundedness of M and the equicontinuity of its elements on $[a,b]$.

The first property means that for the set, M, there exists a positive number, μ, such that $f \in M$ implies $|f(t)| \leq \mu$, $t \in [a,b]$.

The second property means the following: for any $\epsilon > 0$, there exists $\delta = \delta(\epsilon) > 0$, such that $|f(t) - f(s)| < \epsilon$ for $|t-s| < \delta$, $t,s \in [a,b]$, $f \in M$. It is also called equi-uniform continuity.

Proofs can be found in many textbooks, including Corduneanu [123]. The book by Kolmogorov and Fomin [295] contains the criterion but also an interesting proof of Peano's existence for Cauchy's problem, without transforming the problem into an integral equation, as a direct application of Ascoli-Arzela's result.

Let us notice that the compactness result of Ascoli–Arzelà is actually concerned with the concept of relative compactness.

What happens when the interval $[a,b]$ is replaced by $[a,b)$, or even $[a,\infty)$? The supremum norm used in (1.26) cannot be considered on the semi-open interval $[a,b)$ or on the half-axis $[a,\infty)$.

In this case, in order to obtain a distance between maps defined on $[a,\infty)$, the case $[a,b)$, $b<\infty$, being totally similar to the half-axis, we shall make recourse to the semi-norms

$$|x(t)|_k = \sup\{|x(t)|;\, t\in[a,a+k],\, k\geq 1\}, \tag{1.27}$$

which obviously form a sufficient family. On the linear space of continuous maps, from $[a,\infty)$ into R^n, we have the sufficient family of semi-norms defined by (1.27). Therefore, we can apply formula (1.25), which in this case becomes

$$d(x,y) = \sum_{k=1}^{\infty} 2^{-k}\left\{|x(t)-y(t)|_k\,[1+|x(t)-y(t)|_k]^{-1}\right\}, \tag{1.28}$$

with $|x(t)|_k$ defined by (1.27).

Therefore, the linear space of continuous maps, from $[a,\infty)$ into R^n, is a metric space. The property of completeness is the result of the fact that each $|x(t)|_k$ is a norm on the restricted space, $C([a,a+k],R^n)$, and this (plus continuity) implies the completeness of the space under discussion, which shall be denoted by $C([a,\infty),R^n)$ and sometimes by $C_c([a,\infty),R^n)$, the index denoting that the convergence induced by the metric is uniform on compact sets in $[a,\infty)$.

In summary, a metric structure as indicated earlier, is better—and natural—for $C([a,\infty),R^n)$, even though normed space (Banach structures are possible and useful for "parts" of $C([a,\infty),R^n)$.

We shall present now a class of Banach function spaces, denoted $C_g(R_+,R^n)$, where $g: R_+\to(0,\infty)$ is a continuous function, whose role is to serve as a *weight* for the concept of *boundedness*. Namely, $C_g(R_+,R^n)$ is defined by

$$C_g(R_+,R^n) = \{x;\, R_+\to R^n \text{ continuous and such that } |x(t)|\leq A_x g(t),\, t\in R_+,\, A_x>0\}. \tag{1.29}$$

The norm in $C_g(R_+,R^n)$ is given by

$$|x|_{C_g} = \sup\left\{\frac{|x(t)|}{g(t)};\, t\in R_+\right\}. \tag{1.30}$$

Obviously, $|x|_{C_g} = \inf A_x$, with A_x in (1.29). It is shown (see, for instance, Corduneanu [120]) that $C_g(R_+,R^n)$ is a Banach space, with the norm given by (1.30). The special case $g(t)\equiv 1$ on R_+ leads to the space of bounded functions on R_+, with values in R^n, the norm being the supremum norm. This space of bounded continuous functions on R_+ is denoted by $BC(R_+,R^n)$. It contains as subspaces several important Banach spaces, from R_+ into R^n, such as the space of functions with limit at ∞, $\lim_{t\to\infty} x(t) = x_\infty \in R^n$, which is encountered when we deal with the so-called *transient* solutions to FEs. This space is usually denoted by $C_\ell(R_+, R^n)$, and it is isomorphic and isometric to the space $C([0,1], R^n)$. Prove this statement! The subspace of $C_\ell(R_+, R^n)$, for which $\lim_{t\to\infty} x(t) = \theta$ is denoted $C_0(R_+, R^n)$, and it is the space of *asymptotic stability* (each motion, described by elements of $C_0(R_+, R^n)$, tends to the equilibrium point $x_\infty = \theta$). We will also deal with the subspace of space $BC(R, R^n)$, known as the space of almost periodic functions on R, with values in R^n (Bohr almost periodicity). It is denoted by $AP(R, R^n)$ and contains all continuous maps from R into R^n, such that they can be uniformly approximated on R by vector trigonometric polynomials: for each $\epsilon > 0$ and $x \in AP(R, R^n)$, there exists vectors $a_1, a_2, \ldots, a_n \in R^n$ and reals $\lambda_1, \lambda_2, \ldots, \lambda_m$, such that

$$\left| x(t) - \sum_{k=1}^{m} a_k e^{i\lambda_k t} \right| < \epsilon,\ t \in R. \tag{1.31}$$

Inequality (1.31) shows that $x \in AP(R, R^n)$ is as close as we want from oscillatory functions. For the classical types of almost periodic functions, see the books authored by Bohr [72], Besicovitch [61], Favard [208], Levitan [326], Corduneanu [129, 156], Fink [213], Amerio and Prouse [21], Levitan and Zhikov [327], Zaidman [547], and Malkin [355]. Appendix to this book will be dedicated to some new developments (not necessarily continuous functions).

We shall continue to enumerate function spaces, this time, having in mind the *measurable functions/elements*. These spaces of great importance in the development of modern analysis have appeared at the beginning of the past century, primarily due to Lebesgue's discovery of measure theory. Actually, the first function spaces amply investigated in the literature are known as Lebesgue's spaces or L^p-spaces.

The space $L^p(R, R^n)$, $p \geq 1$, is the linear space of all measurable maps from R into R^n, such that $\int_R |x(s)|^p\, ds < \infty$, the ds representing the Lebesgue measure on R. The norm of this space is

$$|x|_{L^p} = \left\{ \int_R |x(s)|^p\, ds \right\}^{\frac{1}{p}},\ 1 \leq p < \infty. \tag{1.32}$$

The case $p = \infty$ is characterized by

$$|x|_{L^\infty} = \operatorname*{ess\text{-}sup}_{t\in R} |x(t)| < \infty. \tag{1.33}$$

The theory of these Banach spaces, whose elements are, in fact, equivalence classes of functions (i.e., two functions are equivalent, if and only if they coincide, except on set of points of Lebesgue measure zero) is largely diffused in many books/textbooks available. Let us indicate only Yosida [541], Lang [320], and Amann and Escher [20].

Let us consider now the problem of compactness (or relative compactness) in the space $L^p([a,b],R^n)$, providing a useful result (Riesz):

Let $M \subset L^p([a,b],R^n)$, $1 < p < \infty$, be a subset; necessary and sufficient conditions for the (relative) compactness of M are as follows:

(1) M is a bounded set in L^p, that is, $|x|_{L^p} \leq A < \infty$, for each $x \in L^p$;
(2) $\lim \int_a^b |x(t+h) - x(t)|^p \, dt = 0$ as $h \to 0$.

We notice the fact that $x(t+h)$ must be extended to be zero, outside $[a,b]$.

There are other criteria of compactness in L^p-spaces, due to Kolmogorov a.o. See, for instance, the items mentioned already, in this section, or Kantorovich and Akilov [274].

Some results concerning L^p-spaces, with a weight function, have been obtained by Milman [399], in connection with stability theory for integral equations.

Also, Kwapisz [305] introduced and applied to integral equations normed spaces of measurable functions, with a mixed norm, such as $x \to \sup\{g(t) \int_0^t |x(s)| \, ds\}$, on finite intervals or on the semiaxis R_+. Such spaces are useful when fixed-point theorems are used for existence of solutions to FEs. (See Kwapisz [304, 305]).

From the L^p-spaces theory, many other classes of measurable functions have been constructed, with important applications to the theory of FEs. An example, frequently appearing in literature, are the spaces L^p_{loc}. These spaces occur naturally when dealing with global existence of solutions.

For instance, the space $L^2_g(R,R^n)$ will consist of all measurable maps from R into R^n, such that $x \in L^2_g$ is determined by

$$\int_R g(t) \, |x(t)|^2 \, dt < \infty, \tag{1.34}$$

where $g(t)$ is measurable from R to R_+. The norm adequate for this space is, obviously,

$$x \to \left\{ \int_R g(t) \, |x(t)|^2 \, dt \right\}^{\frac{1}{2}}. \tag{1.35}$$

By using various weight functions g, one can achieve more generality in regard to the behavior/global properties of the solution.

Finally, we will mention the definition of $L^2_{\text{loc}}(R,R^n)$. This is a linear metric space (Fréchet), which belongs to the larger class of linear locally convex topological spaces.

In order to obtain the linearly invariant distance function for $L^2_{\text{loc}}(R,R^n)$, which consists of all locally integrable maps from R into R^n, that is, such that each integral

$$\int_{|t|\leq k} |x(s)|^2\, dt < +\infty,\ k \geq 1,$$

we will use a formula similar to (1.28).

Since

$$|x_k| = \left\{ \int_{|t|\leq k} |x(s)|^2\, ds \right\}^{\frac{1}{2}},\ k \geq 1,$$

is a semi-norm, the distance function on $L^2_{\text{loc}}(R,R^n)$ will be given (by definition)

$$d(x,y) = \sum_{k=1}^{\infty} 2^{-k} |x-y|_k\, (1 + |x-y|_k)^{-1}. \tag{1.36}$$

The convergence in $L^2_{\text{loc}}(R,R^n)$ is, therefore, the L^2-convergence on each finite interval of R (or, on each compact set in R).

This definition for $L^2_{\text{loc}}(R,R^n)$ can be extended from R^n to Hilbert spaces, or even to Banach spaces.

The variety of function spaces encountered in investigating the solutions of functional differential equations is considerable, and we do not attempt to give a full list. We will mention one more category of function spaces, containing the L^p-spaces, for the sake of their frequent use in the theory of FDE. Apparently, these spaces have been first used by N. Wiener. Their definition and systematic use is given in Massera–Schäffer [374].

The space $M(R_+,R^n)$ consists of all locally measurable functions on R_+, such that

$$|x|_M = \sup_{t\in R_+} \left\{ \int_t^{t+1} |x(s)|\, ds \right\} < \infty.$$

The space $M_0(R_+,R^n) \subset M(R_+,R^n)$ with the same norm contains only those elements for which

$$\sup \left\{ \int_t^{t+1} |x(s)|\, ds \right\} \to 0 \text{ } as \text{ } t \to \infty.$$

Such functions are important in connection with the concept of *almost periodicity* and other applications in the theory of FDE.

1.5 SOME NONLINEAR AUXILIARY TOOLS

The development of Functional Analysis has brought to the investigation of various classes of FEs, such as integral equations, for instance, but later to the theory of partial differential equations, many tools and methods. Volterra, who started the systematic investigation of integral equations in the 1890s, considered the problem of existence for these equations as the inversion of an integral operator (generally nonlinear). At the beginning of the twentieth century, Fredholm created the theory of linear integral equations, with special impact on the spectral side. Hilbert went further in this regard, contributing substantially to the birth of the theory of orthogonal series, related to symmetric/hermitian kernels of the Fredholm type equations. Due to remarkable contributions from Fréchet, Riesz (F and M), and other mathematicians, the new field of functional analysis (i.e., dealing with spaces/classes whose elements are functions) made substantial progress, but it was centered around the linear problems.

Starting with the third decade of the twentieth century, essentially *nonlinear* results have appeared in the literature. One of the first items in nonlinear functional analysis is the theory (or method) of fixed points, generally speaking for *nonlinear* operators/maps. The history takes us back to Poincaré and Brouwer, when only finite dimensional (Euclidean) spaces were involved. The problem of fixed points was considered by mathematicians starting the third decade of the twentieth century, during which period both best known results have been obtained.

The *contraction mapping principle* was the first tool, due to Banach in the case of what we call now Banach spaces. The case of complete metric spaces, as it is usually encountered nowadays, is due to V. V. Nemytskii (*Uspekhi Mat. Nauk*, 1927). This principle, usually encompassing all the results that can be obtained by the iteration method (successive approximations), can be stated as follows:

Let (S,d) denote a complete metric space and $T: S \to S$ a map, such that

$$d(Tx,Ty) \leq ad(x,y),\ x,y \in S, \tag{1.37}$$

for fixed a, $0 \leq a < 1$. Then, there exists a unique $x^* \in S$, such that $Tx^* = x^*$.

The proof of this statement is based on the iterative process

$$x_{k+1} = Tx_k,\ k \geq 0,\ x_0 \in S, \tag{1.38}$$

the initial term x_0 being arbitrarily chosen in S. One easily shows that $\{x_k; k \geq 0\}$ is a Cauchy sequence in (S,d), whose limit is x^*. The uniqueness follows directly by application of (1.37).

Proofs are available in many books on Functional Analysis or FEs (see, for instance, Corduneanu [120, 135]).

Let us mention that an estimate for the "error" is easily obtained, namely

$$d(x^*, x_k) \leq a^k (1-a)^{-1} d(x_0, Tx_0),\ k \geq 1. \tag{1.39}$$

The fixed point theorem due to *Schauder* is formulated for Banach spaces and can be stated as follows:

Let B be a Banach space (over reals) and $T : B \to B$ a continuous map/operator such that

$$TM \subset M, \tag{1.40}$$

with $M \subset B$ a closed convex set, while TM is relatively compact. Then, there exists at least a fixed point of T, that is, $Tx^* = x^* \in M$.

The definition of a *convex* set, say M, is $x, y \in M$ implies $ax + (1-a)y \in M$, for each $a \in (0,1)$.

For a proof of the Schauder fixed-point result, see the proof of a more general result (the next statement) in Corduneanu [120]. *Schauder–Tychonoff* fixed point theorem. Let E be a locally convex Hausdorff space and $x \to Tx$ a continuous map, such that

$$TK \subset A \subset K \subset E, \tag{1.41}$$

where K denotes a convex set, A being compact. Then, there exists at least one $x^* \in A \subset K$, such that $Tx^* = x^*$.

Both fixed point results formulated already imply the relative compactness of the image $TK \subset A$, and we mention the fact that uniqueness does not, generally, hold.

The set of fixed points of a map, under appropriate conditions, satisfies certain interesting and useful properties, such as compactness, convexity, and others.

We invite the readers to check the property of compactness, under aforementioned conditions. Another fixed-point result, known as the *Leray–Schuader Principle* also involves the concept of compactness of the operator, but also the idea of an "a priori" estimate for searched solution. Namely, one considers the equation

$$x = Tx,\ x \in B, \tag{1.42}$$

with B a Banach space and $T : B \to B$ a compact operator (i.e., taking bounded sets in B, into relatively compact sets). One associates to (1.42) the parameterized equation

$$x = \lambda Tx, \ \lambda \in [0,1], \tag{1.43}$$

assuming that (1.43) is solvable in B for each $\lambda \in [0,1)$. Moreover, each solution x satisfies the "a priori" estimate

$$|x|_B \leq K < \infty, \ \lambda \in [0,1), \tag{1.44}$$

where K is a fixed number. Then, equation (1.42), which corresponds to $\lambda = 1$ in (1.43), possesses a solution in B.

The proof of the principle can be found in Brézis [77] and Zeidler [550,551], including also some applications.

Several other methods/principles in functional analysis are known and widely applied in the study (particularly, in existence results) of various classes of FEs. We will mention here the method based on *monotone operators* (see Barbu [45], Deimling [190], and Zeidler [551]).

The definition of a *strongly monotone* operator $A : H \to H$, with H a Hilbert space, is

$$< Ax - Ay, x - y >_H \geq m|x - y|_H^2, \ x, y \in H,$$

for some $m > 0$. A very useful result can be stated as follows:

Consider in H the equation

$$Ax = y, \tag{1.45}$$

with $A : H \to H$ strongly monotone. Further, assume A satisfies on H the Lipschitz condition

$$|Au - Av|_H \leq L|u - v|_H. \tag{1.46}$$

Then, for each $y \in H$, Equation (1.45) has a unique solution $x \in H$.

The proof of this result can be found in Zeidler [550], and it is done by means of Banach fixed point (contraction).

An alternate statement of the solvability property of (1.45), for any $y \in H$, is obviously the property of A to be onto H (or surjective).

A somewhat similar result, still working in a Hilbert space H, can be stated as follows: Equation (1.45) is solvable for each $y \in H$, when A is continuous, monotone, that is,

$$< Ax - Ay, x - y >_H \geq 0, \ x, y \in H, \tag{1.47}$$

and *coercive*

$$\lim < Ax, x >_H |x|_H^{-1} = \infty, \ as \ |x| \to \infty. \tag{1.48}$$

See the proof, under slightly more general conditions, in Deimling [190] or Barbu and Precupanu [46].

All aforementioned references, in regard to monotone operators, contain applications to various types of FEs.

Other methods/procedures leading to the existence of solutions of various classes of FEs are based on diverse form of the *implicit functions* theorem (in Banach spaces and Hilbert Spaces); see Zeidler [550].

When we deal with FDE with finite delay, which we will consider in Chapters 2 and 3, we use another constructive method called the *step method.* We briefly discuss this method, which is frequently used to construct solutions to FDE of the form

$$\dot{x}(t) = F(t,x(t),x(t-h)),\ x \in R^n, \tag{1.49}$$

with $h > 0$ the delay, or time delay. In order to make the first step in constructing the solution, it is necessary, assuming the initial moment is $t = 0$, to know $x(t)$ on the interval $[-h,0]$. In other words, one has to associate to (1.49) the initial condition $x(t) = \phi(t)$, $t \in [-h,0]$. Using the notation $x_t(s) = x(t+s)$, $s \in [-h,0]$, this condition can be written in the form

$$x_0(t) = \phi(t),\ t \in [-h,0]. \tag{1.50}$$

If we assign the initial function ϕ from a certain function space, say $C([-h,0],R^n)$, then equation (1.49) becomes an ODE, on the interval $[0,h]$:

$$\dot{x}(t) = F(t,x(t),\phi(t-h)),\ t \in [0,h], \tag{1.51}$$

and the initial condition at $t = 0$ will be

$$x(0) = \phi(0). \tag{1.52}$$

The *second step*, after finding, from (1.51) to (1.52), $x(t)$ on $[0,h]$ will require to solve the ODE (1.49) on $[h,2h]$, starting at $x(h)$, as found from $x(t)$, on $[0,h]$.

The process continues and, at each step, one finds $x(t)$, on an interval $[mh,(m+1)h]$, $m \geq 1$, solving the ODE (1.49),

$$\dot{x}(t) = F(t,x(t),x(t-h)),\ t \in [mh,(m+1)h], \tag{1.53}$$

under initial condition at *mh*, as determined from the preceding step. In this way, the solution $x(t)$ of (1.49), appears as a chain, say

$$x_1(t), x_2(t), \ldots, x_m(t), \ldots \tag{1.54}$$

each term in (1.54) being found from (1.53), for different values of m, taking as initial value for $x_m(t)$, the final value of $x_{m-1}(t)$, $m \geq 1$. In this way, one obtains a continuous solution on $[-h,T)$ for (1.49), where T denotes the largest value of T, such that the solution is defined ($T = \infty$, when the solution is global on R_+).

The method of integration by steps has several variants, and in order to make it more convenient for numerical purposes, one uses the construction of the chain by means of the recurrent equation

$$\dot{x}_{m+1}(t) = F(t, x_m(t), x_{m-1}(t-h)),\ m \geq 1,$$

which allows making each step by performing a single quadrature.

It is remarkable that the step method allows investigation of global problems, such as stability. See Kalmar-Nagy [273] for illustration.

1.6 FURTHER TYPES OF FEs

The mathematical literature has been enriched, by other types of FEs than those usually encountered in the classical period. We have in mind, particularly, the various classes of *discrete* equations, the *fractional-order differential* equations and the *difference equations*. Discrete equations have been largely investigated. Modern computers have been used to perform computational procedures on those equations, which have represented mathematical models with large number of variables and calculations.

It is true that some FEs, pertaining to the aforementioned types, have been involved in research (may be only sporadically) and we can mention here a cycle of papers due to Bochner [68–71] from 1929 to 1931, who investigated the almost periodicity of some classes of equations with differential, integral and difference operations. A very simple example is, for instance, the FE

$$\dot{x}(t) + ax(t-h) + \int_0^t A(t-s)x(s)\,ds = f(t),$$

for which an initial condition would be of the form $x_0(t) = \phi(t)$, $t \in [-h,0]$, $h > 0$. See also Hale and Lunel [242].

In this section, we will only illustrate some types of equations belonging to those three categories mentioned above and state some results available in the literature.

Let us start by introducing some examples of *discrete* equations, by providing the necessary concepts leading to their solution belonging to various spaces of sequences.

One example is constructed by the space $s = s(N, R^n)$ or $S(Z, R^n)$ of all sequences $\xi = \{x_k; x_k \in R^n, k \geq 1\}$. These sequences form a Fréchet space, either real or complex, with the distance (compare with (1.36)),

$$d(\xi, \eta) = \sum_{k=1}^{\infty} 2^{-k} |\xi_k - \eta_k| (1 + |\xi_k - \eta_k|)^{-1}. \tag{1.55}$$

In case of sequences on Z, the sum must be considered from $-\infty$ to $+\infty$.

Within the space s, one can consider several sequence spaces, such as the space of all *bounded* sequences $s_b(N, R^m)$, respectively, $s_b(Z, R^m)$, with the supremum norm, its subspaces of *almost periodic* sequences or $s_\ell(N, R^m)$ which consists of all elements ξ of $s_\ell(N, R^m)$, such that $\lim \xi_k$ as $n \to \infty$ exists (finite!), (convergent sequences) or the subspace, traditionally denoted by $c_0(N, R^m)$, containing only those elements from $s_\ell(N, R^m)$ for which $\lim \xi_k = \theta \in R^m$, the null vector of R^m. Obviously, these concepts make sense when ξ belongs to a Banach space, in particular to Hilbert space $\ell^2(N, R^m)$.

Let E be a real Banach space and consider the operator equation, in discrete form

$$(Lx)(n) = (Gx)(n),\ n \in N, \tag{1.56}$$

where L denotes a linear operator on E, while G stands for a nonlinear operator on E. Equation (1.56) is a neutral one (it means, not solved in respect to the unknown element x). Under some conditions, we can reduce (1.56) to a normal form, namely

$$x(n) = (Fx)(n),\ n \in N, \tag{1.57}$$

which can be treated easier. Moreover, equations like (1.57) have caught the attention of researchers long time ago, the number of results being considerably higher than in case of (1.56).

If L has a bounded inverse operator on E, a situation warranted by the condition

$$|Lx|_E \geq m\, |x|_E,\ \forall x \in E,\ m > 0, \tag{1.58}$$

then equation (1.56) is equivalent to (1.57), with $F = L^{-1}G$. The term equivalent must be understood in the sense they have the same solutions (if any).

We assume that the "nonlinear" operator G satisfies the Lipschitz continuity:

$$|Gx - Gy|_E \leq K\, |x - y|_E,\ K > 0,\ x, y \in E. \tag{1.59}$$

Since $F = L^{-1}G$, we find easily

$$|Fx - Fy|_E = |L^{-1}(Gx - Gy)|_E \leq m^{-1} K |x - y|_E, \tag{1.60}$$

since $|L^{-1}|_E \leq m^{-1}$ (if in (1.58) we let $x = Ly$, $y \in E$, then we obtain $|Ly|_E \leq m^{-1}|y|_E$). From (1.60) one obtains for $x, y \in E$,

$$|Fx - Fy|_E \leq m^{-1} K |x - y|_E, \tag{1.61}$$

which implies the property of contraction (Banach) when

$$K < m. \tag{1.62}$$

Therefore, equation (1.57) has a unique solution in E. This property is then true for (1.56).

We will now apply this existence result, to obtain conditions for the existence of a solution $x = (x_1, x_2, \ldots, x_m, \ldots)$ in one of the spaces (of sequences) listed in the text, say $\ell^2(N, R^n)$. It is known that this is the space Hilbert used for constructing the infinite dimensional analysis (and geometry!). It is the prototype of separable Hilbert spaces (containing a countable subset, everywhere dense in it). Namely, we consider the infinite system of quasilinear equations

$$x_k = \sum_{j=1}^{\infty} a_{kj} x_j + f_k(x),\ k \geq 1, \tag{1.63}$$

where a_{kj} are real numbers and $x \to f_k(x)$, $k \geq 1$, are maps from E into itself. Obviously, (1.63) can be rewritten in concise form as

$$x = Ax + f(x), \tag{1.64}$$

with A the double infinite matrix

$$A = (a_{ij}),\ i,j = 1,2,\ldots$$

and $f(x) = (f_1(x), f_2(x), \ldots, f_m(x), \ldots)$. It is quite obvious that (1.64) is of the same form as (1.57). Hence, we have to look for conditions assuring the contraction, on ℓ^2, of the operator on the right-hand side of (1.64). First, we shall assume an inequality of the form

$$\sum_{k=1}^{\infty} \sum_{j=1}^{\infty} |a_{kj}|^2 \leq \alpha^2 < \infty. \tag{1.65}$$

This guarantees that the operator A in (1.64) is bounded on $\ell^2(N,R^n)$. Concerning the nonlinear term $f(x)$, we assume that

$$\sum_{k=1}^{\infty} |f_k(x)|^2 < \infty, \; x \in \ell^2(N,R^n), \tag{1.66}$$

and the Lipschitz conditions hold

$$|f_k(x) - f_k(y)| \leq \beta_k |x-y|, \; x,y \in \ell^2,$$

with

$$\sum_{k=1}^{\infty} \beta_k{}^2 = \beta^2 < \infty. \tag{1.67}$$

The assumptions (1.65)–(1.67), allows us to write (with the ℓ^2-norm)

$$\begin{aligned}|Ax+f(x)-Ay-f(y)| &= |A(x-y)+f(x)-f(y)| \\ &\leq |A(x-y)|+|f(x)-f(y)| \\ &\leq (\alpha+\beta)|x-y|,\end{aligned}$$

which implies the contraction of the operator $x \to Ax+f(x)$ on $\ell^2(N,R^n)$, as soon as

$$\alpha+\beta < 1. \tag{1.68}$$

We can conclude that, under assumptions (1.65)–(1.68), the infinite system (1.63) (or (1.64)) has a unique solution in the space $\ell^2(N,R^n)$. This solution consists of a sequence of vectors $x^{(k)}$, $k \geq 1$, in the Hilbert space $\ell^2(N,R^n)$.

This discussion provides an elementary example of the kind of problems that may occur in applications. The discrete equations are adequate for treatment by means of the electronic computers. The literature in this field is very rich and various aspects and problems can be found in Collatz [108].

The following remark is related to condition (1.66). Namely, if (1.66) holds for an element $y \in \ell^2$, then it will be valid for any $x \in \ell^2$. This is a consequence of the obvious inequality

$$|f_j(x)|^2 \leq 2\,(|f_j(y)|^2 + \beta_j^2\,|x-y|^2), \; j \geq 1,$$

relying also on (1.67).

Another remark is that $x \in \ell^2(N,R^n)$ implies $x_m \to \theta \in R^n$, which implies the fact that only a finite number of coordinates of the solution can provide sufficient information about it (with regard to the numerical treatment of the problems).

We invite the readers to obtain the existence of a solution of the equation

$$\Delta f(x,m) = g(x,m),\ m \geq 0, \tag{1.69}$$

where $\Delta f(x,m) = f(x_{m+1}, m+1) - f(x_m, m)$, the sequence x_m, $m \geq 0$, being constructed recurrently with an arbitrary $c_0 \in s_b(N, R^n)$. In other words, we need conditions to obtain the existence of a bounded sequence $\{x_m;\ m \geq 0\} \subset s_b(N, R^n)$ for equation (1.69), which is of a neutral type. One uses the supremum norm.

The following conditions assure the existence of a unique solution:

1) $f : N \times R^n \to R^n$ is such that the equation

$$f(u,m) = v \in R^n$$

has a unique solution, say $u_m \in R^n$, for each v;

2) f satisfies the condition

$$|f(u,m)| \geq h(|u|),\ m \geq 1,$$

with $h(r)$, $0 \leq r < \infty$, monotonically increasing, while $h^{-1}(r)$ has sublinear growth,

$$h^{-1}(r) \leq \alpha r + \beta,\ r \geq 0,\ \alpha, \beta > 0;$$

3) $g : N \times R^n \to R^n$ satisfies

$$|g(u,m)| \leq c_m |u|,\ n \geq 0,\ u \in R^n,$$

where the series $\sum_{k=1}^{\infty} c_k$ is convergent.

Then, equation (1.69) has a unique solution in $s_b(N, R^n)$, for each initial choice $x_0 \in R^n$.

Details of the proof can be found in the paper by Corduneanu [146], together with other existence results.

Let us notice, with regard to the discrete FEs, that a rich literature is available. We mention some of the best known sources/books, in which a large variety of such equations (frequently including recurrent ones) is treated, with both classical or modern tools. Mickens [395], Kelley and Peterson [284], and Pinney [451].

The *difference equations* are adequate to be treated as discrete ones, as in above shown examples, or they can be considered in the class of equations with continuous argument. Let us look at the very elementary example of the equation $x(t+1) = 2x(t)$, on the real axis R. Then, looking for an exponential solution $x(t) = a^t$, $a > 0$, one finds that each function of the form $x(t) = c\,2^t$, with $c \in R$, is a solution depending on the continuous argument $t \in R$.

This second manner of looking at solutions depending on a continuous argument had enjoyed early attention in the literature, and we shall now dwell on these types of solutions. Following Shaikhet [493], we shall consider the scalar difference equation, somewhat simplified, namely

$$x(t+h_0) = F(t, x(t), x(t-h_1), \ldots, x(t-h_m)), \tag{1.70}$$

on the semiaxis $t > t_0 - h_0$, under initial conditions

$$x(s) = \phi(s),\ s \in S = [t_0 - h_0 - \max_{1 \le k \le m} (h_k, t_0)]. \tag{1.71}$$

In (1.71), $x \in R$, $h_k > 0$, $1 \le k \le m$, $\phi : S \to R$ being the initial datum, usually continuous. The following growth condition is imposed on F:

$$|F(t, x_0, x_1, \ldots, x_m)| \le \sum_{j=0}^{m} a_j\,|x_j|, \tag{1.72}$$

with $a_j \in R_+ \setminus \{0\}, j = 0, 1, \ldots, m$.

We shall denote by $x(t; t_0, \phi)$ the solution of (1.70), (1.71). It is obtained by the step method, first on $[t_0, t_0 + h_0]$, by substituting ϕ to x on the right-hand side, then on $[t_0 + h_0, t_0 + 2h_0]$, substituting $x(t)$ already found on $[t_0, t_0 + h_0]$ on the right-hand side, and so on. This process uniquely determines $x(t)$, for $t > t_0 - h_0$.

Now we will anticipate on the concept of stability, which will be defined and developed in Chapter 3. More precisely, we shall formulate as follows the definitions of stability and asymptotic stability of the solution $x = 0$ of (1.70).

We notice first, on behalf of assumption (1.72), that $x = 0$ is a solution of (1.70).

Stability of $x = 0$ means that to each $\epsilon > 0$, $t_0 \ge 0$, there corresponds $\delta = \delta(\epsilon, t_0) > 0$, such that

$$|x(t; t_0, \phi)| < \epsilon \text{ for } t \ge t_0, \tag{1.73}$$

provided $|\phi|_c = \sup_{s \in S} |\phi(s)| < \delta(\epsilon, t_0)$.

Asymptotic stability means stability, and additionally,

$$\lim_{t \to \infty} x(t : t_0, \phi) = 0, \tag{1.74}$$

for all ϕ, such that $|\phi|_c < \eta(t_0)$.

We are now able to formulate a stability result for the zero solution of the difference equation (1.70), which is an example of an equation with difference and continuous argument.

Theorem 1.1 *Assume the existence of a functional (Liapunov type) $V(t) = V(t, x(t), x(t-t_1), \ldots, x(t-t_m))$, nonnegative and such that, for some positive c_1, c_2, p, one has*

$$V(t) \leq c_1 \sup_{s \leq t} |x(s)|^p, \; t \in [t_0, t_0 + h_0), \tag{1.75}$$

$$\Delta V(t) = V(t + h_0) - V(t) \leq c_2 |x(t)|^p, \; t \geq t_0. \tag{1.76}$$

Then $x = 0$ is a stable solution for equation (1.70).

Proof. The conditions (1.75), (1.76) and the nonnegativity of $V(t)$ imply

$$V(t) \geq c_2 |x(t)|^p, \; t \geq t_0, \tag{1.77}$$

$$V(t) \leq V(t - h_0) \leq V(t - 2h_0) \ldots \leq V(s), \; t \geq t_0, \tag{1.78}$$

with $s = t - \left[\frac{t - t_0}{h_0}\right] h_0 \in [t_0, t_0 + h]$, which implies

$$\sup_{s \in [t_0, t_0 + h_0]} V(s) \leq c_1 \sup_{t \leq t_0 + h_0} |x(t)|^p. \tag{1.79}$$

After taking into account (1.70)–(1.72), we obtain for $t \leq t_0 + h_0$,

$$\begin{aligned} |x(t)| = |F(t, x(t), \ldots)| &\leq a_0 |\phi(t - t_0)| + \sum_{j=1}^{m} a_j |\phi(t - t_0 - h_j)| \\ &\leq \left(\sum_{j=0}^{m} a_j \right) |\phi|_c \\ &= A |\phi|_c, \end{aligned} \tag{1.80}$$

where $A = \sum_{j=0}^{m} |a_j| > 0$, which leads easily to the inequality

$$c_2 |x(t)|^p \leq c_1 A^p |\phi|_c^p, \; t \geq t_0, \tag{1.81}$$

relying also on (1.77)–(1.80).

The inequality (1.81), obtained for the solution $x(t) = x(t;t_0,\phi)$, shows that the stability is assured.

Moreover, imposing a rather restrictive condition, namely $A < 1$, the solution $x = 0$ of equation (1.70) is asymptotically stable. To obtain this conclusion, one should rely on (1.81) and $A < 1$, which leads to the inequality

$$|x(t)| \leq A^{[\frac{t-t_0}{h_0}]} |\phi|_c, \; t \geq t_0, \tag{1.82}$$

which assures (1.74) because $[\frac{t-t_0}{h_0}] \to \infty$ as $t \to \infty$. Of course $[.]$ denotes the integer part function. This inequality is obtained following the same steps as in case of deriving (1.80).

It is adequate to notice the fact that, in Shaikhet [493], one finds further similar results of stability, or integrability, as well as some procedures to construct Liapunov functions for difference equations. See also Kolmanovskii and Shaikhet [294], for related results.

In concluding this section, we will briefly illustrate some results and methods in the case of another class of FEs that came relatively recently under the scrutiny of researchers. We have in mind the class of *fractional differential equations*. The literature is rather vast, and we notice the apparition of a fractional calculus, for instance the one based on Caputo's concept of a fractional order derivative. This concept, similar to that known as Riemann–Liouville fractional-order derivative, implies the use of the operation of integration. Therefore, there is a rather close connection between the fractional differential equations and the (singular) integral or IDEs. It is worth mentioning that this new direction of research has been brought into actuality by its engineering applications (see, for instance, Caputo [88]).

An important distinction should be made when we deal with difference equations and look for solutions, which depend on a continuous variable or a discrete variable. Let us look again at the very simple equation, $x(t+1) = 2x(t)$, we considered earlier, in this section, for which $x(t) = c\,2^t$ is a family of continuous solutions on R, while regarded as a discrete one, the result is that each geometric progression $\{2c, 2^2c, \ldots, 2^mc, \ldots\}$ is a solution, defined on the positive integers. Apparently, the discrete variable is preferred in many papers, due to the fact that its results are most adequately handled by digital machines. And this aspect is stressed by the new tendency in mathematical modeling, with emphasis on discrete models.

Returning now to the class of fractional differential equations, based on the concept of fractional-order derivative, we will notice that its roots are in the work of the nineteenth century well known mathematicians Liouville and Riemann. Probably attracted by the beauty and usefulness of Euler's Γ (gamma) function, they discovered the fractional derivative of a function (also the fractional integral), which can be defined by means of Euler's function.

The Riemann–Liouville derivative, of (fractional) order q, of a function $f : R_+ \backslash \{0\} \to R$ is given by the formula

$$D_{0+}^{q} f(t) = \frac{1}{\Gamma(n-q)} \left(\frac{d}{dt}\right)^{n} \int_0^t \frac{f(s)\,ds}{(t-s)^{q+1-n}},\ n = [q]+1, \tag{1.83}$$

where $[q]$ means the integer part of q.

The fractional derivative, introduced more recently by Caputo, is represented by the formula

$$D_c^{\alpha} f(t) = \frac{1}{\Gamma(n-\alpha)} \int_0^t (t-s)^{-\alpha+n-1} f^{(n)}(s)\,ds, \tag{1.84}$$

with $n = [\alpha]+1$. It is assumed that $f(t)$ is n times differentiable on $R_+ \setminus \{0\}$.

There are other types of fractional derivatives, with frequent use, especially in applications. See the books by Miller and Ross [397] and Das [187], in which details and applications are provided. We shall illustrate the type of results obtained in the literature (both mathematical, i.e., theoretic and applied). The following result is taken from recent papers of Cernea [95, 96] stating the existence of continuous selections for initial value problems of the form

$$D_c^{\alpha} x(t) \in F(t, x(t)),\ a.e. \text{ on } [0,T], \tag{1.85}$$

under initial conditions

$$x(0) = x_0,\ x'(0) = x_1,$$

with $\alpha \in (1,2]$, $F : [0,T] \times R \to \mathcal{P}(R)$ a set valued map whose values are parts/subsets of R. Moreover, it is assumed that $x_0, x_1 \neq 0$. The following hypotheses are assumed:

H_1. The map $F : [0,T] \times R \to \mathcal{P}(R)$ has nonempty closed set-values and is measurable in $L([0,T]) \times B(R)$, where B denotes Borel measurability.

H_2. There exists $L(\cdot) \in L^1([0,T], R_+$, such that $a.e.$ on $t \in [0,T]$, the Lipschitz type condition

$$d_H(F(t,x), F(t,y)) \leq L(t)\,|x-y|, \tag{1.86}$$

for $(t,x), (t,y) \in [0,T] \times R$, where d_H denotes the Pompeiu–Hausdorff distance in $\mathcal{P}(R)$.

H_3. Let S be a separable metric space, and $a,b : S \to R$, $c : S \to (0,\infty)$ continuous maps. There exist continuous mappings $g,p : S \to L^1([0,T],R)$, and $y : S \to C([0,T],R)$, such that

$$(Dy(s))_c^{\alpha}(t) = g(s)(t),\, d(g(s)(t), F(t,y(s)(t)) \leq p(s), \tag{1.87}$$

$a.e.$ for $t \in [0,T]$, $s \in S$.

In order to state the existence result, we need to introduce the following notation

$$I^{\alpha} f(t) = \int_0^t \frac{(t-s)^{\alpha-1}}{\Gamma(\alpha)} f(s)\, ds,\ \alpha > 0, \tag{1.88}$$

where $f : [0,\infty) \to R$ is locally Lebesgue integrable. This integral is known as the fractional integral of order $\alpha > 0$.

Let us now define the auxiliary function

$$\xi(s) = (1 - |I^{\alpha} L(t)|_c^{-1})\,[|a(s) - y(s)(0)| + T|b(s) - y'(s)(0)| \\ + c(s) + |I^{\alpha} p(s)|_c],\ s \in S. \tag{1.89}$$

The following result is an answer to the existence problem, formulated already, for (1.85):

Theorem 1.2 *Under assumptions H_1, H_2, and H_3, the existence of a solution to problem (1.85), under assigned initial conditions, is assured, in a sense to be specified below, if we impose the extra condition*

$$|I^{\alpha} L(t)|_c < 1. \tag{1.90}$$

The meaning of solution is as follows: There exists a mapping $x(\cdot) : S \to C([0,T],R)$, such that for any $s \in S$, $x(s)(\cdot)$ is a solution of the problem

$$D_c^{\alpha} z(t) \in F(t,z(t)),\ z(0) = a(s),\ z'(0) = b(s),$$

satisfying

$$|x(s)(t) - y(s)(t)| \leq \xi(s),\ (t,s) \in [0,T] \times S, \tag{1.91}$$

with y and ξ defined by (1.87) and (1.89), respectively.

The proof can be found in Cernea [95] and relies on a rather sophisticated construction. It is just a sample on how demanding this kind of problems are.

We are concluding this introductory Chapter, whose main purpose is to familiarize the reader with some concepts and methods we will use in the book, as well as summarily approaching some topics that will not be covered in the coming chapters (mainly, types of FEs, old or new, that are in the attention of many researchers).

At the same time, we want to emphasize the fact that other concepts, like equations on time scale, or equations with several independent variables, will not be represented in our presentation. The vastness of the mathematical and science and engineering production in this field is certainly one of the reasons we had to limit ourselves to a number of topics, namely existence, estimates, stability, and oscillations.

Some pertinent references for this section of the introductory chapter are Miller and Ross [397], Das [187], Kilbas, Srivastava and Trujillo [285], Pinney [451], Kelley and Peterson [284], Mickens [395], Abbas, Benchohra and N'Guérékata [2], and Diethelm [196].

More references/sources, concerning the topics summarily discussed in Section 1.6, can be found in those quoted the aforementioned references.

2

EXISTENCE THEORY FOR FUNCTIONAL EQUATIONS

Traditional functional equations are either differential or integral. Integro-differential equations represent the by-products of the differential and the integral equations and constitute a large class of equations with many applications.

To be more accurate, we should also mention the equations of Cauchy's type and their generalizations, the prototype being the simple but famous equation $f(x+y) = f(x) + f(y)$. This last type of equation has also known a real flourishing development. However, in this book, we are not concerned with these equations due to rather different procedures in their investigations, which are somewhat different from those we encounter in our exposition. See the list of references at the end of this book under the name Aczel, who is a leading promoter of this type of functional equation.

The main objective of this chapter is to provide various results on the local existence for functional equations within the classes of continuous or measurable functions (real-valued or with values in R^n). The problem of global existence (i.e., on a preassigned interval) will be also discussed. We will also deal with the existence of solutions in spaces of measurable functions like Lebesgue spaces L^p, $1 \le p \le \infty$.

Functional Differential Equations: Advances and Applications, First Edition.
Constantin Corduneanu, Yizeng Li and Mehran Mahdavi

Our approach will consist of applying general existence results, in most cases obtained by fixed-point methods, to various classes of functional equations (ordinary differential equations, integral equations in a single variable, integro-differential equations, finite or infinite delay equations, and equations involving operators acting on function spaces, such as causal operators).

Some results will be concerned with the existence of solutions with special properties such as positiveness, or constant sign, boundedness on unbounded intervals, asymptotic behavior, belonging to various function spaces, uniqueness, and dependence on data.

2.1 LOCAL EXISTENCE FOR CONTINUOUS OR MEASURABLE SOLUTIONS

We shall start with a general functional equation of the form

$$x(t) = (Vx)(t), \quad t \in [t_0, t_0 + a], \tag{2.1}$$

where $t_0 \in R$ and $a > 0$, while V stands for an operator defined on the space of continuous functions $C([t_0, t_0 + a], R^n)$, $n \geq 1$. We shall further assume that V is a causal operator, which means that $(Vx)(t)$, $t \in (t_0, t_0 + a]$, is determined by the values of x on the interval $[t_0, t]$.

Formal definition of the causality property of an operator is the following: $V : C([t_0, t_0 + a], R^n) \to C([t_0, t_0 + a], R^n)$ will be called *causal* if for each $s \in (t_0, t_0 + a]$, from $x(t) = y(t)$ on $[t_0, s]$, one obtains $(Vx)(t) = (Vy)(t)$ on the same interval.

Another property we will require for the operator V is that of *fixed initial value*, that is, $(Vx)(t_0) = \text{constant} \in R^n$, with respect to x in $C([t_0, t_0 + a], R^n)$.

Obviously, the property of fixed initial value is verified by the classical Volterra integral operator

$$(Vx)(t) = f(t) + \int_{t_0}^{t} F(t, s, x(s))\, ds, \quad t \in [t_0, t_0 + a]. \tag{2.2}$$

The following existence result is proven in Corduneanu [135].

Theorem 2.1 *Assume the following conditions hold for the operator V in equation (2.1):*

(1) *The operator V is causal on $C([t_0, t_0 + a], R^n)$.*

(2) *V is continuous and compact on the same space.*

(3) *V satisfies the fixed initial value property.*

Then, there exists a number δ, $0 < \delta \leq a$, such that equation (2.1) has a solution $x(t) \in C([t_0, t_0 + \delta], R^n)$.

Proof. The result is based on the Schauder fixed point theorem, relying on Ascoli–Arzelà criterion of compactness for sets of continuous functions on a compact interval (see Chapter 1).

Let $x^0 \in R^n$ denote the fixed initial value of the operator V, that is, $(Vx)(t_0) = x^0$ for any $x \in C([t_0, t_0 + a], R^n)$.

Let us consider now the equation

$$x(t) - x^0 = (Vx)(t) - x^0, \tag{2.1$'$}$$

which is obviously equivalent to (2.1). The compactness of V allows us to state the following: for a fixed $r > 0$, $r < a$, we can write $|(Vx)(t) - x^0| < \epsilon$ for $t - t_0 < \delta$ and all $x \in C$ with $|x(t) - x^0| \leq r$, $t \in [t_0, t_0 + a]$, $\delta = \delta(\epsilon)$. We assume now $\epsilon \leq r$, then from (2.1)$'$ we obtain $|(Vx)(t) - x^0| < \epsilon \leq r$ for any $x(t)$ such that $|x(t) - x^0| \leq r$ on the interval $[t_0, t_0 + \delta]$.

These considerations lead to the conclusion that the ball of radius r with center at x^0, in the space $C([t_0, t_0 + \delta], R^n)$, is taken by the operator V into itself. Since V is continuous and compact and the ball $|x(t) - x^0| \leq r$ is convex and closed, we can apply the Schauder's fixed-point theorem in the Banach space $C([t_0, t_0 + \delta], R^n)$. Therefore, there exists at least one solution of Equation (2.1)$'$ in the space $C([t_0, t_0 + \delta], R^n)$.

Remark 2.1 *Theorem 2.1 provides a local existence result in an interval $[t_0, t_0 + \delta]$ which belongs to the interval $[t_0, t_0 + a]$. It is possible to obtain a global existence result when $x(t)$ is defined on the whole interval $[t_0, t_0 + a]$. We invite the readers to show that this result can be obtained when $|(Vx)(t)| \leq M < \infty$, for $t \in [t_0, t_0 + a]$ and any $x \in C([t_0, t_0 + a], R^n)$.*

Corollary 2.1 *Consider the functional differential equation*

$$\dot{x}(t) = (Vx)(t), \quad t \in [t_0, t_0 + a], \tag{2.3}$$

with the initial condition

$$x(t_0) = x^0 \in R^n, \tag{2.4}$$

under the assumption (1) from Theorem 2.1. Instead of condition (2) in Theorem 2.1, we consider the weaker condition.

(2') The operator V is continuous and takes bounded sets into bounded sets of $C([t_0, t_0 + a], R^n)$.

Then, there exists a number δ, $0 < \delta \leq a$, *such that equation (2.3) with the initial condition (2.4) has a continuously differentiable solution defined on* $[t_0, t_0 + \delta]$.

The proof of this corollary follows directly from the proof of Theorem 2.1 if we notice that the auxiliary operator

$$(Ux)(t) = x^0 + \int_{t_0}^{t} (Vx)(s)\, ds, \tag{2.5}$$

satisfies the conditions of Theorem 2.1. We have to prove that U is compact under (2'), while the fixed initial value property is obviously verified from (2.5).

Concerning the compactness of U, we observe that $x(t) \in B =$ bounded set in $C([t_0, t_0 + a], R^n)$ implies $(Vx)(t) \in B_1 \subset C([t_0, t_0 + a], R^n)$, with B_1 also bounded. Hence, one can find $K > 0$ such that $|(Ux)(t) - (Ux)(\tau)| \leq K|t - \tau|$, $t, \tau \in [t_0, t_0 + a]$, $x(t) - x^0 \in C([t_0, t_0 + a], R^n)$ and $|x(t) - x^0| \leq A$. Therefore, the image of a bounded set is equicontinuous.

This ends the proof of the Corollary 2.1.

The results stated in Theorem 2.1 and Corollary 2.1 contain as special cases several local existence results, and we will mention here the cases of Peano's existence theorem for the classical initial value problem for ordinary differential equations

$$\dot{x}(t) = f(t, x(t)), \quad x(t_0) = x^0 \in R^n, \tag{2.6}$$

and the existence for the functional differential equation with finite delay

$$\dot{x}(t) = f(t, x_t), \quad t \in [0, T),\ x_0(t) = \phi(t),\ t \in [-h, 0], \tag{2.7}$$

where $x_t(s) = x(t + s)$, $s \in [-h, 0]$, $h > 0$.

In case of Peano's theorem, one has

$$(Vx)(t) = f(t, x(t)), \quad t \in [t_0, t_0 + a]. \tag{2.8}$$

Such kind of causal operators are known, in the literature, under the name of Niemytskii's operators. The compactness of this type of operators is the consequence of uniform continuity of the function $f(t, x)$, on any bounded closed set of the space R^n, which follows from the continuity of f.

With respect to the initial value problem (2.7), the following conditions will be assumed:

(1) The operator

$$(Vx)(t) = f(t, x_t), \quad t \in [0, T), \tag{2.9}$$

is continuous and compact on the space $C([-h, 0], R^n)$ for each $t \in [0, T)$.

(2) $\phi(t) \in C([-h, 0], R^n)$.

Under conditions (1) and (2), the operator defined by (2.9) obviously verifies the requirements of Corollary 2.1, which allows us to conclude the existence of a solution to problem (2.7).

This type of initial value problems is engaging function spaces which are infinite dimensional, as initial source for the initial value problem.

Remark 2.2 *The compactness of the operator V can be secured by assuming the continuity of the map $(t, \phi) \to f(t, \phi)$, from $[0, T) \times C([-h, 0], R^n)$ into R^n.*

From Theorem 2.1, an existence result can be easily obtained for the classical Volterra nonlinear equation

$$x(t) = f(t) + \int_0^t k(t, s, x(s))\, ds, \quad t \in [0, T], \tag{2.10}$$

under continuity assumption on the functions f and k, with the operator V defined by

$$(Vx)(t) = f(t) + \int_0^t k(t, s, x(s))\, ds. \tag{2.11}$$

The readers are invited to formulate the details of the proof of this existence result.

All the aforementioned results have been concerned with the continuous functions–solutions. We will now consider some cases in which measurable functions/solutions are involved.

Theorem 2.2 *Consider the functional equation (2.1), $x(t) = (Vx)(t)$, $t \in [t_0, t_0 + a]$, $a > 0$, under the following conditions:*

(1) *The operator $V : L^p([t_0, t_0 + a], R^n) \to L^p([t_0, t_0 + a], R^n)$, $1 \leq p < \infty$, is causal (hence equalities must be understood* a.e.*).*

(2) *The operator V is continuous and compact on the space $L^p([t_0, t_0+a], R^n)$.*

Then, equation (2.1) has a solution $x(t) \in L^p([t_0, t_0+\delta], R^n)$, for some δ, $0 < \delta < a$.

The proof of theorem 2.2 is provided in Corduneanu [149], based on the Schauder's fixed-point theorem and the compactness criterion for subsets of the space $L^p([t_0, t_0+a], R^n)$ (known as Kolmogorov's criterion).

We notice the fact that we have omitted the limit case $p = \infty$. The continuous case discussed in Theorem 2.1 represents a substitute for this situation, due to the fact that $C([t_0, t_0+a], R^n)$ is a subset of $L^\infty([t_0, t_0+a], R^n)$.

We shall also observe that the property of fixed initial value has not been required for the operator V, a circumstance caused by the fact that each element of L^p is actually a class of measurable functions such that each function in the class differs from another on a set of measure zero.

Corollary 2.2 *Consider the functional differential equation (2.3), $\dot{x}(t) = (Vx)(t)$, $t \in [t_0, t_0+a]$, with the initial condition (2.4), $x(t_0) = x^0 \in R^n$, under the following assumptions:*

(1) *V is a causal continuous operator on the space $L^p([t_0, t_0+a], R^n)$, $1 \leq p < \infty$.*
(2) *V takes bounded sets in $L^p([t_0, t_0+a], R^n)$ into bounded sets.*

Then, there exists a solution of problem (2.3) and (2.4) which is absolutely continuous on an interval $[t_0, t_0+\delta]$, $0 < \delta \leq a$. The derivative of the solution satisfies $\dot{x} \in L^p([t_0, t_0+\delta], R^n)$.

The proof of Corollary 2.2 is given in Corduneanu [149]. The underlying space can be chosen as the space $AC([t_0, t_0+a], R^n)$ of absolutely continuous functions with the norm of $|x|_{AC} = |x(t_0)| + |\dot{x}(t)|_{L^p}$.

We shall conclude this section with an illustration of Corollary 2.2 to the functional differential equation of the form

$$\dot{x}(t) = f\left(t, x(t), \int_{t_0}^{t} k(t,s)x(s)\,ds\right), \quad t \in [t_0, T], \tag{2.12}$$

with the initial condition $x(t_0) = x^0 \in R^n$.

In order to satisfy the conditions of Corollary 2.2, we will assume

(1) $f : [t_0, T] \times R^n \times R^n \to R^n$ satisfies Carathéodory's condition of measurability for each fixed x and y and continuity in (x, y) for almost all t in $[t_0, T]$.

(2) The following estimate holds for f:

$$|f(t,x,y)| \leq M(|x| + |y|) + g(t),$$

where $M > 0$ is a constant and $g \in L^p$.

(3) The matrix kernel $k(t,s)$ is measurable in (t,s) for $t_0 \leq s \leq t \leq T$ and satisfies the condition

$$\left\{ \int_{t_0}^{T} \left(\int_{t_0}^{t} |k(t,s)|^p \, ds \right)^{\frac{q}{p}} dt \right\}^{\frac{1}{q}} = K_0 < \infty,$$

where q is the conjugate index to p (i.e., $p^{-1} + q^{-1} = 1$) and $K_0 > 0$.

Based on conditions (1), (2), and (3), there exists an absolutely continuous solution of the problem (2.12), $x(t_0) = x^0 \in R^n$, defined on some interval $[t_0, t_0 + \delta] \subset [t_0, T]$.

2.2 GLOBAL EXISTENCE FOR SOME CLASSES OF FUNCTIONAL DIFFERENTIAL EQUATIONS

This section will start with some results of existence and uniqueness based on a method introduced by Bielecki [65]. This method uses function spaces consisting of elements, which are bounded with respect to a weighted norm. We have dealt with such spaces, denoted by $C_g([0,T), R^n)$, where g represents the weighting function (see Chapter 1). Bielecki's method has been widely used in the theory of functional equations. In Corduneanu [125] a survey has been presented for the case of integral equations until 1984. Kwapisz [304, 305] has extended the applicability of the method by introducing a new kind of weight function, as well as new types of functional equations.

To begin, we shall consider an ordinary differential equation in a real Banach space E:

$$\dot{x}(t) = f(t, x(t)), \quad t \in [0,T),\ T \leq \infty, \tag{2.13}$$

where x, f take values in E. We shall also associate the initial condition

$$x(0) = x_0 \in E, \tag{2.14}$$

to (2.13) and will be looking for a solution defined on the whole interval $[0,T)$.

The following assumptions will be made for the function $f : [0,T) \times E \to E$:

(1) The map f is continuous.
(2) There exists a nonnegative locally integrable function $\lambda : [0,T) \to R_+$, such that

$$|f(t,x) - f(t,y)|_E \leq \lambda(t)\, |x-y|_E, \tag{2.15}$$

for any $t \in [0,T)$ and $x, y \in E$.

The existence and uniqueness result for problem (2.13) and (2.14) can be stated as follows:

Theorem 2.3 *Equation (2.13), with the initial condition (2.14) has a unique continuously differentiable solution $x(t)$, $t \in [0,T)$, provided conditions (1) and (2) formulated earlier hold, while*

$$|f(t,\theta)|_E \leq K\lambda(t) \exp\left\{\alpha \int_0^t \lambda(s)\, ds\right\}, \quad t \in [0,T), \tag{2.16}$$

$K > 0$, a constant and $\alpha > 1$.

Proof. We shall actually deal with the integral equation equivalent to problem (2.13) and (2.14)

$$x(t) = x_0 + \int_0^t f(s, x(s))\, ds, \quad t \in [0,T). \tag{2.17}$$

On the space $C_g([0,T), E)$, with $g(t) = \exp\{\alpha \int_0^t \lambda(s)\, ds\}$, we consider the operator

$$(Ux)(t) = x_0 + \int_0^t f(s, x(s))\, ds, \quad t \in [0,T). \tag{2.18}$$

Let us now show that U is, indeed, acting on the space $C_g([0,T), E)$, with g chosen as before. On the interval $[0,T)$, we can write the inequalities

$$\begin{aligned} |(Ux)(t)|_E &\leq |x_0|_E + \int_0^t |f(s,x(s))|_E\, ds \\ &\leq |x_0|_E + \int_0^t \{|f(s,x(s)) - f(s,\theta)|_E + |f(s,\theta)|_E\}\, ds \\ &\leq |x_0|_E + \int_0^t \lambda(s)\, |x(s)|_E\, ds + \int_0^t |f(s,\theta)|_E\, ds. \end{aligned} \tag{2.19}$$

Taking into account (2.16) and the fact that $x(t) \in C_g$, which amounts to

$$|x(t)|_E \leq K_0 \exp\left\{\alpha \int_0^t \lambda(s)\,ds\right\}, \quad t \in [0,T),$$

from (2.19), we obtain after neglecting negative terms on the right-hand side, for $t \in [0,T)$,

$$|(Ux)(t)|_E \leq \left[|x_0|_E + \alpha^{-1}(K_0 + K)\right] \exp\left\{\alpha \int_0^t \lambda(s)\,ds\right\}.$$

This inequality proves that the operator U is acting on the space $C_g([0,T),E)$.

It remains to prove that the operator U is a contraction on C_g. Since

$$\begin{aligned}
|(Ux)(t) - (Uy)(t)|_E &\leq \int_0^t |f(s,x(s)) - f(s,y(s)|\,ds \\
&\leq \int_0^t \lambda(s)|x(s) - y(s)|_E\,ds \\
&= \int_0^t \lambda(s) \underbrace{|x(s) - y(s)|_E\, e^{-\alpha \int_0^s \lambda(u)\,du}}\, e^{\alpha \int_0^s \lambda(u)\,du}\,ds \\
&\leq |x(t) - y(t)|_{C_g} \int_0^t \lambda(s)\, e^{\alpha \int_0^s \lambda(u)\,du}\,ds \\
&\leq \alpha^{-1}\, |x(t) - y(t)|_{C_g}\, e^{\alpha \int_0^t \lambda(s)\,ds}, \quad t \in [0,T).
\end{aligned}$$

From this sequence of inequalities we obtain, taking supremum on $[0,T)$, after multiplying by $\exp\{-\alpha \int_0^t \lambda(s)\,ds\}$,

$$|Ux - Uy|_{C_g} \leq \alpha^{-1}\, |x - y|_{C_g}. \tag{2.20}$$

From (2.20) we see that the operator U is a contraction on C_g.

This concludes the proof of Theorem 2.3, according to the Banach's fixed-point theorem.

Remark 2.3 *Condition (2.16) represents a restriction on the growth of $f(t,\theta)$ as $t \to T$. We point out the fact that this restriction is imposed only for the reason to assure that the operator U, defined on the whole space C_g, $g(t) = \exp\{\alpha \int_0^t \lambda(s)\,ds\}$, will be acting on C_g. If instead of (2.16) we would have assumed*

$$|f(t,\theta)|_E \leq \mu(t) \in C([0,T),E),$$

then the proof should have been conducted using the function $\bar{\lambda}(t) = \max\{\lambda(t), \mu(t)\}$, $t \in [0,T)$.

Another functional equation we shall treat in this section is the classical Volterra equation

$$u(t) = g(t) + \int_0^t f(t,s,u(s))\,ds, \quad t \in [0,T],\ T > 0. \tag{2.21}$$

The underlying space will be the Lebesgue space $L^p([0,T],R^n)$, $p > 1$. The following assumptions will be made on equation (2.21):

(1) $f : [0,T] \to R^n$ is a measurable function, such that

$$\int_0^T \left(\int_0^S |f(t,s,0)|^p\,ds \right) dt < +\infty; \tag{2.22}$$

(2) A Lipschitz-type condition of the form

$$|f(t,s,u) - f(t,s,v)| \leq L(t,s)\,|u-v|, \tag{2.23}$$

for u and v in R^n, almost everywhere in $[0,T] \times [0,T]$ is verified, with $L(t,s)$ satisfying

$$\int_0^T \left[\int_0^t L^q(t,s)\,ds \right]^{\frac{p}{q}} \mathrm{dt} < \infty, \quad p^{-1} + q^{-1} = 1. \tag{2.24}$$

In order to formulate and prove the existence and uniqueness of the solution of equation (2.21), we need to introduce in $L^p([0,T],R^n)$ an equivalent norm to the classical one. This will constitute a change in the manner used by Bielecki. The new norm has been introduced by Kwapisz [304].

Namely, with $\omega : [0,T] \to R_+$, a continuous and positive function, one denotes

$$|u|_{p,\omega} = \left[\sup \left\{ \omega^{-1}(t) \int_0^t |u(s)|^p\,ds;\, t \in [0,T] \right\} \right]^{\frac{1}{p}}. \tag{2.25}$$

The following double inequality can be easily obtained for $|u|_{p,\omega}$,

$$c_1\,|u|_p \leq |u|_{p,\omega} \leq c_2\,|u|_p, \tag{2.26}$$

where

$$c_1 = [\sup\{\omega(t) \colon t \in [0,T]\}]^{-\frac{1}{p}},$$

and

$$c_2 = [\inf\{\omega(t) \colon t \in [0,T]\}]^{-\frac{1}{p}}.$$

Theorem 2.4 *Consider the integral equation of Volterra type (2.21) under assumptions (1) and (2).*

Then, there exists a unique solution $u(t) \in L^p([0,T],R^n)$ of equation (2.21).

Proof. Let us consider on the space $L^p([0,T],R^n)$, endowed with the norm $|\cdot|_{p,\omega}$, the operator

$$(Vu)(t) = g(t) + \int_0^t f(t,s,u(s))\,ds, \quad t \in [0,T]. \tag{2.27}$$

We have to prove now that the operator V, defined by (2.27) takes its values in $L^p([0,T],R^n)$. Since $g \in L^p$, the only remaining fact to be proven is that $\int_0^t f(t,s,u(s))\,ds \in L^p([0,T],R^n)$. This property is secured by assumptions (1) and (2), more specifically by the inequalities (2.22)–(2.24), applying Hölder's inequality.

We have to prove now that the operator V is a contraction on the space $L^p([0,T],R^n)$, with the norm $|.|_{p,\omega}$. We obtain from (2.27),

$$\begin{aligned}
|(Vu_1)(t) - (Vu_2)(t)|^p &\le \left[\int_0^t |f(t,s,u_1(s)) - f(t,s,u_2(s)|\,ds\right]^p \\
&\le \left[\int_0^t L(t,s)\,|u_1(s) - u_2(s)|\,ds\right]^p \\
&\le \left[\int_0^t L^q(t,s)\,ds\right]^{\frac{p}{q}} \cdot \int_0^t |u_1(s) - u_2(s)|^p\,ds \\
&\le M(t)\int_0^t |u_1(s) - u_2(s)|^p\,ds,
\end{aligned}$$

where

$$M(t) = \left[\int_0^t L^q(t,s)\,ds\right]^{\frac{p}{q}}, \quad t \in [0,T].$$

In what follows, we choose $\omega(t) = \exp\left(\lambda \int_0^t M(s)\,ds\right)$, $t \in [0,T]$, $\lambda > 1$. Integrating first and last terms from the sequence of inequalities, with respect to t, from 0 to $\tau \le T$, we obtain

$$\begin{aligned}
&\int_0^\tau |(Vu_1)(t) - (Vu_2)(t)|^p\,dt \\
&\quad \le \int_0^\tau \left[M(t)\int_0^t |u_1(s) - u_2(s)|^p\,ds\right] dt \\
&\quad = \int_0^\tau \left[M(t)\exp\left(\lambda\int_0^t M(s)\,ds\right)\exp\left(-\lambda\int_0^t M(s)\,ds\right)\right.
\end{aligned}$$

$$\times \int_0^t |u_1(s) - u_2(s)|^p \, ds \Bigg] \, dt$$
$$\leq |u_1 - u_2|_{p,\omega}^p \int_0^\tau M(t) \exp\left(\lambda \int_0^t M(s)\, ds\right) dt$$
$$\leq \lambda^{-1} |u_1 - u_2|_{p,\omega}^p \exp\left(\lambda \int_0^\tau M(s)\, ds\right).$$

This means that

$$\exp\left(-\lambda \int_0^\tau M(s)\, ds\right) \int_0^\tau |(Vu_1)(t) - (Vu_2)(t)|^p \, dt \leq \lambda^{-1} |u_1 - u_2|_{p,\omega}^p,$$

$u_1, u_2 \in L^p([0,T], R^n)$.

Taking supremum on the left-hand side, with respect to $\tau \in [0,T]$, one obtains

$$|Vu_1 - Vu_2|_{p,\omega}^p \leq \lambda^{-1} |u_1 - u_2|_{p,\omega}^p, \tag{2.28}$$

which is equivalent to

$$|Vu_1 - Vu_2|_{p,\omega} \leq \lambda^{-\frac{1}{p}} |u_1 - u_2|_{p,\omega}, \tag{2.29}$$

for $u_1, u_2 \in L^p([0,T], R^n)$.

Inequality 2.29 proves that the operator V, defined by (2.27) is a contraction mapping of the Banach space $L^p([0,T], R^n)$ into itself.

This ends the proof of Theorem 2.4.

In concluding this section, we will illustrate another case of global existence, the result being obtained by means of the fixed theorem of Tychonoff. The statement and proof of the existence result are given in Corduneanu [120].

Theorem 2.5 *Let us consider the functional differential equation*

$$\dot{x}(t) = (Vx)(t), \quad t \in [0,T), \tag{2.30}$$

with the initial condition $x(0) = x^0 \in R^n$, *under the following assumptions:*

(1) *V is a causal, continuous operator from $C([0,T), R^n)$ into $L^1_{\text{loc}}([0,T), R^n)$.*
(2) *There exists real-valued functions $A(t)$ and $B(t)$, defined on $[0,T)$, with $A(t)$ continuous and positive and $B(t)$ locally integrable, such that*

$$\int_0^t B(s)\, ds \leq A(t) - A(0), \quad t \in [0,T), \tag{2.31}$$

while $x(t) \in C([0,T),R^n)$ and $|x(t)| \leq A(t)$ imply

$$|(Vx)(t)| \leq B(t), \quad a.e.\ on\ [0,T). \tag{2.32}$$

Then, under assumptions (1) and (2), there exists a solution $x(t) \in C([0,T),R^n)$ of our problem, provided $|x^0| \leq A(0)$. This solution will satisfy the estimate $|x(t)| \leq A(t)$, $t \in [0,T)$.

We will now consider an illustration of Theorem 2.5, to the case of linear functional differential equations of the form

$$\dot{x}(t) = (Lx)(t) + f(t), \quad t \in [0,T), \tag{2.33}$$

under the following assumptions:

(1) $L : C([0,T),R^n) \to C([0,T),R^n)$ is a linear, causal and continuous operator.
(2) $f(t) \in L^1_{\text{loc}}([0,T),R^n)$.

We will now prove the existence, on the whole interval $[0,T)$, of a solution to (2.33) by checking the validity of the assumptions of Theorem 2.5.

Since assumption (1) in the theorem is obviously satisfied, there remains to construct the functions $A(t)$ and $(B(t)$.

Since L is a continuous operator, we can write this condition in the form $|Lx|_{L^1} \leq \alpha(t)\,|x|_C$, $t \in [0,T)$, with $\alpha(t) > 0$ a nondecreasing function on $[0,T)$. Hence, with $B(t) = \alpha(t)A(t) + |f(t)|$, one should have

$$\int_0^t [\alpha(s)A(s) + |f(s)|]\,ds \leq A(t) - A(0), \quad t \in [0,T],$$

choosing $x^0 = \theta \in R^n$. The general case, with $|x^0| < A(0)$, does not present any difficulty. This inequality can be rewritten as follows:

$$A(t) \geq A(0) + \int_0^t |f(s)|\,ds + \int_0^t \alpha(s)A(s)\,ds. \tag{2.34}$$

Inequality 2.34 will be the consequence of the similar inequality of the form

$$A(t) \geq c(t) + \int_0^t \alpha(s)A(s)\,ds, \quad t \in [0,T), \tag{2.35}$$

with $c(t) = A(0) + \int_0^t |f(s)|\,ds > 0$. Equation (2.35) can be reduced to an equivalent form by the substitution

$$y(t) = \int_0^t \alpha(s)A(s)\,ds, \quad t \in [0,T). \tag{2.36}$$

One obtains $\dot{y}(t) = \alpha(t)A(t)$, $t \in [0,T)$, which leads, together with (2.35) to the inequality

$$\dot{y}(t) \geq \alpha(t)y(t) + \alpha(t)c(t), \quad t \in [0,T). \tag{2.37}$$

Since $y(0) = 0$, according to (2.36), we observe that the solution of the equation $\dot{y}(t) = \alpha(t)y(t) + \alpha(t)c(t)$, with $y(0) = 0$, given by

$$y(t) = e^{\int_0^t \alpha(s)\,ds} \int_0^t e^{-\int_0^s \alpha(u)\,du} \alpha(s)c(s)\,ds,$$

is positive on $[0,T)$. Therefore, according to (2.37), $\dot{y}(t)$ is positive on $[0,T)$ because both terms on the right-hand side are positive. Since (2.36) implies $A(t) = \frac{\dot{y}(t)}{\alpha(t)} > 0$ on $[0,T)$, the constructed functions $A(t)$ and $B(t)$, $t \in [0,T)$, satisfy all requirements of Theorem 2.5.

This conclusion ends the discussion of equation (2.33), having proven the global existence of solutions.

We ask the readers to apply Theorem 2.5 in case of integro-differential equations of the form

$$\dot{x}(t) = f\left(t, x(t), \int_0^t k(t,s,x(s))\,ds\right), \quad t \in [0,T),$$

under initial condition $x(0) = x^0 \in R^n$.

2.3 EXISTENCE FOR A SECOND-ORDER FUNCTIONAL DIFFERENTIAL EQUATION

In this section we will consider the second-order functional differential equation, which involves a linear causal operator, namely

$$\ddot{x}(t) + (L\dot{x})(t) = (Nx)(t), \quad t \in [0,T), \tag{2.38}$$

with the initial conditions

$$x(0) = x^0 \in R^n, \quad \dot{x}(0) = v^0 \in R^n, \tag{2.39}$$

under the following hypotheses on the operators L and N. Let us point out that by the substitution

$$\dot{x} = y \Longrightarrow x(t) = x^0 + \int_0^t y(s)\,ds, \quad t \in [0,T), \tag{2.40}$$

equation (2.38) becomes of the first order,

$$\dot{y}(t) + (Ly)(t) = (\widetilde{N}y)(t), \quad t \in [0,T), \tag{2.41}$$

with $\widetilde{N}$ determined by N, on behalf of (2.40).

While $L : L^2_{\text{loc}}([0,T),R^n) \to L^2_{\text{loc}}([0,T),R^n)$ is assumed linear, continuous, and causal, N is generally assumed to be nonlinear, acting also on $L^2_{\text{loc}}(R_+,R^n)$. One may consider N as a perturbing operator of the linear functional differential equation $\ddot{x}(t) + (L\dot{x})(t) = f(t)$.

The theory of linear equations of the form (2.41) is presented, for instance, in Corduneanu [149]. In particular, a variation of parameters formula is valid. In the case of equation (2.41), by applying this formula one obtains an equivalent integral equation to problem (2.38) and (2.39) on $[0,T)$:

$$x(t) = X(t,0)x^0 + \int_0^t X(t,s)v^0\,ds + \int_0^t X(t,s)\int_0^s (Nx)(u)\,du\,ds, \tag{2.42}$$

with $X(t,s)$ the fundamental matrix defined by the formula

$$X(t,s) = I + \int_s^t \tilde{k}(t,u)\,du, \tag{2.43}$$

$0 \le s \le t < T$, with $\tilde{k}$ the resolvent kernel associated to $k(t,s)$ from the relationship

$$\int_0^t (Lx)(s)\,ds = \int_0^t k(t,s)x(s)\,ds, \quad t \in [0,T), \tag{2.44}$$

which is valid for any $x \in L^2_{\text{loc}}([0,T),R^n)$, as shown in the previously quoted reference, or in the joint paper by Corduneanu and Mahdavi [181].

It is now obvious that we have to concentrate on the integral equation (2.42) and prove existence results.

A first case which can be relatively simply clarified corresponds to the situation when the operator N is also linear. Then, we can write a formula similar to (2.44):

$$\int_0^t (Nx)(s)\,ds = \int_0^t k_0(t,s)x(s)\,ds, \quad t \in [0,T). \tag{2.45}$$

Substituting in (2.44), and using the notation

$$k_1(t,s) = \int_s^t X(t,u)k_0(u,s)\,du, \quad 0 \le s \le t < T, \tag{2.46}$$

we can rewrite (2.42) as

$$x(t) = f(t) + \int_0^t k_1(t,s)x(s)\,ds, \quad t \in [0,T), \tag{2.47}$$

where

$$f(t) = X(t,0)x^0 + \int_0^t X(t,s)v^0\,ds, \quad t \in [0,T). \tag{2.48}$$

We shall make one more assumption, namely,

$$k_0(t,s) \in L^2_{\text{loc}}(\Delta, \mathcal{L}(R^n,R^n)), \tag{2.49}$$

with $\Delta = \{0 \le s \le t < T\}$.

Returning to the linear integral equation (2.47), we derive from (2.46) the property $k_1(t,s) \in L^2_{\text{loc}}(\Delta, \mathcal{L}(R^n,R^n))$. This property of $k_1(t,s)$ allows us to place ourselves in the classical framework of the theory of integral equations (see, e.g., Tricomi [520]).

The discussion carried above about problem (2.38) and (2.39), equivalent to equation (2.47), leads us to the following existence result.

Proposition 2.1 *Consider the integral equation (2.42), equivalent to problem (2.38) and (2.39), under the following assumptions:*

(1) *The operators L and N are linear, causal, and continuous, acting on the space $L^2_{\text{loc}}(R_+,R^n)$.*
(2) *The associated kernels $k(t,s)$ and $k_0(t,s)$ from formulas (2.44) and (2.45) satisfy the conditions*

$$k(t,s) \in L^\infty_{\text{loc}}(\Delta, \mathcal{L}(R^n,R^n)),$$
$$k_0(t,s) \in L^2_{\text{loc}}(\Delta, \mathcal{L}(R^n,R^n)),$$

and

$$|(Nx)(t) - (Ny)(t)| \le \lambda(t)\,|x(t) - y(t)|, \text{ a.e. on } R_+,$$

where $\lambda : R_+ \to R_+$ is a nondecreasing map. Also, $X(t,s)$ satisfies

$$\int_0^t |X(t,s)|\, ds \leq M < \infty, \quad t \in R_+.$$

Then, there exists a unique solution to problem (2.38) and (2.39), or equivalently to the integral equation (2.42). This solution is defined on R_+, takes its values in R^n and is locally absolutely continuous on R_+.

Proof. We shall begin by constructing the sequence of successive approximations, starting from equation (2.42) and taking into account the notation (2.48). This sequence is defined by the recurrent formula

$$x_{k+1}(t) = f(t) + \int_0^t X(t,s) \int_0^s (Nx_k)(u)\, du\, ds, \quad k \geq 0, \tag{2.50}$$

the first approximation being $x_0(t) \in L^2_{\text{loc}}(R_+, R^n)$, chosen arbitrarily. We shall choose a continuous function.

An important observation we have to make is that all the approximants $x_k(t)$, $k \geq 1$, are absolutely continuous (hence, continuous) functions on R_+. Therefore, instead of dealing with the convergence of $L^2_{\text{loc}}(R_+, R^n)$, we can use the more common convergence of the space $C(R_+, R^n)$, which means the uniform convergence on any interval $[0,T] \subset R_+$.

We point out the fact that the third property listed in assumption (2), in the statement of Proposition 2.1, is stronger than the continuity of N on $L^2_{\text{loc}}(R_+, R^n)$. We prefer to impose this condition, to achieve the convergence of successive approximations in $C(R_+, R^n)$, instead of $L^2_{\text{loc}}(R_+, R^n)$.

From equation (2.50) we obtain, for $k \geq 1$,

$$x_{k+1}(t) - x_k(t) = \int_0^t X(t,s) \int_0^s [N(x_k - x_{k-1})(u)]\, du\, ds, \quad t \in R_+. \tag{2.51}$$

This implies the inequality

$$|x_{k+1}(t) - x_k(t)| \leq \int_0^t |X(t,s)| \int_0^s |(Nx_k)(u) - (Nx_{k-1})(u)|\, du\, ds, \quad t \in R_+,$$

and after taking into account the third condition in (2), one can write

$$|x_{k+1}(t) - x_k(t)| \leq \int_0^t |X(t,s)| \int_0^s |\lambda(u)|\, |x_k(u) - x_{k-1}(u)|\, du\, ds, \quad k \geq 1.$$

This inequality can be further processed, leading, for $k \geq 1$, on behalf of assumption (2) to,

$$|x_{k+1}(t) - x_k(t)| \leq M \int_0^t \lambda(u)| \, |x_k(u) - x_{k-1}(u)| \, du, \quad t \in R_+. \tag{2.52}$$

From (2.52) we derive

$$|x_{k+1}(t) - x_k(t)| \leq M \int_0^t \lambda(s) \sup_{u \in [0,s]} |x_k(u) - x_{k-1}(u)| \, ds, \quad k \geq 1,$$

and denoting

$$y_k(t) = \sup_{u \in [0,t]} |x_k(u) - x_{k-1}(u)|, \quad k \geq 1,$$

one obtains

$$y_{k+1}(t) \leq M \int_0^t \lambda(s) y_k(s) \, ds, \quad t \in R_+, \, k \geq 1. \tag{2.53}$$

(2.53) leads, by induction, to the inequality

$$y_{k+1}(t) \leq A \frac{M^k}{k!} \left(\int_0^t \lambda(s) \, ds \right)^k t \in [0,T], \quad T < \infty, \tag{2.54}$$

where $A = \sup |x_1(t) - x_0(t)|, 0 \leq t \leq T$.

In conclusion, the sequence of successive approximations $\{x_k(t); \, k \geq 1\}$ is uniformly convergent on any finite segment $[0,T] \subset R_+$.

Now we return to equation (2.42), and notice that it is obtained from (2.50), letting $k \to \infty$. Of course, we have to keep in mind that the convergence in $C(R_+, R^n)$ is the uniform convergence on each $[0,T] \subset R_+$.

This ends the proof of Proposition 2.1, which regards the linear case.

In regard to the nonlinear case, which means that N is nonlinear on $L^2_{\mathrm{loc}}(R_+, R^n)$, we notice that the Lipschitz condition imposed on N, namely

$$|(Nx)(t) - (Ny)(t)| \leq \lambda(t) \, |x(t) - y(t)|, \tag{2.55}$$

is all we have used in the linear case, in Proposition 2.1, which helps us to avoid repeating the proof for nonlinear N.

More precisely, the proof of Proposition 2.1, allows us to state the following result without proof.

Theorem 2.6 *Let us consider problem (2.38) and (2.39), which is the same as solving the integral equation (2.42). Assume that the operator L is linear, causal and continuous on $L^2_{\text{loc}}(R_+, R^n)$, with N nonlinear, causal, and satisfying (2.55),* a.e. *on R_+. Moreover, assume that first inclusion in condition (2) of Proposition 2.1 holds true, as well as the inequality listed at the end of (2), with regard to $X(t,s)$.*

Then, there exists a unique solution of the problem, which can be constructed by the method of successive approximations defined by

$$x_{k+1}(t) = f(t) + \int_0^t X(t,s) \int_0^s (Nx_k)(s)\,ds, \quad k \geq 0. \tag{2.56}$$

Remark 2.4 *The uniqueness is proven in the classical manner, starting with a solution of (2.42), say $y(t)$, and estimating the differences $y(t) - x_k(t)$, $k \geq 1$. One obtains $y(t) = \lim x_k(t)$ as $k \to \infty$, uniformly on each $[0,T] \subset R_+$. Since the limit of a convergent sequence is unique, there results the uniqueness of the solution.*

The readers are invited to perform the adequate operations in estimating the differences $y(t) - x_k(t)$, with $y(t)$ a solution whose existence is assumed.

2.4 THE COMPARISON METHOD IN OBTAINING GLOBAL EXISTENCE RESULTS

The comparison method, in the study of ordinary differential equations, has been used extensively in the theory of existence, stability, and almost all chapters of this field. Later on, it also served in the investigation of various classes of functional differential equations.

In the framework adopted by Conti [110] (see also Sansone and Conti [489]), the comparison method is built up by the systematic use, under *the most general conditions* imposed on the equation and auxiliary items, of two concepts: the Liapunov's function/functional and the differential inequalities (Chaplyguin).

In this section we shall present this method in its original form, as it appears in the previously quoted references. We will also apply the comparison method to stability (Chapter 3). The method has been used extensively in the past 50 years. We invite the readers to search the list of references at the end of this book for the names Lakshmikantham and Leela, Matrosov, Brauer, Martynyuk. See also Corduneanu [123].

We will not adopt here the most general formulations encountered in the literature on this subject; instead, we present the results in a classical framework.

Consider the system of ordinary differential equations

$$\dot{x}(t) = f(t,x(t)), \quad t \in [t_0,T),\ T \leq \infty \tag{2.57}$$

where $f(t,x(t))$ is defined in a domain $\Omega \in R^{n+1}$, is continuous there, and further properties will be specified later (usually, local Lipschitz condition).

Let $V(t,x)$ be a real-valued map defined on Ω. We assume the differentiability of $V(t,x)$, at each point $(t,x) \in \Omega$.

It $v(t) = V(t,x(t))$, with $x(t)$ a solution of (2.57), defined on some interval (t_1,t_2), then

$$\dot{v}(t) = \frac{\partial V}{\partial t} + \left(\frac{\partial V}{\partial x}, f\right), \tag{2.58}$$

where $\frac{\partial V}{\partial x}$ denotes the gradient of V with respect to the x-variable, and

$$\left(\frac{\partial V}{\partial x}, f\right) = \frac{\partial V}{\partial x_1} f_1 + \cdots + \frac{\partial V}{\partial x_n} f_n, \quad (t,x) \in \Omega.$$

One denotes the right-hand side of (2.58) by

$$V^{\cdot}(t,x) = \frac{\partial V}{\partial t} + \left(\frac{\partial V}{\partial x}, f\right), \quad (t,x) \in \Omega, \tag{2.59}$$

and call it the derivative of $V(t,x)$ with respect to the system (2.57). This term is justified if we take into account formula (2.58), in which $v(t) = V(t,x(t))$, with $x(t)$ a solution of (2.57).

The key element of the *comparison method* is the differential inequality

$$\frac{\partial V}{\partial t} + \left(\frac{\partial V}{\partial x}, f\right) \leq \omega(t,V), \quad (t,x) \in \Omega, \tag{2.60}$$

where $\omega(t,y)$ is defined and continuous in a domain $\Delta \subset R^2$, satisfying also a local Lipschitz condition with respect to y.

These assumptions will guarantee local existence and uniqueness of a solution of the comparison equation

$$\dot{y} = \omega(t,y), \tag{2.61}$$

through any point $(t_0,y_0) \in \Delta$.

In order to be able to state the theorem of global existence for the system (2.57), we need the following elementary result on ordinary differential inequalities.

Proposition 2.2 *Consider the differential inequality*

$$\dot{y}(t) \leq \omega(t, y(t)), \tag{2.62}$$

under aforementioned assumptions of continuity in Δ *and locally Lipschitz. If* $y(t)$ *is a differentiable solution of inequality (2.62), such that* $y(t_0) = y_0$ *and* $\widetilde{y}(t)$ *is the solution of the comparison equation (2.61) with* $\widetilde{y}(t_0) = y_0 \geq y(t_0)$, *then*

$$y(t) \leq \widetilde{y}(t), \tag{2.63}$$

for all $t \geq t_0$ *for which both sides of (2.63) are defined.*

Proof. Assume that inequality (2.63) does not hold. Then we can find $t_1 > t_0$, such that

$$y(t_1) > \widetilde{y}(t_1). \tag{2.64}$$

Let us now consider the function $z(t) = y(t) - \widetilde{y}(t)$ and notice that $z(t_0) \leq 0$ while $z(t_1) > 0$. If τ is the greatest value in $[t_0, t_1]$, such that $z(\tau) = 0$, then we can find a compact subset $K \subset \Delta$ such that $(t, y(t))$ as well as $(t, \widetilde{y}(t))$ belong to K, for $t \in [\tau, t_1]$. Hence, $\dot{z}(t) = \dot{y}(t) - \dot{\widetilde{y}}(t) \leq \omega(t, y(t)) - \omega(t, \widetilde{y}(t)) \leq L[y(t) - \widetilde{y}(t)]$, $t \in [\tau, t_1]$, which means

$$\dot{z}(t) \leq Lz(t), \quad t \in [\tau, t_1], \tag{2.65}$$

where L is the Lipschitz constant. But (2.65) and $z(\tau) = 0$ imply $z(t) \leq z(\tau) exp\{L(t - \tau)\}$, $t \in [\tau, t_1]$, that is, $z(t) = 0$ for $t \in [\tau, t_1]$. This fact is in contradiction with $z(t_1) > 0$.

Proposition 2.2 is thus proven.

Corollary 2.3 *Consider system (2.57) and let* $V(t,x)$ *be a real-valued map defined on* Ω *and differentiable there. Assume that the derivative of* V *with respect to system (2.57) satisfies the differential inequality (2.60), with* $\omega(t,y)$ *satisfying the conditions of Proposition 2.2.*

Then, the solution $x(t) = x(t; t_0, x^0)$ *of system (2.57), defined for* $t \in [t_0, T)$ *and corresponding to the initial condition* $x(t_0; t_0, x^0) = x^0 \in R^n$, *satisfies the inequality*

$$V(t, x(t; t_0, x^0)) \leq y(t; t_0, y_0), \quad t \geq t_0, \tag{2.66}$$

for all t *such that both* $y(t; t_0, y_0)$ *and* $x(t; t_0, x^0)$ *are defined, provided* $V(t_0, x^0) \leq y_0$.

The proof of Corollary 2.3 follows from Proposition 2.2, due to the fact that denoting $v(t) = V(t,x(t;t_0,x^0))$ satisfies the comparison inequality (2.62).

We are now prepared to prove a general existence result for system (2.57). The domain of definition of the function $f(t,x)$ will be the half-space $\Omega = \{(t,x);\ t \geq 0,\ x \in R^n\}$. The condition imposed on the comparison function $\omega(t,y)$ will assure the existence of all its solutions on the positive semiaxis. In this manner, we shall obtain only solutions to (2.57), defined on the positive semiaxis.

Theorem 2.7 *Let us consider the ordinary differential system (2.57) in the half-space Ω, mentioned above, with f continuous and locally Lipschitz. Assume that $V(t,x)$ is differentiable from Ω into R_+ such that $V(t,x) \to \infty$ as $|x| \to \infty$, uniformly in t, on any compact set. Suppose also that the comparison inequality (2.60) is verified, with ω defined on $\Delta = \{(t,y);\ t \geq 0,\ y \geq 0\}$, continuous and locally Lipschitz, $\omega(t,0) = 0$, $t \in R_+$, whose solutions are defined on R_+.*

Then, each solution of (2.57), say $x(t;0,x^0)$, is defined on R_+.

Proof. One has to show that the function $x(t;0,x^0)$ is defined for all $t \geq 0$, and all $x^0 \in R^n$. As seen in Corollary 2.3, we will have

$$V(t,x(t;0,x^0)) \leq y(t;0,y_0), \tag{2.67}$$

for all $t > 0$, such that both sides of (2.67) are defined. Since by hypothesis $y(t;0,y_0)$ is defined for all $t \geq 0$, there remains to be proven that (2.67) cannot be violated for any $x^0 \in R^n$. From the conditions on f, we know that equation (2.57) has a solution on some interval $[0,T)$, $T < \infty$. If the graph $\{t,x(t;0,x^0), t \in [0,T)\}$ is bounded in R^{n+1}, then from the obvious inequality

$$|x(t;0,x^0) - x(s;0,x^0)| \leq \int_s^t |f(u,x(u;0,x^0))|\,du,$$

with $t,s \to T$, we derive on behalf of Cauchy criterion the existence of the finite limit $\lim_{t\to T} x(t;0,x^0) = \bar{x} \in R^n$.

Then, it would be possible to extend the solution $x(t;0,x^0)$ beyond T. One should simply apply the local existence result at the point $(T,\bar{x}) \in R^{n+1}$.

The conclusion of this construction is that no finite T can be such that $[0,T)$ is the maximal existence interval for the solution $x(t;0,x^0)$. Consequently, the solution $x(t;0,x^0)$, considered on its maximal interval of existence, is defined on R_+.

This ends the proof of Theorem 2.7.

Remark 2.5 *In many sources on ordinary differential equations, one finds different conditions on the function V, renouncing its differentiability. Instead, one assumes that $V(t,x)$ satisfies a local Lipschitz condition, which allows to deal with Dini's derivative numbers along the trajectories/solutions of the system under consideration. This approach can be found, for instance, in the papers by Corduneanu [114]. See also Sansone and Conti [489] and Yoshizawa [539].*

Corollary 2.4 *Consider system (2.57) in the half-space $\Omega = \{(t,x);\ t \geq 0,\ x \in R^n\}$. One assumes that*

$$|f(t,x)| \leq \omega(t,|x|), \quad t \geq 0,\ |x| > a,\ a > 0, \tag{2.68}$$

besides the properties of continuity and local Lipschitz. We also assume that $\omega(t,y)$ verifies the same conditions in Δ.

Then, if the comparison equation (2.62) has all its solutions, defined on R_+, system (2.57) enjoys the same property.

In such situations, one says that the solutions are continuable in the future.

We will conclude this section by deriving from Corollary 2.4, the classical criterion of Wintner for existence in the future.

Namely, if one chooses $\omega(t,y) = g(y)$, with $g(y)$ satisfying

$$\int^{\infty} \frac{du}{g(u)} = +\infty, \tag{2.69}$$

then (2.68) becomes $|f(t,x)| \leq g(|x|)$, $t \geq 0$, $x \in R^n$. The comparison equation (2.61) being $\dot{y} = g(y)$ has separable variables. One obtains

$$\int_{y_0>0}^{y(t)} \frac{du}{g(u)} = t,$$

a condition assuring that $y(t)$ cannot become unbounded for any finite t. Therefore, all solutions exist in the future.

2.5 A FUNCTIONAL DIFFERENTIAL EQUATION WITH BOUNDED SOLUTIONS ON THE POSITIVE SEMIAXIS

In this section we will investigate the existence of bounded solutions to the first-order equation

$$\dot{x}(t) + (Lx)(t) = f(x(t)) + g(t), \quad t \in R_+, \tag{2.70}$$

with $L: C(R_+,R^n) \to L^1_{\rm loc}(R_+,R^n)$ a linear, causal, and continuous operator satisfying other conditions, to be specified in the following text, as well as those related to f and g.

The existence can be reduced to the similar problem for integral equations. For instance, getting the inverse of the operator

$$x(t) \to \dot{x}(t) + (Lx)(t), \quad t \in R_+,$$

from $AC_{loc}(R_+,R^n)$, into $L^1_{loc}(R_+,R^n)$, which means to write the variation of constants formula

$$x(t) = X(t,0)x^0 + \int_0^t X(t,s)[-f(x(s)) + g(s)]\,ds, \quad t \in R_+, \tag{2.71}$$

assuming, of course, that f and g are giving sense to the integral in (2.71). Also, the initial condition

$$x(0) = x^0 \in R^n, \tag{2.72}$$

must be considered.

For the validity of the formula of variation of constants, see Corduneanu [149]. Hence, problem (2.70) and (2.72) is equivalent to the integral equation (2.71).

We will now state the result of existence for problem (2.70) and (2.72).

Theorem 2.8 *Consider system (2.70), with condition (2.72), under the following assumptions:*

(1) *The operator $L : C(R_+,R^n) \to L^1_{\rm loc}(R_+,R^n)$ is linear, causal, and continuous.*
(2) *L and $f : R^n \to R^n$ continuously are satisfying the positiveness condition*

$$\int_0^t < (Lx)(s) - f(x(s)), \quad x(s) > ds \geq 0, \ t \in R_+. \tag{2.73}$$

for any $x \in C(R_+,R^n)$, where $< \cdot, \cdot >$ denotes the scalar product (R^n).
(3) *The operator L satisfies the condition*

$$L[BC(R_+,R^n)] \subset M_0(R_+,R^n),$$

where BC denotes bounded and continuous, while $M_0(R_+,R^n)$ consists of those functions in $L^1_{\text{loc}}(R_+,R^n)$, such that

$$\int_t^{t+1} |x(s)|\, ds \to 0 \ \text{ as } \ t \to \infty, \tag{2.74}$$

the norm being $\sup \int_t^{t+1} |x(s)|\, ds$, $t \in R_+$.

(4) $g \in L^1(R_+,R^n)$.

Then, any solution of the problem is bounded on R_+, and its ω-limit set belongs to the ω-limit set of the solutions of the ordinary differential equation $\dot{y}(t) = f(y(t))$, $t \in R_+$.

Proof. Let $x(t) = x(t;t_0,x^0)$ be the solution of (2.70) and (2.72), which is defined on some interval $[0,T)$, $T \leq \infty$ (possibly, for small T only). On $[0,T)$, we multiply both sides of (2.70) scalarly by $x(t)$, and integrate on $[0,t)$, $t < T$. We can write the result of these operations as follows:

$$\begin{aligned} &\frac{1}{2}\frac{d}{dt}|x(t)|^2 + \int_0^t < (Lx)(s) - f(x(s)), x(s) > ds \\ &\quad = \frac{1}{2}|x^0|^2 + \int_0^t < g(s), x(s) > ds, \ t \in R_+. \end{aligned} \tag{2.75}$$

If we take into account equation (2.73), one obtains from (2.75)

$$|x(t)|^2 \leq |x^0|^2 + 2\int_0^t < g(x), x(s) > ds, \tag{2.76}$$

on $[0,T)$. Denoting $\sup\{|x(s)|; 0 \leq s \leq t < T\} = X(t)$, we derive from (2.76) the following inequality

$$X^2(t) \leq |x^0|^2 + 2X(t)\int_0^\infty |g(s)|\, ds, \tag{2.77}$$

on $[0,T)$. From (2.77), there results the estimate

$$X(t) \leq \int_0^\infty |g(s)|\, ds + \left[|x^0|^2 + \left(\int_0^\infty |g(s)|\, ds\right)^2\right]^{\frac{1}{2}},$$

also on $[0,T)$, from which follows

$$|x(t)| \leq \int_0^\infty |g(s)|\, ds + \left[|x^0|^2 + \left(\int_0^\infty |g(s)|\, ds\right)^2\right]^{\frac{1}{2}}, \tag{2.78}$$

on $[0,T)$. Because the right-hand side in (2.78) is independent of $t \in [0,T)$, there results (see the proof of Theorem 2.7) that the solution exists on R_+ and the estimate in (2.78) is valid on R_+.

Therefore, the boundedness of all solutions of problem (2.70) and (2.72), or equivalently to (2.71), is proven.

In order to get information with respect to the ultimate behavior of the solutions, more precisely to identify as much as possible the ω-limit sets of the solutions of (2.70) and (2.72), we need one more common property of all solutions to this problem.

Namely, we will show that each solution is actually uniformly continuous on R_+.

Let $x(t) = x(t;0,x^0)$, $t \in R_+$, $x^0 \in R^n$, be an arbitrary solution of (2.70) and (2.72). We first notice the fact that each solution has a nonempty ω-limit set, because of its boundedness on R_+.

For an arbitrary solution $x(t)$ of (2.70), we see that, according to our assumptions and because of its boundedness on R_+, one has

$$\dot{x}(t) \in L^\infty(R_+,R^n) \oplus M_0(R_+,R^n). \tag{2.79}$$

From (2.79), we derive a formula of the form $\dot{x}(t) = u(t) + v(t)$, $t \in R_+$, where $u \in L^\infty$ and $v \in M_0$. With regard to the first term, it is obvious from

$$x(t) - x(s) = \int_s^t u(\tau)\,d\tau + \int_s^t v(\tau)\,d\tau \tag{2.80}$$

that it is guaranteeing an estimate of the form $\left|\int_s^t u(\tau)\,d\tau\right| \leq A|t-s|$. With the second term in (2.80), we notice that the sequence $\{v_m(t);\, m \geq 1\}$, where $v_m(t) = v(t+m-1)$, $t \in [0,1]$, is a convergent sequence to $\theta \in L^1([0,1],R^n)$. Hence, it is a compact set in $L^1([0,1],R^n)$, and according to a compactness criterion due to Dunford and Pettis (see Theorem 2.2.3 in Corduneanu [135]) the integrals of the functions form a set which is equi-uniformly continuous. This condition is sufficient to assure the uniform continuity of the integrals $\left\{\int_0^t v(s)\,ds;\ v \in L^1([0,1],R^n)\right\}$, for all $v \in M_0$ occurring in the representation: $x(t) = \int_0^t [u(\tau)+v(\tau)]\,d\tau$. This argument concludes the discussion about the uniform continuity, on R_+, of each solution of (2.70).

The last step in the proof of Theorem 2.8 is proving that any ω-limit point of an arbitrary solution of (2.70) is an ω-limit point of a solution of the ordinary differential equation $\dot{y} = f(y(t))$, $t \in R_+$. We observe from (2.70), that the family of translates $\{x(t+h);\, h \in R_+\}$ of the solution $x(t)$ is compact in $C(R_+,R^n)$, that is, with respect to the uniform convergence on each compact interval of R_+. This follows from the fact, which is obvious from equation (2.70), that on each compact interval of R_+, the derivatives $\dot{x}(t)$, $\dot{x}(t+h)$,

$t+h \in R_+$, are uniformly bounded by a function which can be represented as the sum of two functions in M_0 and in L^∞. The proof of uniform continuity (in the text) relied on similar circumstances. Therefore, we can find a sequence $\{\tau_m; m \geq 1\} \subset R_+$, with $\tau_m \to \infty$ as $m \to \infty$, such that

$$\lim_{m \to \infty} x(t+\tau_m) = y(t), \quad t \in R_+. \tag{2.81}$$

We shall now prove that $y(t)$ is a solution of the ordinary differential equation/system $\dot{y}(t) = f(y(t))$, $t \in R_+$.

Integrating both sides of equation (2.70), with $s+\tau_m$ as argument (instead of s), from t to $t+h$, $h > 0$ small, one obtains

$$\int_t^{t+h} \dot{x}(s+\tau_m)\, ds + \int_t^{t+h} (Lx)(s+\tau_m)\, ds$$
$$= \int_t^{t+h} f(x(s+\tau_m))\, ds + \int_t^{t+h} g(s+\tau_m)\, ds.$$

We notice that the first integral equals $x(t+\tau_m-h) - x(t+\tau_m)$, while the second tends to zero $= \theta$, as $m \to \infty$, because $Lx \in M_0$. The same is true for the second integral on the right-hand side, due to the fact that $L^1 \subset M_0$. What remains from above equality as $m \to \infty$, is the equation

$$y(t+h) - y(t) = \int_t^{t+h} f(y(s))\, ds, \quad h > 0 \text{ small},$$

which is equivalent to (2.81), due to the arbitrariness of h.

This ends the proof of Theorem 2.8.

Remark 2.6 *Condition (2.73) is obviously verified when L is weakly monotone, that is,* $\int_0^t < (Lx)(s), x(s) > ds \geq 0$, $t \in R_+$, *which is less restrictive than* L^2*-monotonicity, while f satisfies* $\int_0^t < (f(x(s)), x(s) > ds \geq 0$, $t \in R_+$. *Let us mention that the strict monotonicity* $\int_0^t < (Lx)(s), x(s) > ds \geq \lambda \int_0^t |x(s)|^2\, ds$, $\lambda > 0$, *is acceptable, and such a condition can take place, in the scalar case, for a truly nonlinear function,* $f(x) = \lambda x - \alpha x^3$, $\alpha > 0$.

These considerations, leading to the result in Theorem 2.8, can be applied in case of other functional equations, similar to (2.70).

We shall state a result, regarding the equation

$$\dot{x}(t) + (L\dot{x})(t) = f(x(t)) + g(t), \quad t \in R_+. \tag{2.82}$$

Theorem 2.9 *Let us consider equation (2.82), under the following assumptions:*

(1) *L is a linear, continuous, and causal operator on the space* $L^2_{\text{loc}}(R_+,R^n)$.

(2) *The positiveness condition*

$$\int_0^t < (Lx)(s) + \eta x(s), \ \ x(s) > ds \geq 0, \ t \in R_+,$$

is valid for any $x \in L^2_{\text{loc}}(R_+,R^n)$, *where* $\eta \in R$ *is such that* $\eta < 1$.

(3) *The operator L takes the space* $L^2(R_+,R^n)$ *into* $M_0(R_+,R^n)$.

(4) $f : R^n \to R^n$ *is of gradient type, that is,*

$$f(x) = \nabla U(x), \quad U \subset C^{(1)}(R^n,R),$$

with

$$U(x) \to \infty \ as \ |x| \to \infty.$$

(5) $g \in L^2(R_+,R^n)$.

Then, equation/system (2.82) has all its solutions defined on R_+*; they are bounded there, while the* ω*-limit set of any solution belongs to the* ω*-limit set of solutions of the ordinary differential equation* $\dot{y}(t) + f(y(t)) = 0$, $t \in R_+$.

The proof is omitted, being similar to that of Theorem 2.8. We invite the readers to conduct this process as an exercise.

We conclude this section by noting that the theorems can be applied to integro-differential equations of the form

$$\dot{x}(t) + \int_0^t k(t,s)x(s)\,ds = f(x(t)) + g(t), \quad t \in R_+.$$

Several examples of this kind can be found in Cassago, Jr. and Corduneanu [92], or in Corduneanu [135, 149].

2.6 AN EXISTENCE RESULT FOR FUNCTIONAL DIFFERENTIAL EQUATIONS WITH RETARDED ARGUMENT

The functional differential equations with retarded/delayed argument have been treated in the book by Myshkis [411], which is the first book in the literature dedicated to a wide class of functional equations, as different with respect to the ordinary differential, integral or integro-differential ones.

As a short historical note, we point out that Krasovskii [299] has contributed to the theory of equations with delay by bringing the methods known for ordinary differential equations, particularly the Liapunov's functions and the ensuing stability theory. Hale [240] has also substantially contributed to this branch of investigation by introducing methods of functional analysis. Some of these methods and results will find their place in the forthcoming chapters of this book.

Returning to the case of functional equations with delay, we will use the notations introduced in the work of Hale and his followers. Namely, we shall deal with the equation

$$\dot{x}(t) = f(t, x_t), \text{ a.e.,} \quad t \in [0,T),\ T > 0, \tag{2.83}$$

with $x_t(s) = x(t+s), s \in [-h,0], h > 0, t \in [0,T)$, in the underlying real Banach space B, under the functional initial data

$$x(s) = x_0(s) \in C([-h,0],B). \tag{2.84}$$

Let us point out the fact that the notation, used in (2.83), is equivalent to the notation

$$\dot{x}(t) = (Fx)(t), \quad t \in [0,T), \tag*{(2.83)'}$$

provided F is a *causal operator*. Both notations are widely used in the literature and we shall choose, in what follows, the notation (2.83).

The existence theorem we shall prove below is like that in Corduneanu [135], and has features similar to those encountered in Lakshmikantham et al. [311], which provides several results.

Theorem 2.10 *Let us consider problem (2.83) and (2.84) in the real Banach space B, under the following assumptions:*

(1) *For each $t \in [0,T)$, the map $t \to f(t,\phi)$, from $[0,T) \times C([-h,0],B)$ into B, is continuous in (t,ϕ).*
(2) *The operator $F : C([0,T),B) \to C([0,T),B)$, defined by*

$$(Fx)(t) = f(t, x_t), \quad x \in C([0,T),B) \tag{2.85}$$

satisfies the Lipschitz-type condition

$$|(Fx)(t) - (Fy)(t)|_B \leq \lambda(t)\, |x(t) - y(t)|_B, \tag{2.86}$$

for $t \in [0,T)$, and $x, y \in C([-h,0],B)$, where $\lambda(t)$ is a positive nondecreasing function on $[0,T)$.

Then, there exists a unique solution $x(t) \in C([0,T),B)$ of problem (2.83) and (2.84).

Proof. A few explanations on the notations used before. By $C([0,T),B)$, one denotes the space of continuous maps from $[0,T)$ into B, with $T \leq \infty$, endowed with the topology of uniform convergence on any $[0,T_1] \subset [0,T)$. This is a Fréchet space, whose topology is determined by the family of semi-norm

$$|x|_m = \sup\{|x(t)|_B, t \in [0,t_m]\}, \tag{2.87}$$

where $t_m \to T$ as $m \to \infty$.

The initial condition (2.84) is used to determine $x_t(s) = x(t+s)$, with $s \in [-h,0]$.

We can now basically repeat other feature of the proof of Theorem 2.10 by using the Bielecki's approach from the proof of Theorem 2.3, which also relates to the case of Banach spaces.

Remark 2.7 *In Remark 2.3 to Theorem 2.3, we noticed the feature that an extra condition (3) appearing in the proof is used only to obtain the solution in the space C_g defined there. In Theorem 2.10, we did not specify the space to which the solution belongs, because we did not impose any growth condition on $f(t,x_t)$, as $t \to \infty$. So, the only information provided by Theorem 2.10 consists in the inclusion $x(t) \in C([0,T),B)$.*

Theorem 2.10 is a global existence result obtained under rather restrictive conditions. In the book by Corduneanu [135], one can find a local existence (not uniqueness) result by using the method of singular perturbations.

Consider again equation (2.83)′, which in case of causality of the operator F, can also be written in the form (2.83). Of course, the initial condition (2.84) is associated with the equation (2.83)′, in order to construct a (local) solution to this equation.

The following result is taken from the book mentioned above.

Theorem 2.11 *Consider equation (2.83)′, under the following conditions on the operator F:*

(1) *$F : C([t_0,T],B) \to C([t_0,T],B)$ is a causal continuous operator.*
(2) *F takes bounded sets in $C([t_0,T],B)$ into bounded sets.*
(3) *$x(t_0) = x^0 \in B$.*

Then, there exists $\delta > 0$, $t_0 + \delta \leq T$, such that equation (2.83)′, with the initial condition (3), has a solution defined on $[t_0, t_0+\delta]$.

This result, concerned with causal operators and with initial condition (3), though stated for point-wise initial datum, is actually covering the case of causal operators with delay, like those encountered in Theorem 2.10. The equivalent integral equation is of the form $(2.83)'$, with the operator F defined by (2.85).

More information about this type of problems can be found in Corduneanu [135, 149], Lakshmikantham et al. [311], Azbelev et al. [33], and in a large number of journal papers. See our list of references.

To conclude this section, we will mention a few types of functional differential equations involving the class of causal operators.

A first example is offered by the ordinary differential equations $\dot{x}(t) = f(t,x(t))$, with x, f belonging to a Banach space (including R^n, C^n, $n \geq 1$).

A second example is given by the "traditional" equations with delayed argument, of the form $\dot{x}(t) = f(t,x(t),x(t-h))$, $h > 0$, the initial condition having the form (2.84).

Another example is found in the theory of integro-differential equations, having the form $\dot{x}(t) = f(t,x(t)) + \int_{t_0}^{t} k(t,s,x(s))\,ds$, or, more general, $\dot{x}(t) = f\left(t,x(t),\int_{t_0}^{t} k(t,s,x(s))\,ds\right)$, in both cases the initial condition having the form $x(t_0) = x^0$. Of course, the theory depends on the choice of the underlying space.

All these types of equations belong to the type "causal," the procedures and results stated earlier being largely applicable.

An example of an equation with "infinite delay," is

$$\dot{x}(t) = f\left(t,x(t),\int_{-\infty}^{t} k(t,s)x(s)\,ds\right), \quad t \geq t_0,$$

the initial conditions being

$$x(t) = x^0(t), \quad t \leq t_0,\ x(t_0) = x^0.$$

For results regarding existence, uniqueness, estimates, or asymptotic behavior, see Lakshmikantham et al. [312], Lakshmikantham and Rao [316], as well as the references included in these books.

The history of such functional or functional differential equations began with Volterra [528], and has been successfully developed, including their applications in various fields, ever since.

The nonlinear functional equations of Volterra type, including the abstract Volterra equations (causal), constitute a conspicuous chapter and have many applications.

Concerning the nonlinear Fredholm equations of the form

$$x(t) = f(t) + \int_a^b k(t,s,x(s))\,ds,$$

with wide occurrence in applications, the theory is still far from offering handy results.

2.7 A SECOND-ORDER FUNCTIONAL DIFFERENTIAL EQUATION WITH BOUNDED SOLUTIONS ON THE POSITIVE SEMIAXIS

In this section we will consider the functional differential equation

$$\ddot{x}(t) + (Lx)\dot{x}(t) + \nabla F(x) = f(t), \quad t \in R_+, \tag{2.88}$$

with $x : R_+ \to R^n$, $t \in R_+$ and L a linear operator satisfying properties to be described below.

The problem we shall consider consists in finding solutions to (2.88), if possible defined on R_+, under initial conditions

$$x(0) = x^0 \in R^n, \quad \dot{x}(0) = v^0 \in R^n. \tag{2.89}$$

We shall look for such hypotheses on the data that will assure the existence of solutions defined on R_+.

Let us notice that equation (2.88) represents a multidimensional generalization of the classical Lieńard's equation of the nonlinear oscillations.

Before getting into details, let us point out the fact that a local existence result, say for $t \in [0,T)$, can be obtained if one relies on already known results. We shall apply Corollary 2.1.

Let us transform equation/system (2.88) into a first-order system, with $2n$ scalar equations. Denoting $\dot{x} = v$, and taking (2.88) into account, we can write the system

$$\dot{x}(t) = v(t), \quad \dot{v}(t) = -(Lx)(t)v(t) - \nabla F(x)(t) + f(t), \tag{2.90}$$

with the initial conditions derived from (2.89),

$$x(0) = x^0 \in R^n, \quad v(0) = v^0 \in R^n. \tag{2.91}$$

From (2.90), we obviously have a causal system, if $(Lx)(t)$ is causal. Hence, if we apply Corollary 2.1 to the system (2.90), we obtain the existence of a local solution to this system, satisfying also (2.91).

Therefore, with regard to equation (2.88), we can rely on the *existence of a local solution* $x(t)$, of class $C^{(2)}$ or $C^{(1)}$, depending on the other hypotheses. Under the second situation, when the solution is in $C^{(1)}$, it will be also *a.e.* differentiable.

More precisely, the local existence will be assumed by the following hypotheses:

(a) $L : C(R_+, R^n) \to C(R_+, R^n)$ is causal, continuous, and takes bounded sets into bounded sets.

(b) $F : R^n \to R$ is differentiable and coercive, that is, satisfies

$$F(x) \to \infty \text{ as } |x| \to \infty. \tag{2.92}$$

(c) $f \in C(R_+, R^n)$.

Remark 2.8 *We notice the fact that the local solution* $x(t)$, $t \in [0,T)$, $T > 0$, *is such that* $\dot{x}(t) \in AC([0,T], R^n)$, *while* $\ddot{x}(t) \in L^1([0,T), R^n)$.

In order to deal with the global existence problem and boundedness of the solution to (2.88), we will make two more assumptions on the data.

(d) $L : C(R_+, R^n) \to C(R_+, R^n)$ satisfies the positivity condition

$$\int_0^t < (Lx)y(s), y(s) > ds \geq \delta \int_0^t |y(s)|^2 ds,$$

for all $t > 0$, and $x, y \in C(R_+, R^n)$, where $\delta > 0$ is constant.

(e) $f \in L^2(R_+, R^n)$.

We observe that (e) and (c) are independent of each other. We need (e) in order to obtain the boundedness on R_+.

The following result can now be proven.

Theorem 2.12 *Assume that hypotheses (a), (b), (d), and (e) hold true.*

Then, there exists a solution $x(t)$ *of equation (2.88), with the initial condition (2.89), for any* $x^0, v^0 \in R^n$, *defined on* R_+, *and possessing the following properties:*

(1) $x(t) \in BC(R_+, R^n)$ *is continuously differentiable and uniformly continuous on* R_+.

(2) $\dot{x}(t) \in BC(R_+, R^n) \cap L^2(R_+, R^n)$ *is locally absolutely continuous on* R_+.

(3) $\ddot{x}(t)$ *is locally integrable, and* $\ddot{x}(t) \in L^\infty(R_+, R^n) \cup L^2(R_+, R^n)$, *provided* $BC(R_+, R^n)$ *is invariant for L. In this case,* $\dot{x}(t) \to \theta$ *as* $t \to \infty$.

Proof. Let $x(t)$, $t \in [0,T)$ be a local solution of our problem. In order to show that any local solution can be extended on the whole semiaxis R_+, we need to obtain an estimate from which we can derive the possibility of continuing such a solution, on R_+. To this aim, let us multiply scalarly both sides of equation (2.88) by $\dot{x}(t)$, then we integrate the result from 0 to $t > 0$,

$$\frac{1}{2}\frac{d}{dt}|\dot{x}(t)|^2 + < (Lx)(s)\dot{x}(s), \dot{x}(s) > ds + F(x(t))$$
$$= \frac{1}{2}|v_0|^2 + F(x^0) + \int_0^t < f(s), \dot{x}(s) > ds.$$

Taking hypothesis (d) into account, we obtain the following from the above inequality

$$\frac{1}{2}|\dot{x}(t)|^2 + \delta \int_0^t |\dot{x}(s)|^2 ds + F(x(t))$$
$$\leq \frac{1}{2}|v_0|^2 + F(x^0) + \frac{\epsilon}{2}\int_0^t |\dot{x}(s)|^2 ds + \frac{1}{2\epsilon}\int_0^t |f(s)|^2 ds,$$

where $\epsilon > 0$ is arbitrary. If we choose $\epsilon < 2\delta$, then the last inequality above implies

$$\frac{1}{2}|\dot{x}(t)|^2 + \left(\delta - \frac{\epsilon}{2}\right)\int_0^t |\dot{x}(s)|^2 ds + F(x(t))$$
$$\leq \frac{1}{2}|v_0|^2 + F(x^0) + \frac{1}{2\epsilon}\int_0^t |f(s)|^2 ds, \quad t \in [0,T). \tag{2.93}$$

Let us denote

$$C = C(x^0, v^0, f) = \frac{1}{2}|v_0|^2 + F(x^0) + \frac{1}{2\epsilon}\int_0^\infty |f(s)|^2 ds.$$

Then (2.93) becomes

$$\frac{1}{2}|\dot{x}(t)|^2 + \left(\delta - \frac{\epsilon}{2}\right)\int_0^t |\dot{x}(s)|^2 ds + F(x(t)) \leq C, \quad t \in [0,T). \tag{2.94}$$

The following consequences of inequality (2.94) shall be stated in the following text.

First, from (2.94) we derive the fact that the solution $x(t)$ can be extended beyond T. This fact is due to the existence of the $\lim x(t)$, as $t \uparrow T$, which follows from $|\dot{x}(t)|^2 \leq 2C$, whence $|x(t) - x(\tau)| \leq (2C)^{\frac{1}{2}}|t-\tau|$, with $t,\tau \in [0,T)$. Consequently, $x(t)$ can be extended to the right, on $[0,T')$, $T' > T$. As we have seen in the proof of Theorem 2.7, the extension is possible, on the whole R_+, which is needed to obtain global existence.

Second, the boundedness of $x(t)$ on R_+ follows from (2.94), due to the fact that it implies, on each $[0,T)$, $T>0$, $F(x(t)) \le C = constant$, and on behalf of (2.92) we obtain the boundedness of $x(t)$, $x(t) \in BC(R_+,R^n)$. The uniform continuity of $x(t)$, on R_+, follows from the inequality

$$|x(t)-x(\tau)| = \left|\int_\tau^t \dot{x}(s)\,ds\right| \le \left(\int_0^\infty |\dot{x}(s)|^2\,ds\right)^{\frac{1}{2}} |t-\tau|,$$

because (2.94) is valid for $t \in R_+$ and implies $|\dot{x}(t)|^2 \in L^2(R_+,R^n)$.

Third, we notice from (2.88) and properties (1) and (2), that the inclusion shown for $\ddot{x}(t)$ in property (3) takes place *when* $BC(R_+,R^n)$ *is invariant* for *the operator* L. We need only to prove that $\dot{x}(t)$ tends to $\theta \in R^n$, as $t \to \infty$. We shall achieve this by applying a frequently used lemma due to Barbălat [42]. In order to perform the last step in the proof, let us notice that $\dot{x}(t) \in BC(R_+,R^n)$, $\dot{x}(t) \in L^2(R_+,R^n)$ and L leaves invariant the space $BC(R_+,R^n)$. This last property, of $\dot{x}(t)$, combined with equation (2.88), show that the inclusion in assumption (3) of Theorem 2.12, is valid.

Indeed, the term $(Lx)(t)\dot{x}(t) \in BC(R_+,R^n)$, because $\dot{x} \in BC(R_+,R^n)$ and L leaves this space invariant. The third term in the left hand side of (2.88) is bounded on R_+, due to the continuity of the first order derivatives of $F(x)$, and the boundedness of $x(t)$, $t \in R_+$. Hence, the second and third term in (2.88) are bounded on R_+, that is, belong to $BC \subset L^\infty$, while $f \in L^2(R_+,R^n)$, according to (e).

Then, there results the inclusion in property (3), from which we can derive $\ddot{x}(t) = u(t) + v(t)$, with $u \in L^\infty(R_+,R^n)$, and $v \in L^2(R_+,R^n)$. One has $\dot{x}(t) - \dot{x}(\tau) = \int_\tau^t \ddot{x}(s)\,ds$, and

$$|\dot{x}(t)-\dot{x}(\tau)| \le \left|\int_\tau^t |\ddot{x}(s)|\,ds\right| \le \left|\int_\tau^t (|u(s)|+|v(s)|)\,ds\right|.$$

But $u \in L^\infty(R_+,R^n)$ leads to the estimate $\left|\int_\tau^t |u(s)|\,ds\right| \le c_1|t-\tau|$, while $v \in L^2(R_+,R^n)$ implies $\left|\int_\tau^t |v(s)|\,ds\right| \le \left(\int_0^\infty |v(s)|^2\,ds\right)^{\frac{1}{2}} |t-\tau|^{\frac{1}{2}}$. Hence, $|\dot{x}(t) - \dot{x}(\tau)| \le c_1|t-\tau| + c_2|t-\tau|^{\frac{1}{2}}$, where c_1, c_2 are constants, $c_2 = |v|_{L^2}$. The last inequality proves the uniform continuity, on R_+, of the derivative $\dot{x}(t)$. Further on, $|\dot{x}(t)|$ is uniformly continuous on R_+, because $||\dot{x}(t)-\dot{x}(\tau)|| \le |\dot{x}(t)-\dot{x}(\tau)|$. Taking into account the boundedness of $\dot{x}(t)$ on R_+, we find out that the function $|\dot{x}(t)|^2$ is uniformly continuous on R_+.

This statement follows from the following relationships:

$$\begin{aligned}\left||\dot{x}(t)|^2-|\dot{x}(\tau)|^2\right| &= (|\dot{x}(t)|+|\dot{x}(\tau)|)(|\dot{x}(t)|-|\dot{x}(\tau)|)\\ &\le 2\sup_{t\in R_+}|\dot{x}(t)|\,|\dot{x}(t)-\dot{x}(\tau)|.\end{aligned}$$

We are finally able to conclude that $|\dot{x}(t)|^2$ is integrable on R_+ and is uniformly continuous. These two conditions/properties satisfied by $|\dot{x}(t)|^2$, imply by Barbălat's lemma, $|\dot{x}(t)|^2 \to 0$, as $t \to \infty$, which is the same as $\dot{x}(t) \to \theta$, as $t \to \infty$.

All properties stipulated in Theorem 2.12 being established, we can conclude the proof of Theorem 2.12.

2.8 A GLOBAL EXISTENCE RESULT FOR A CLASS OF FIRST-ORDER FUNCTIONAL DIFFERENTIAL EQUATIONS

This section is concerned with the existence of solutions for the equation

$$\dot{x}(t) = (Lx)(t) + (Nx)(t), \quad t \in [0,T), \tag{2.95}$$

which is a special form of the general equation $\dot{x}(t) = (Vx)(t)$, $t \in [0,T)$, when we want to emphasize the fact that we start with a linear equation $\dot{x}(t) = (Lx)(t)$, and perturb the right-hand side by, in general, a nonlinear term.

Let us point out that a similar equation to (2.95), but of second order, has been considered in Section 2.3, when global results were obtained under assumptions leading to existence and uniqueness. This time we shall apply a result from Corduneanu [135; Theorem 3.2.2], which concerns the functional equation $x(t) = (Vx)(t)$. Namely, we shall rely on

Theorem 2.13 *Consider the functional equation*

$$x(t) = (Vx)(t), \quad t \in [0,T), \tag{2.96}$$

with $L^2_{\text{loc}}([0,T),R^n)$ as underlying space. Assume the following conditions on V:

(1) *$V : L^2_{\text{loc}}([0,T),R^n) \to L^2_{\text{loc}}([0,T),R^n)$ is a causal, continuous, and compact operator.*

(2) *There exists a continuous map $A : [0,T) \to R_+ \setminus 0$, such that for $x(t) \in L^2_{\text{loc}}([0,T),R^n)$, with*

$$\int_0^t |x(s)|^2 ds \le A(t), \quad t \in [0,T), \tag{2.97}$$

one has

$$\int_0^t |(Vx)(s)|^2 ds \le A(t), \quad t \in [0,T). \tag{2.98}$$

Then, equation (2.96) has a solution $x(t) \in L^2_{\text{loc}}([0,T),R^n)$, satisfying (2.97).

The proof of Theorem 2.13 is provided in the aforementioned reference.

We shall apply Theorem 2.13 to obtain a result related to the problem (2.95), with the initial condition $x(0) = x^0 \in R^n$, to the equivalent integro-functional equation

$$x(t) = X(t,0)x^0 + \int_0^t X(t,s)(Nx)(s)\,ds, \quad t \in [0,T). \tag{2.99}$$

In (2.99), $X(t,s)$, $0 \leq s \leq t < T$, is the fundamental matrix associated with the equation $\dot{x}(t) = (Lx)(t)$, $t \in [0,T)$. For details, see the beginning of the Section 2.3.

Consequently, in order to work in the space $L^2_{\text{loc}}([0,T),R^n)$, we will consider as V in (2.96) the operator

$$(Vx)(t) = X(t,0)x^0 + \int_0^t X(t,s)\,(Nx)(s)\,ds, \quad t \in [0,T). \tag{2.100}$$

It is obvious from (2.100) that V is defined on $L^2_{\text{loc}}([0,T),R^n)$. The values of this operator are taken in the space $AC_{\text{loc}}([0,T),R^n) \subset C([0,T),R^n)$. In order to satisfy the conditions required for the operator V in Theorem 2.13, we need to impose adequate conditions on the operators L and N as follows:

(1) The operator L is continuous and causal on the whole space $L^2_{\text{loc}}([0,T),R^n)$.
(2) The operator N acting on $L^2_{\text{loc}}([0,T),R^n)$ is continuous, casual, and satisfies the growth condition

$$\int_0^t |(Nx)(s)|^2\,ds \leq \lambda(t)\int_0^t |x(s)|^2\,ds + \mu(t), \quad t \in [0,T), \tag{2.101}$$

with $\lambda(t)$ and $\mu(t)$ positive nondecreasing on $[0,T)$.

The following global existence result can be stated:

Theorem 2.14 *Consider the integro-functional equation (2.99) under assumptions (1) and (2).*

Then, there exists a solution of equation (2.99); therefore, also a solution for problem (2.95), $x(0) = x^0 \in R^n$, defined on the interval $[0,T)$, with values in R^n, such that the following properties hold:

1. *The solution is locally absolutely continuous on $[0,T)$ and satisfies (2.99) and (2.95) only* a.e. *on $[0,T)$.*
2. *The solution satisfies the inequality $|x(t)| \leq A(t)$ on $[0,T)$ with a function $A(t)$ to be specified in the course of the proof.*

Proof. We consider equation (2.99), for which the associated operator V is given by (2.100). We have to check the validity of the conditions required by Theorem 2.13 for the operator V given by (2.100).

It is obvious from (2.100) that V is a causal operator on $L^2_{\text{loc}}([0,T),R^n)$, because it appears that it is the sum of two causal operators, taking into account the assumed causality of N. The continuity of V, which means continuity on each interval $[0,T_1] \subset [0,T)$, follows from the continuity of N, combined with that of the linear integral operator

$$y(t) \to \int_0^t X(t,s)\,y(s)\,ds, \tag{2.102}$$

on $L^2_{\text{loc}}([0,T),R^n)$, which follows from the square integrability, on each triangle $0 \leq s \leq t \leq T_1$, of $X(t,s)$,

$$\int_0^{T_1} dt \int_0^t |X(t,s)|^2\,ds = k(T_1) < +\infty. \tag{2.103}$$

We know that $X(t,s)$ is locally absolutely continuous for almost all $s \in [0,t]$, while the formula (2.43), with $\tilde{k} \in L^2(\Delta_1, \mathcal{L}(R^n,R^n))$, $\Delta_1 = \{(s,t); 0 \leq s \leq t \leq T_1, T_1 < T\}$, proves its continuity in s.

From (2.100), taking into account the growth condition (2.101), we obtain the following sequence of inequalities

$$\begin{aligned} |(Vx)(t)|^2 &\leq 2|X(t,0)|^2\,|x^0|^2 + 2\int_0^t |X(t,s)|^2\,ds \int_0^t |(Nx)(s)|^2\,ds \\ &\leq 2|X(t,0)|^2\,|x^0|^2 + 2\gamma(t)\left[\lambda(t)\int_0^t |x(s)|^2\,ds + \mu(t)\right], \end{aligned}$$

where $\gamma(t) = \int_0^t |X(t,s)|^2\,ds$, $t \in [0,T)$. By integrating with respect to t, we obtain the following inequality

$$\begin{aligned} \int_0^t |(Vx)(t)|^2 \leq 2|x^0|^2 &\int_0^t |X(s,0)|^2\,ds \\ &+ 2\int_0^t \gamma(s)\,[\lambda(s)A(s) + \mu(s)]\,ds, \end{aligned} \tag{2.104}$$

taking only those $x(t)$ satisfying (2.97).

In order to satisfy (2.98) in condition (2) of Theorem 2.13, we need to impose the following inequality:

$$2|x^0|^2 \int_0^t |X(s,0)|^2\,ds + 2\int_0^t \gamma(s)\,\mu(s)\,ds + 2\int_0^t \gamma(s)\,\lambda(s)\,A(s)\,ds \le A(t), \quad t \in [0,T). \tag{2.105}$$

Inequality (2.105), for the unknown function $A(t)$, has the form

$$A(t) \ge \int_0^t \alpha(s)A(s)\,ds + \beta(t), \quad t \in [0,T), \tag{2.106}$$

with $\alpha(t)$ and $\beta(t)$ nonnegative locally integrable functions on $[0,T)$.

Such integral inequalities have been discussed in Corduneanu [149], where it has been shown to possess positive solutions on $[0,T)$.

Therefore, we can conclude the proof, observing that under the conditions of Theorem 2.13 there exists a positive function $A(t)$, as required in the theorem.

Remark 2.9 *The equation $\dot{x}(t) = (Vx)(t)$, $t \in [0,T)$ can be reduced to the form considered in Theorem 2.13, by integrating both sides from 0 to $t < T$, taking also into account the initial condition $x(0) = x^0 \in R^n$. The conditions to be imposed on the operator V are milder than those of Theorem 2.13. In the book by Corduneanu [149], one finds a set of conditions assuring the existence of the solutions. These conditions require that $V : C([0,T),R^n) \to L^1_{loc}([0,T),R^n)$ be causal and continuous, also satisfying the following condition: $|x(t)| \le A(t)$, $t \in [0,T)$ implies $|(Vx)(t)| \le B(t)$ a.e. on $[0,T)$, with $A(t)$ and $B(t)$ positive functions on $[0,T)$, verifying the inequality $A(t) - A(0) \ge \int_0^t B(s)\,ds$, $t \in [0,T)$. One also has to assume $|x^0| < A(0)$.*

Remark 2.10 *The formulation of the assumptions in Theorem 2.13 was such that $x^0 \in R^n$ was arbitrary, not having the restriction we mentioned at the end of Remark 2.9. We have though to point out that the construction of the auxiliary function $A(t)$ depends on x^0.*

We propose to the readers to formulate adequate conditions on the functions involved in the equation

$$\dot{x}(t) = f\left(t, x(t), \int_0^t k(t,s)x(s)\,ds\right), \quad x(0) = x^0.$$

This functional differential equation obviously constitutes a special case of the equation we dealt with in Remark 2.9.

2.9 A GLOBAL EXISTENCE RESULT IN A SPECIAL FUNCTION SPACE AND A POSITIVITY RESULT

First, we shall be concerned, in this section, with the functional equation

$$x(t) = (Vx)(t), \quad x : [0,t) \to B, \tag{2.107}$$

where B is a real Banach space, while the underlying function space is

$$E_{p,\rho}([0,T),B), \quad T \leq \infty, \tag{2.108}$$

consisting of those maps $x(t)$ of the interval $[0,T)$ into B, locally integrable and such that

$$\sup_{t\in[0,T)} \left\{ \rho(t) \int_0^t |x(s)|^p \, ds \right\} = |x|_{p,\rho}^p < \infty. \tag{2.109}$$

It will be always assumed in this section that $p \geq 1$, and $\rho(t) > 0$ is continuous on $[0,T)$, even though the continuity could be dropped.

Let us mention that such spaces were first considered by Kwapisz [304, 305], on closed intervals, primarily in connection with Bielecki's method (see Section 2.2).

We will try to carry out some work by using the spaces $E_{p,\rho}([0,T),B)$, which we defined in Corduneanu and Mahdavi [179], where proof has been presented to show that they are Banach spaces. We omit here the details of this proof, which rely on results in the Banach space theory of common knowledge.

Instead, we would like to find out what kind of convergence is implied by the norm $x \to |x|_{p,\rho}$.

It turns out that the convergence in the space $E_{p,\rho}([0,T),B)$ is the L^p-convergence on any $[0,\tau] \subset [0,T)$.

Indeed, if we restrict the functions in $E_{p,\rho}$ to an interval $[0,T_1] \subset [0,T)$, we can apply what we had in Section 2.2, namely

$$\begin{aligned} \inf \rho(t) \int_0^{T_1} |x(s)|^p \, ds &\leq \int_0^{T_1} \rho(s) \, |x(s)|^p \, ds \\ &\leq \sup \rho(t) \int_0^{T_1} |x(s)|^p \, ds, \end{aligned} \tag{2.110}$$

with inf and sup taken on $[0,T_1]$. Since both $\inf \rho(t)$ and $\sup \rho(t)$ are strictly positive on $[0,T_1]$, we derive the fact mentioned before about the L^p-convergence on each closed interval in $[0,T)$.

We can state that the $E_{p,\rho}$-convergence implies the L^p-convergence on each compact interval of $[0,T)$.

We remark that the spaces $E_{p,\rho}$, as defined earlier, contain functions enjoying a certain kind of *behavior* for $t \to T$; this behavior depends intimately on the weight function $\rho(t)$. So, by extending the definition of Kwapisz to semi-open intervals, we achieve a variety of asymptotic behaviors at the open extremity of the interval of definition of the functions in $E_{p,\rho}([0,T),R^n)$.

Let us consider now the functional equation (2.107) under the main assumption of continuity of the operator V, from $E_{p,\rho}([0,T),R^n)$ into itself. This property is a consequence of the following assumption

$$\int_0^t |(Vx)(s)-(Vy)(s)|^p \, ds \leq \lambda \int_0^t |x(s)-y(s)|^p \, ds, \quad t \in [0,T), \tag{2.111}$$

where $\lambda > 0$, for $x, y \in E_{p,\rho}([0,T),B)$. If we multiply both sides of (2.111) by $\rho(t) > 0$, we obtain

$$\rho(t) \int_0^t |(Vx)(s)-(Vy)(s)|^p \, ds \leq \lambda \rho(t) \int_0^t |x(s)-y(s)|^p \, ds, \quad t \in [0,T).$$

Taking the sup of both sides and then the pth-root of both sides, we obtain

$$|Vx - Vy|_{p,\rho} \leq \lambda^{\frac{1}{p}} |x-y|_{p,\rho}, \tag{2.112}$$

and since $\lambda^{\frac{1}{p}} > 0$, we can conclude that $V : E_{p,\rho}([0,T),B) \to E_{p,\rho}([0,T),B)$ is a Lipschitz continuous operator.

Moreover, if one assumes $\lambda \in (0,1)$, then V is a contraction map on $E_{p,\rho}([0,T),B)$. Hence, Equation (2.107) has a unique solution in $E_{p,\rho}([0,T),B)$.

Summarizing the discussion carried earlier on equation (2.107), we can state the following result.

Theorem 2.15 *Consider the functional equation (2.107), under the assumption (2.111). If $\lambda \in (0,1)$, then equation (2.107) has a unique solution in $E_{p,\rho}([0,T),B)$, for any $p \geq 1$ and $\rho(t) : [0,T) \to R_+$ continuous.*

This unique solution is obtained by iteration, according to the formula $x_{k+1}(t) = (Vx_k)(t)$, $t \in [0,T)$, $k \geq 0$, starting with an arbitrary $x_0(t) \in E_{p,\rho}([0,T),B)$. The convergence is that given by the norm $x \to |x|_{p,\rho}$, as defined by (2.109).

Remark 2.11 *We did not state explicitly that V must be causal. It is evident from (2.111) that this property holds true, as well as the continuity.*

The second part of this section will be dedicated to a problem of existence for the integro-functional equation

$$x(t) = f(t) + \int_0^t Q(t,s)\,(Fx)(s)\,ds, \quad t \in [0,T), \tag{2.113}$$

under conditions to be specified later, the operator F being *causal*, in a sense somewhat different than this term has been used in this book. More precisely, we shall look for a "positive" solution, which actually means that it will belong to a cone, in the appropriate function space.

Equation (2.113) is a scalar one, the functions involved being real-valued.

The proof of the existence result will be based on Theorem 34.2, which is in the book by Krasnoselskii and Zabreiko [298]. In order to state this theorem, we need to introduce some definitions and notations.

The underlying space will be the space $L^\infty([0,T],R)$, consisting of the maps which are essentially bounded on $[0,T)$. In $L^\infty([0,T],R)$, we will consider the cone of positive functions, which shall be denoted by K_+. The notation

$$K_G = \{w;\, w \in K_+,\, 0 \le w(t) \le G,\, t \in [0,T]\}$$

will be used below.

An operator $F : L^\infty([0,T],R) \to L^\infty([0,T],R)$ will be called *causal* if it verifies the condition

$$P_\tau F P_\tau = P_\tau F, \quad \tau \in [0,T],$$

where P_τ is the projection operator given by

$$(P_\tau w)(t) = \begin{cases} w(t), & t \in [0,\tau], \\ 0, & t \in (\tau, T]. \end{cases}$$

Theorem 2.16 *Let us consider equation (2.113), under the following assumptions:*

(1) $Q : \{0 \le s \le t \le T, R_+\}$ *is nonnegative, measurable kernel.*
(2) $f \in K_+$ *is given.*
(3) $F : L^\infty([0,T],R) \to L^\infty([0,T],R)$ *is an operator dominated by a linear casual operator A on* $L^\infty([0,T],R)$*, that is,* $Fv \le Av$*,* $v \in K_R$.
(4) $b(s) = ess - \sup_{s \le t \le T} Q(t,s) \in L^1([0,T],R)$.

(5) *The inequality*

$$\left(\exp(|A|_{L^\infty} \int_0^T b(s)\,ds)\right) |f|_{L^\infty} \leq R, \tag{2.114}$$

holds true.

Then, equation (2.113) has a solution $x \in K_R$, satisfying the inequality

$$|x|_{L^\infty([0,T),R)} \leq \left(\exp(|A|_{L^\infty} \int_0^T b(s)\,ds)\right) |f|_{L^\infty([0,T),R)}.$$

The proof of Theorem 2.16 is relying on the following result from the previously quoted book by Krasnoselskii and Zabreiko. It can be stated as follows:

(KZ) Let $F : K_R \to K_+$ be a continuous map. Assume there is a linear positive operator M, while $f \in K_+$, such that

$$Fv \leq Mv + f, \quad v \in K_R. \tag{2.115}$$

Moreover, let the spectral radius of M be less than 1, and

$$|(I-M)^{-1} f|_{L^\infty} \leq R.$$

Then F has a fixed point $x \in K_R$, and

$$|x|_{L^\infty} \leq |(I-M)^{-1} f|_{L^\infty}. \tag{2.116}$$

As mentioned before, the *proof* of the statement (KZ) is contained in the quoted reference.

For the proof of Theorem 2.16, we need the following property for the linear casual operator A, on $L^\infty([0,T),R)$:

For any $\tau \in (0,T)$, the inequality

$$|A|_{L^\infty([0,\tau),R)} \leq |A|_{L^\infty([0,T),R)}, \tag{2.117}$$

holds true.

Indeed, for each $v \in L^\infty([0,T),R)$ we have the sequence of relations

$$\begin{aligned} |Av|_{L^\infty([0,\tau),R)} &= |P_\tau Av|_{L^\infty([0,T),R)} \\ &= |P_\tau A P_\tau v|_{L^\infty([0,T),R)} \\ &\leq |A P_\tau v|_{L^\infty([0,T),R)} \\ &\leq |A|_{L^\infty([0,T),R)} |P_\tau v|_{L^\infty([0,T),R)} \\ &= |A|_{L^\infty([0,T),R)} |v|_{L^\infty([0,\tau),R)}, \end{aligned}$$

which implies (2.117), due to arbitrariness of $v \in L^\infty([0,T),R)$.

Another auxiliary property needed in the proof of Theorem 2.16 is the following, related to the operator V, from $L^\infty([0,T),R)$ into K_R; namely,

$$v \to (Vv)(t) = \int_0^t Q(t,s)\,(Av)(s)\,ds,$$

which is a majorant for the operator

$$v \to \int_0^t Q(t,s)\,(Fv)(s)\,ds, \quad t \in [0,T),$$

and satisfies the inequalities

$$|V^k|_{L^\infty} \le \frac{|A|_{L^\infty}^k}{k!}\left(\int_0^T b(u)\,du\right)^k, \quad k \ge 1, \tag{2.118}$$

the L^∞ sign standing for $L^\infty([0,T),R)$.

The inequality (2.118) is obtained by induction on k, as follows. For an arbitrary $w \in L^\infty([0,T),R)$, denoting $w_k = V^k w$, $k \ge 1$, and taking into account

$$|w_{k+1}|_{L^\infty([0,T),R)} \le \int_0^t b(s)\,|(Aw_k)(s)|\,ds, \quad t \in [0,T),$$

as well as (2.117). One obtains for $k > 1$,

$$|w_{k+1}|_{L^\infty([0,t),R)} \le |A|_{L^\infty([0,T),R)}^k \int_0^t b(s)\,|w_k(s)|_{L^\infty([0,s),R)}\,ds, \tag{2.119}$$

which routinely leads to (2.118). Therefore, the process of successive approximations, for the equation $x(t) = (Vx)(t)$, $t \in [0,T)$, is convergent. But if one takes into account that equation (2.113) can be written, by means of V, as $v = (I-V)^{-1}f$, then (2.118) guarantees the convergence in L^∞ of the successive approximations (iteration method).

This ends the proof of Theorem 2.16.

As an *application* of this result, Giĺ indicates, among other examples, the equation

$$\dot{x}(t) + \int_0^a x(t-s)\,dg(s) = (Fx)(t), \quad t \in [0,T], \tag{2.120}$$

in which F is a causal operator on $L^\infty([0,T],R)$, $a > 0$, $g \in BV([0,T],R)$ is nondecreasing, and an initial condition of the form $x(t) = \phi(t)$, $t \in [-a,0]$, is associated. The existence of solution is obtained under the assumption $a\,e\,Var g|_{[0,a]} < 1$.

2.10 SOLUTION SETS FOR CAUSAL FUNCTIONAL DIFFERENTIAL EQUATIONS

This section is dedicated to the study of functional differential equations of the form

$$\dot{x}(t) = (Vx)(t), \ a.e., \quad t \in [0,T], \tag{2.121}$$

under various conditions on the operator V, with the initial condition

$$x(0) = x^0 \in R^n. \tag{2.122}$$

Since the uniqueness of the solution of problem (2.121) and (2.122) is not taking place, in general, it is interesting to investigate the solution set corresponding to this problem.

In case $(Vx)(t) = f(t, x(t))$, with f continuous, the problem of existence of the solution set, known as the funnel with its vertex in (t_0, x^0), has been widely investigated after Peano gave his existence theorem (for at least, one solution). The geometric study of the funnel leads to interesting results. Many books on ordinary differential equations contain such results (Peano's phenomenon).

We shall give here some results, with regard to the topological structure of solution set for problem (2.121) and (2.122).

Theorem 2.17 *Consider equation (2.121), with the initial condition (2.122), under the following hypotheses:*

(1) $V : C([0,T], R^n) \to L^1([0,T], R^n)$ *is a causal continuous operator.*
(2) *There exists a function* $\mu(t) \in L^1([0,T], R)$, *such that for any* $x \in C([0,T], R^n)$ *we have* $|(Vx)(t)| \leq \mu(t)$, *a.e. for* $t \in [0,T]$.

Then, the solution set, denoted by $S_V(x^0; R^n)$, *is nonempty, compact, and connected in the space* $C([0,T], R^n)$.

Proof. We shall rely on several results from real analysis and elementary topology, which can be found in any textbook treating these subjects.

Let us notice that condition (2), in Theorem 2.17, implies conditions (2) and (3) of Theorem 2.1, while Corollary 2.1 is guaranteeing the existence of a solution to (2.121) and (2.122), on some interval $[0,a]$, $a \leq T$. Therefore, $S_V(x^0; R^n)$ is nonempty. Further, the compactness of $S_V(x^0; R^n)$ is obtained as follows: each solution of (2.121) and (2.122), defined on $[0,a]$, $a \leq T$, satisfies the equation

$$x(t) = x^0 + \int_0^t (Vx)(s)\, ds, \quad t \in [0,a]. \tag{2.123}$$

On behalf of assumption (2) in Theorem 2.17, we obtain from (2.123) the estimate

$$|x(t)| \leq |x^0| + \int_0^t \mu(s)\, ds, \quad t \in [0,a]. \tag{2.124}$$

Hence, the solution set of our problem is uniformly bounded in the space $C([0,a],R^n)$. But (2.123) and assumption (2) also imply

$$|x(t) - x(\tau)| \leq \left| \int_\tau^t \mu(s)\, ds \right|, \quad t,\tau \in [0,a], \tag{2.125}$$

for each $x(t) \in S_V(x^0;R^n)$. This means that $S_V(x^0;R^n)$ is an equicontinuous subset of the space $C([0,a],R^n)$, and according to Ascoli–Arzelà criterion this assures the compactness of $S_V(x^0;R^n)$ in $C([0,a],R^n)$.

Now we shall consider the property of *connectedness* of $S_V(x^0;R^n)$. Assume this property does not hold. Then, we can represent the set $S_V(x^0;R^n)$ in the form

$$S_V(x^0;R^n) = A \cup B, \quad A \cap B = \emptyset, \tag{2.126}$$

with A and B closed in $C([0,a],R^n)$. Since both A and B are compact, as closed subsets of the compact set $S_V(x^0;R^n)$, there results that

$$d(A,B) = \rho > 0, \tag{2.127}$$

because A and B are disjoint sets in the space $C([0,a],R^n)$.

Let us denote $P = (\tau,x)$ the point corresponding to $\tau \in [0,a]$, and $|x - x^0| \leq b$, with

$$b = \int_0^a \mu(s)\, ds, \tag{2.128}$$

and consider the function of P

$$f(P) = d(P,B) - d(P,A). \tag{2.129}$$

The following two properties are immediate:

$$f(P) \geq \rho, \quad P \in A \quad \text{and} \quad f(P) \leq -\rho, \quad P \in B. \tag{2.130}$$

We shall denote by $y(t)$, $t \in [0,\tau]$, a solution of (2.123), reaching the point $y(\tau) \in A$. Similarly, let $z(t)$ be a solution of (2.123), reaching the point $z(\tau) \in B$. Then (2.130) implies

$$f(\tau, y(\tau)) \geq \rho, \quad f(\tau, z(\tau)) \leq -\rho. \tag{2.131}$$

Let us construct, first, a sequence of continuous functions, associated with $y(t)$, say $\{y_m(t), m \geq 1\}$, with $y_m(t) \to y(t)$ as $m \to \infty$, uniformly on $[0,\tau]$, according to Euler's approximation procedure by polygonal functions. This means that the graph of $y_m(t)$, in the (t,y) plane, is obtained by joining with a line segment the consecutive points, on the graph $(t, y(t))$, $t \in [0,\tau]$, corresponding to the equidistant subdivision given by the points $km^{-1}\tau$, $k = 0, 1, \ldots, m$. The uniform continuity of $y(t)$ assures the continuity of $y_m(t)$ on $[0,\tau]$.

Second, one constructs a similar sequence $\{z_m(t), m \geq 1\}$, associated with the solution $z(t)$.

We introduce now the sequences $(m \geq 1)$

$$\alpha_m(t) = \int_0^t [(Vy)(s) - (Vy_m)(s)]\, ds, \quad t \in [0,\tau], \tag{2.132}$$
$$\beta_m(t) = \int_0^t [(Vz)(s) - (Vz_m)(s)]\, ds,$$

and notice the relations

$$\lim \alpha_m(t) = \lim \beta_m(t) = 0 \text{ as } m \to \infty, \tag{2.133}$$

uniformly for $t \in [0,\tau]$. The statement (2.133) is a consequence of the continuity of the operator V on $C([0,\tau], R^n)$ and of the Lebesgue dominated convergence in the space $L^1([0,\tau], R^n)$.

From (2.130), (2.132), and (2.123), also taking into account the construction of the sequences $\{y_m(t), m \geq 1\}$ and $\{z_m(t), m \geq 1\}$, one obtains for $m \geq N_0$

$$f(\tau, y_m(\tau)) > 0, \quad f(\tau, z_m(\tau)) < 0. \tag{2.134}$$

Without loss of generality, we can assume that (2.134) holds for $m \geq 1$.
Define now on $[0,\tau] \times [0,1]$, for $m \geq 1$,

$$\gamma_m(t,u) = (1-u)\,\alpha_m(t) + u\,\beta_m(t), \tag{2.135}$$

and notice that $|\gamma_m(t,u)| \leq |\alpha_m(t)| + |\beta_m(t)|$, $m \geq 1$, which implies

$$\lim_{m\to\infty} \gamma_m(t,u) = 0, \quad (t,u) \in [0,\tau] \times [0,1], \tag{2.136}$$

uniformly in (t,u). Further, let us construct the sequence $\{x_m(t,u); m \geq 1\}$, such that

$$x_m(t,u) = \begin{cases} x^0, & 0 \leq t < \frac{\tau}{m}, \\ x^0 + \int_0^{t-\frac{\tau}{m}} (Vx_m)(s,u)\,ds + \gamma_m(t,u), & \frac{\tau}{m} \leq t \leq \tau. \end{cases} \tag{2.137}$$

This procedure, due to Tonelli, for constructing approximate solutions must be understood in the following manner. After using the initial value on $[0, \frac{\tau}{m}]$, one looks at the second row in (2.137), first on the interval $[\frac{\tau}{m}, \frac{2\tau}{m}]$. The integration will be performed on $[0, \frac{\tau}{m}]$, where $x_m(s,u)$ is already known. In this step, one determines $x_m(t,u)$ on $[\frac{\tau}{m}, \frac{2\tau}{m}]$, then on $[\frac{2\tau}{m}, \frac{3\tau}{m}]$, and so on, until $x_m(t,u)$ is obtained for $t \in [0,\tau]$, in m steps. These functions are continuous in (t,u), on $[0,\tau] \times [0,1]$.

Let us consider now the auxiliary functions

$$f_{m,\tau}(u) = f(\tau, x_m(\tau,u)), \quad u \in [0,1]. \tag{2.138}$$

$f_{m,\tau}(u)$ is continuous on $[0,1]$, and one obtains from (2.137)

$$x_m(\tau,0) = y_m(\tau), \quad x_m(\tau,1) = z_m(\tau), \tag{2.139}$$

and taking into account (2.134)

$$\begin{aligned} f_{m,\tau}(0) &= f(\tau, y_m(\tau)) > 0, \\ f_{m,\tau}(1) &= f(\tau, z_m(\tau)) < 0. \end{aligned} \tag{2.140}$$

From (2.140) and the continuity of $f_{m,\tau}(u)$, $u \in [0,1]$, one derives the existence of numbers $u_m \in (0,1)$, $m \geq 1$, such that $f_{m,\tau}(u_m) = 0$.

We will now focus on the sequence $\{x_m(t,u_m); m \geq 1\}$, consisting of approximate solutions to the integral equation (2.123), and show that it has a uniformly convergent subsequence on $[0,\tau]$. In order to achieve this objective, we shall prove that the sequence is uniformly bounded and equicontinuous on $[0,\tau]$. The existence of the subsequence, which is uniformly convergent, will result from Ascoli–Arzelà criterion of compactness in the space $C([0,\tau],R^n)$.

Let us consider the sequence of functions in $C([0,\tau],R^n)$, given by

$$\tilde{x}_m(t) = x^0 + \int_0^t [(Vx_m)(s,u_m) + \gamma_m(s,u_m)]\,ds, \tag{2.141}$$

for $m \geq 1$, whose uniform boundedness on $[0,\tau]$ is easy to check, based on property (2) of the operator V and (2.136). Also, the equicontinuity of the sequence $\{\tilde{x}_m(t);\, m \geq 1\}$ is obtained from

$$|\tilde{x}_m(t) - \tilde{x}_m(\bar{t})| = \left| \int_{\bar{t}}^{t} [(Vx_m)(s,u_m) + \gamma_m(s,u_m)]\, ds \right|$$
$$\leq \left| \int_{\bar{t}}^{t} \mu(s)\, ds \right| + K|t-\bar{t}|, \quad t,\bar{t} \in [0,\tau],$$

with $K \geq |\gamma_m(t,u)|$, $m \geq 1$, an estimate assured by the property given in (2.136).

Therefore, there exists a subsequence of $\{\tilde{x}_m(t);\, m \geq 1\}$, which converges uniformly on $[0,\tau]$ to a continuous function $x(t)$. One can assume that the sequence itself converges uniformly.

Let us now estimate the difference

$$\tilde{x}_m(t) - x_m(t,u_m) = \int_{t_{m,k}}^{t} [(Vx_m)(s,u_m) + \gamma_m(s,u_m)]\, ds, \tag{2.142}$$

where $t_{m,k}$ is chosen, in relationship with t, in formula (2.142), such as $km^{-1}\tau \leq t < (k+1)m^{-1}\tau$. Obviously, there is only one integer k to provide validity to the above inequalities. From (2.142), we obtain, based on property (2) for V and (2.136) for γ_m, the inequality

$$|\tilde{x}_m(t) - x_m(t,u_m)| \leq \int_{km^{-1}\tau}^{t} (\mu(s) + K)\, ds, \tag{2.143}$$

which implies

$$|\tilde{x}_m(t) - x_m(t,u_m)| \leq \int_{km^{-1}\tau}^{(k+1)m^{-1}\tau} \mu(s)\, ds + Km^{-1}\tau. \tag{2.144}$$

Inequality (2.144) shows that for $m \to \infty$, one has $\tilde{x}_m(t) - x_m(t,u_m) \to 0$, uniformly on $[0,\tau]$. Since we have proven that $\{\tilde{x}_m(t);\, m \geq 1\}$ is uniformly convergent on $[0,\tau]$, there results from (2.144) that $\{x_m(t,u_m);\, m \geq 1\}$ is also uniformly convergent on $[0,\tau]$, while

$$x(t) = \lim_{m \to \infty} x_m(t,u_m), \quad t \in [0,\tau], \tag{2.145}$$

is continuous on $[0,\tau]$. Taking the limit in (2.137), as $m \to \infty$, one obtains for $x(t)$ equation (2.123), which is equivalent to our initial value problem (2.121), (2.122).

On the other hand, returning to (2.138), and letting $m \to \infty$, we obtain for $u = u_m$, $m \geq 1$, $f_{m,\tau}(u_m) = 0 = f(\tau, x_m(\tau,u_m))$, $m \geq 1$, which implies

$f(\tau,x(\tau)) = 0$ for $m \to \infty$. But $f(\tau,x(\tau)) = 0$ contradicts the inequality (2.130).

Therefore, our assumption about the representation of $S_V(x^0;R^n) = A \cup B$, with A and B closed in $C([0,a],R^n)$ and disjoint, leads to a contradiction. This ends the proof of Theorem 2.17.

Remark 2.12 *The result of Theorem 2.17, originally given by Kneser in the special case $(Vx)(t) = f(t,x(t))$, has caught the attention of numerous mathematicians interested in the topological properties of differential equations. The proof provided before is taken, with some small modifications, from Corduneanu [123].*

Variants and improvements of the results in Theorem 2.17 have been obtained by various researchers using advanced knowledge from topology and functional analysis.

In concluding this section, we will state some of the results related to the properties, one may say geometric properties, of the solution set $S_V(x^0;R^n)$.

1) Hukuhara property. In the classical case $(Vx)(t) = f(t,x)$, this property states:

   Let $(\tau,\xi) \in \partial S_V(x^0;R^n)$. There exists a solution of the equation $\dot{x}(t) = f(t,x(t))$ connecting the points $(0,x^0)$ and (τ,ξ), such that its graph lies on $\partial S_V(x^0;R^n)$. See for the proof Corduneanu [123].

2) Agarwal–Górniewicz–O'Regan property [8,429]. This property, in case of equation (2.121), with the initial condition (2.122), can be stated as follows:

   (1) The same as in Theorem 2.17.

   (2) For each $r > 0$, there exists $M_r > 0$, such that $|x|_C \le r \Longrightarrow |(Vx)(t)| \le M_r$, *a.e.* on $[0,T]$.

   (3) Or for each $r > 0$, there exists $\mu_r \in L^1([0,T],R^n)$, such that $|x|_C \le r \Longrightarrow |(Vx)(t)| \le \mu_r(t)$, *a.e.* on $[0,T]$.

   Conditions (1) and (2) or (1) and (3) imply the compactness and the connectedness of $S_V(x^0;R^n)$.

3) O'Regan property [429]. Consider problem (2.121) and (2.122), under the following assumptions:

   (1) $V : C([0,T],R^n) \to C([0,T],R^n)$ is continuous, causal, and compact operator.

   (2) $(Vx)(0) = x^0$ for $x \in C([0,T],R^n)$, that is V has the fixed initial value property (Neustadt). Then $S_V(x^0;R^n)$ is a set of the type R_δ.

Properties 1), 2), and 3) all require some special results from functional analysis. Some of them are contained in the quoted papers, while others are indicated as references.

2.11 AN APPLICATION TO OPTIMAL CONTROL THEORY

We now consider the following problem of optimization described by the input–output equation

$$\dot{x}(t) = (Ax)(t) + (Bu)(t), \quad t \in [t_0, T], \tag{2.146}$$

where $x : [t_0, T] \to R^n$ is the output, $u : [t_0, T] \to R^m$ is the input, and A, B are causal operators $A : L^2([0,T], R^n) \to L^2([0,T], R^n)$, $B : L^2([0,T], R^m) \to L^2([0,T], R^n)$, $n, m \in N$.

To equation (2.146), one attaches the initial conditions

$$x(t) = \phi(t), \quad t \in [0, t_0),\ x(t_0) = \theta, \tag{2.147}$$

with $\phi \in L^2([0,t_0), R^n)$, $\theta \in R^n$.

The following quadratic cost functional

$$\begin{aligned} C(x;\phi,u) = \int_0^{t_0} < (P\phi)(t), \phi(t) > dt + \int_{t_0}^{T} (< (Qx)(t), x(t) > \\ + < (Ru)(t), u(t) >) dt \end{aligned} \tag{2.148}$$

will be considered, under some conditions to be specified later.

The main interest related to the already stated optimization problem is to prove the existence of an optimal triplet $(\bar{x}; \bar{\phi}, \bar{u})$, such that

$$C(\bar{x}; \bar{\phi}, \bar{u}) = \min C(x; \phi, u), \tag{2.149}$$

the minimum being taken for $\phi \in \Phi \subset L^2([0,t_0], R^n)$ and $u \in U \subset L^2([t_0,T], R^m)$, where Φ and U denote admissible sets for ϕ and u.

Under certain conditions, we shall formulate later, the following main result in this section can be stated:

Theorem 2.18 *Consider the modified optimal control problem, defined by the equations (2.146)–(2.149), under the following conditions:*

a) *The operators A, B in equation (2.146) are linear, causal, and continuous on the spaces* $L^2([0,T], R^n)$, *respectively* $L^2([t_0,T], R^m) \to L^2([t_0,T], R^n)$.

b) The linear operators P, Q and R, appearing in the cost functional $C(x;\phi,u)$ are bounded, self-adjoint, with P and R positive definite while Q is nonnegative definite;

c) The initial set $\Phi \subset L^2([0,t_0],R^n)$, and the admissible control set $U \subset L^2([t_0,T],R^m)$ are closed, convex sets.

Then, there exists a unique optimal triplet $(\bar{x};\bar{\phi},\bar{u})$, that is, such that (2.149) is satisfied.

Proof. The method of proof is rather elementary and relies on general results from linear functional analysis.

Namely, the first needed result can be stated: *In a Hilbert space H, any closed, convex set contains a unique element of minimal norm.*

A second necessary result, which holds in the more general case of Banach spaces, states: *In a Banach space B, with norm $x \to |x|_1$, a second norm $x \to |x|_2$ is equivalent to the first, if and only if there exists two positive constants a and b, such that*

$$a|x|_1 \leq |x|_2 \leq b|x|_1, \quad x \in B. \tag{2.150}$$

First we need to show that the cost functional $C(x;\phi,u)$, defined by (2.148), makes sense for $\phi \in \Phi$ and $u \in U$. This fact requires that $x(t)$, as determined by (2.146) and (2.147), is defined on the interval $[t_0,T]$. Since $x(t)$ verifies on (t_0,T) the equation

$$\dot{x}(t) = (Ax)(t) + f(t), \tag{2.151}$$

with

$$f(t) = (Bu)(t), \quad t \in [t_0,T], \tag{2.152}$$

which implies

$$f \in L^2([t_0,T],R^n), \tag{2.153}$$

we can apply the representation formula from Corduneanu [149], that is, for $t \in [t_0,T]$,

$$x(t) = \int_{t_0}^{t} X(t,s)f(s)\,ds + \int_0^{t_0} \widetilde{X}(t,s;t_0)\,\phi(s)\,ds, \tag{2.154}$$

with $X(t,s)$ the Cauchy matrix of the homogeneous system $\dot{x}(t) = (Ax)(t)$, $t \in [t_0, T]$, while $\widetilde{X}(t,s;t_0)$ is a matrix related to the same system (see Corduneanu [149] for details). We notice that the second integral in (2.154) represents the solution of homogeneous system, under initial conditions (2.147).

The above considerations show that the cost functional $C(x;\phi,u)$ is defined, under conditions given in Theorem 2.18.

Let us consider now the Hilbert space

$$H = L^2([0,t_0],R^n) \times L^2([t_0,T],R^m), \tag{2.155}$$

in which the scalar product is the sum of scalar products in each factor. This has as consequence the fact that the norm in H is the square root of the sum of squares of the norms in the spaces factor. In the space H, with the topology described above (derived from the norm), the set $\Phi \times U$ is closed and convex, a fact that can be easily checked.

We will now introduce a different scalar product on H, suggested by the form of the cost functional $C(x;\phi,u)$, such that the following relationship holds:

$$C(x;\phi,u) = |(x;\phi,u)|^2_{\widetilde{H}}, \tag{2.156}$$

where by $(x;\phi,u)$ we understand $x(t;\phi,u)$, the solution of the system (2.146), under the initial conditions (2.147), while $\widetilde{H}$ means the space H with the new topology, generated by the scalar product

$$\begin{aligned} < (x;\phi,u),(y;\psi,v) >_{\tilde{H}} = & \int_0^{t_0} < (P\phi)(t),\psi(t) > dt \\ & + \int_{t_0}^{T} [< (Qx)(t),y(t) > + < (Ru)(t),v(t) >]\, dt. \end{aligned} \tag{2.157}$$

Indeed, with this scalar product, in the space $\widetilde{H}$, condition (2.156) is verified. In order to be able to proceed further with the application of the existence and uniqueness of the element of minimal norm in a closed convex set, in this case we mean $\Phi \times U \subset \widetilde{H}$, we need only to prove that $\Phi \times U$ is closed. The convexity is defined algebraically only, while the closedness must be proven for the topology of $\widetilde{H}$.

Actually, we shall prove that the topologies of H and $\widetilde{H}$ are equivalent, which will be sufficient for our aim (because $\Phi \times U$ is closed in H, a statement easy to prove). This step will be done by showing that the following inequalities are satisfied:

$$\alpha\left(\int_0^{t_0}|\phi(t)|^2\,dt+\int_{t_0}^{T}|u(t)|^2\,dt\right)\le|(x;\phi,u)|_{\widetilde{H}}^2$$
$$\le\beta\left(\int_0^{t_0}|\phi(t)|^2\,dt+\int_{t_0}^{T}|u(t)|^2\,dt\right),\tag{2.158}$$

where $0<\alpha<\beta$ are constants, while $\phi\in\Phi$ and $u\in U$.

We will now rely on the representation (2.154), which is taken from Corduneanu [149] and take into account our hypotheses *a*), *b*), and *c*) in Theorem 2.18. From (2.154), with $f(t)=(Bu)(t)$, we obtain

$$\int_{t_0}^{T}|x(t)|^2\,dt\le C_1\int_{t_0}^{T}|u(t)|^2\,dt+C_2\int_0^{t_0}|\phi(t)|^2\,dt,\tag{2.159}$$

for the solution $x=x(t;\phi,u)$, whose existence is assured by our hypotheses (see details in Corduneanu quoted earlier). The constants C_1, C_2, both positive, are valid for any $\phi\in L^2([0,t_0],R^n)$ and $u\in L^2([t_0,T],R^m)$, and they can easily be determined from (2.154). Moreover, for obtaining the first inequality in (2.158), one can use directly the formula for the definition of the scalar product in $\widetilde{H}$, and taking into consideration assumption *b*) from Theorem 2.18.

On behalf of the earlier discussion, we conclude the existence and uniqueness of the optimal triplet $(\bar{x};\bar{\phi},\bar{u})$ with $\bar{x}$ absolutely continuous on $[t_0,T]$, and $\bar{\phi}\in L^2([0,t_0],R^n)$, $u\in L^2([t_0,T],R^m)$.

Theorem 2.18 is thereby proven.

Remark 2.13 *The connection between the elements of the optimal triplet is given by the equation, derived from (2.154), namely*

$$\bar{x}(t)=\int_{t_0}^{t}X(t,s)\,(B\bar{u})(s)\,ds+\int_0^{t_0}\widetilde{X}(t,s;t_0)\,\phi(s)\,ds.\tag{2.160}$$

Since one can always write

$$\int_{t_0}^{t}X(t,s)\,(B\bar{u})(s)\,ds=\int_{t_0}^{t}X_1(t,s)\,\bar{u}(s)\,ds,$$

for a convenient kernel $X_1(t,s)$, as shown for instance in Corduneanu [149], we can rewrite (2.160) in the form

$$\bar{x}(t)=\int_{t_0}^{t}X_1(t,s)\,\bar{u}(s)\,ds+\int_0^{t_0}\widetilde{X}(t,s;t_0)\,\bar{\phi}(s)\,ds.\tag{2.161}$$

Equation (2.161) invites the readers to consider the problem of *synthesis* of the control system (2.146). This means to represent the control $\bar{u}$ which leads to a given trajectory (i.e., $\bar{x}(t)$), in terms of the data. From (2.161), we derive for $\bar{u}(t)$ an integral equation of the form

$$\int_{t_0}^{t} X_1(t,s)\,\bar{u}(s)\,ds = A(t,x(t)), \quad t \in [t_0,T], \tag{2.162}$$

with $x(t)$ assigned (hence, $A(t,x(t)) = \bar{A}(t)$).

Equation (2.162) is a first kind of Volterra linear equation, when $x(t)$ is known. As it is well-known, this type of equation may be deprived of solutions or may admit infinitely many. For instance, if a solution $\bar{u}(t)$ does exist, and $u_0(t)$ is orthogonal to the kernel $X_1(t,s)$, then $\bar{u}(t) + u_0(t)$ is also a solution to (2.162).

The difficulty of getting the feedback equation in this problem stems from the fact that the existence of $X_1(t,s)$ is generally not proven constructively.

Therefore, the problem of synthesis for systems of the form (2.146) remains open.

In concluding this section, we point out that a more general problem, similar to the *LQ*-problem discussed earlier, can be formulated as follows: One considers the system

$$\dot{x}(t) = (Fx)(t) + (Gu)(t), \quad t \in [t_0,T],$$

with F and G nonlinear causal operators on the same function spaces as above and under same initial conditions (2.147). The cost functional should be taken of the form

$$C(x;\phi,u) = \int_0^{t_0} (K\phi)(t)\,dt + \int_{t_0}^{T} L(x;u)(t)\,dt,$$

with K and L nonlinear functionals, defined on the spaces $L^2([0,t_0],R^n)$, respectively $L^2([t_0,T],R^n) \times U$.

The literature is very rich and varied with regard to this problem. We mention here an interesting paper by Prichard and You [463], in the framework of Hilbert spaces, in which the input–output equation has the form

$$x(t) = f(t) + \int_{t_0}^{t} K(t,s)\,x(s)\,ds + \int_{t_0}^{t} N(t,s)\,u(s)\,ds,$$

which constitutes an integral equation for $x(t)$, of Volterra type, but of the second kind (which is, generally, uniquely solvable). Also, see the references in the paper quoted already.

2.12 FLOW INVARIANCE

The concept of flow invariance for systems of ordinary differential equations came to the attention of researchers toward the mid-twentieth century. One of the first contributions is due to Nagumo, who gave the mathematical formulation and the analytic conditions (necessary and sufficient) related to the concept of flow invariance.

Roughly speaking, the property of flow invariance means that the trajectories of solutions of the system must remain inside a close set of R^n (or $B =$ Banach space), a kind of a "cage." Such a feature is very important in applications because the man-made systems (engineering or from other branches of applied science) are acting in a finite part of the space.

Problems related to flow invariance pertain naturally to the problem of finding various estimates for the solutions of the dynamical system. Let us notice that the investigation of such problems is still seldom conducted, particularly in cases when the dynamics are described by means of functional equations other than ODE. Of course, when problems are treated in Banach space instead of R^n, the results can be applied to some PDE. There are still many possibilities to develop the theory in new directions. These possibilities involve new classes of dynamical systems.

We shall provide now the definitions, largely accepted by authors dealing with problems of flow invariance, considering the case of ODE

$$\dot{x}(t) = f(t, x(t)), \quad t \in (a,b), \tag{2.163}$$

with $x, f \in R^n$. More precisely, we shall assume

$$f : (a,b) \times K \to R^n, \tag{2.164}$$

where $K = \bar{K} \subset R^n$.

Definition 2.1 *The closed set K is called invariant (or flow invariant, with respect to system (2.163)) if for every $(t_0, x^0) \in (a,b) \times K$, there exists a solution $x(t; t_0, x^0)$, defined for $t \in [t_0, t_0 + h]$, $h > 0$, such that*

$$x(t; t_0, x^0) \in K, \quad t \in [t_0, t_0 + h]. \tag{2.165}$$

Remark 2.14 *This kind of invariance is called invariance to the right, or invariance in the future. Similarly, one defines invariance to the left.*

Since $h > 0$ can be arbitrarily small, Definition 2.1 should be regarded as defining a local concept of invariance. If instead, we ask for $x(t; t_0, x^0)$ to be considered on its maximal interval of existence (at the right), say (t_0, T), $T \leq b$, then we deal with a stronger property, related to the so-called *viability*.

In this section, which serves as a brief introduction to the subject of flow invariance, we are only concerned with the local aspect. The proofs of a few basic results we include in this section will be omitted. Instead, we refer the readers to adequate sources, as these proofs can be found in existing literature.

The Nagumo's condition of local flow invariance has been considered and discussed in detail by several authors. We mention here Brézis [77] and Motreanu and Pavel [405], where more information is provided.

In Motreanu and Pavel [405], the necessary and sufficient condition for flow invariance is called Nagumo–Brézis condition and it has the following intuitive formulation.

Theorem 2.19 *Let us consider the closed set $K \subset R^n$, as in Definition 2.1. A necessary and sufficient condition for the flow invariance of the set K, with respect to system (2.163) is given by*

$$\lim_{h \to 0+} h^{-1} d(x + hf(t,x); K) = 0, \tag{2.166}$$

for any $(t,x) \in (a,b) \times K$, when $d(x,S)$ is the (Euclidean) distance from x to $S \subset R^n$.

The *proof* of Theorem 2.19 requires a lengthy analysis, especially in regard to the sufficiency of condition (2.166). In the book Motreanu and Pavel [405], the contribution of Brézis [77] is emphasized, leading to the formulation of Theorem 2.19 and its proof. We send the readers to this reference for the full proof of this basic result.

Remark 2.15 *From geometric point of view, condition (2.166) tells us that the vector $f(t,x)$ is tangent to the set K, at any point $(t,x) \in (a,b) \times K$.*

A vector $v \in R^n$ is called tangent to a set $S \subset R^n$, at a point $x \in S$, if the condition

$$\lim_{h \to 0+} h^{-1} d(x + hv; S) = 0.$$

When S is a regular curve in R^n, one obtains the definition of the tangent vector. In general, there are several tangent vectors to a set, forming a closed cone in R^n.

We shall now consider a simple example illustrating the property of flow invariance, namely, the scalar differential equation

$$\dot{x}(t) = x(t)[x(t) - 1], \quad t \geq 0. \tag{2.167}$$

The solution of (2.167), under initial condition $x(t_0) = x^0$, is given by

$$x(t;t_0,x^0) = x^0 \left[x^0 + (1 - x^0)\, e^{(t-t_0)}\right]^{-1}, \tag{2.168}$$

valid, for $t \geq t_0 \geq 0$, $x^0 \in R \setminus (1, \infty)$. In this case, each solution is defined in the future, that is, from t_0 to ∞. This remark helps us conclude that the closed sets $[0,1] \subset R$ and $(-\infty, 0] \subset R$ are flow invariant to the right, with respect to equation (2.167). For instance, it is apparent that $x = 0$ and $x = 1$ are solutions and the local Lipschitz condition for the right-hand side of (2.167) allows us to conclude that any solution, starting in $[0,1]$, remains there forever. On the other hand, each solution, starting in $(-\infty, 0]$, remains there in the future. These statements are simple consequences of (2.168), which allows us to conclude that both sets $[0,1]$ and $(-\infty,0]$ are flow invariant. It is also obvious that the closed set $(-\infty, 1]$ is flow invariant to the right, with respect to (2.167).

For the case $x^0 > 1$, by using formula (2.168), we conclude that $x(t;t_0,x^0)$ ceases to exist (be defined) for $t = t_0 + \ln \frac{x^0}{x^0-1}$. In other words, each solution corresponding to an $x^0 > 1$ has what we call a "finite escape time." Yet this feature does not mean that the property of flow invariance in the future does not hold. Indeed, one can choose h, in Definition 2.1, such that $h < \ln \frac{x^0}{x^0-1}$, with the latter term being positive.

In the remaining part of this section, we briefly discuss, following the paper by Cârjă, Necula, and Vrabie [89], a substitute equivalent of the Nagumo–Brézis condition (2.166). This task will be carried out by using the comparison method presented in Section 2.4, which we applied to global existence. See also Remark 2.5, which indicates sources where Dini's derivative numbers can be used in the comparison inequality instead of the derivative.

The quantity to be estimated being the same as in condition (2.166), that is, $d(x + hf(t,x); K)$, the comparison inequality will look

$$\liminf_{h \to 0+} h^{-1}[d(x + hf(t,x), K) - d(x,K)] \leq \omega(t, d(\dot{x}, K)), \tag{2.169}$$

which is telling us that lower right Dini's number of the distance function, in direction $f(t,x)$, satisfies a comparison inequality, with a function $\omega(t,d)$

which is constructed in the proof. One shows that a possible choice for the comparison function $\omega(t,r)$, corresponding to system (2.163) is

$$\omega(t,r) = \sup_{\substack{t \in V \\ d(x,K)=r}} \liminf_{h \to 0+} h^{-1} [d(x+hf(t,x), K - d(x,K)],$$

where V is a neighborhood of the set K in R^n, or in $D \subset R^n$ when $f : [a,b] \times D \to R^n$. The function $\omega(t,r)$ is well defined, under adequate conditions on f, including one condition similar to the Nagumo's condition of uniqueness, as well as a certain Lipschitz-type condition on f.

In Ref. [91] by Cârjă, Necula and Vrabie, the case of semilinear systems (in Banach spaces) is discussed in detail. An application to control problems, for systems of the form

$$u'(t) = A\,u(t) + g(u(t)) + c(t), \quad u(0) = x,$$

$x \in B =$ Banach space, is discussed. A minimum time problem is also considered, connections with Bellman's problems of Dynamic Programming being emphasized.

2.13 FURTHER EXAMPLES/APPLICATIONS/COMMENTS

In connection with topics discussed in Sections 2.4 and 2.5, the following considerations play an important role in applications. Namely, we want to provide a result due to Burton [84], which guarantees conditions leading to global existence of solutions to ordinary differential equations. Our aim is to signal the possibility of extending this result to the case of FDE with causal operators, certainly more general than the case of ODE.

First, let us consider the equation $\dot{x} = f(t,x)$, and let $f(t,x)$ be continuous on $R_+ \times R_+$, such that $|f(t,x)| \leq a(t)\,g(|x|)$, where a and g are satisfying the conditions, $g(s) \geq 1$,

$$|f(t,x)| \leq a(t)\,g(|x|), \quad \int_0^\infty \frac{ds}{g(s)} = \infty. \tag{2.170}$$

Then the Liapunov-type function

$$V(t,x) = \left[\int_0^{|x|} \frac{ds}{g(s)} + 1 \right] e^{-\int_0^t a(s)\,ds}, \tag{2.171}$$

has the derivative (with respect to the equation $\dot{x} = f(t,x)$),

$$V^{\cdot}(t,x) = -a(t)\,V(t,x(t)) + e^{-a(t)}\,\frac{|x(t)|^{\cdot}}{g(|x(t)|)},$$

on any trajectory $x(t)$. But $|x(t)|^{\cdot} \leq |\dot{x}(t)|$, being the dot for derivative or a Dini number. Hence, we obtain the following from the above expression of $V^{\cdot}(t,x)$:

$$V^{\cdot}(t,x) \leq \frac{|x(t)|^{\cdot}}{g(|x(t)|)}\,e^{-\int_0^t a(s)\,ds} - a(t)\,V(t,x(t)) \leq 0,$$

when we rely on $|x(t)|^{\cdot} \leq |\dot{x}(t)| \leq a(t)\,g(|x(t)|)$.

The inequalities show that $V(t,x(t))$, with $x(t)$ trajectory of $\dot{x}(t) = f(t,x(t))$ is non-increasing. It means that $V(t,x(t))$ is non-increasing along of each trajectory of $\dot{x}(t) = f(t,x(t))$. Therefore, one concludes the existence of the

$$\lim_{t\to\infty} V(t,x(t)) = V_\infty = \text{finite},$$

which implies the existence of $x(t)$ on R_+.

Remark 2.16 *Actually, the proof of the existence of a solution beyond an interval* $[0,T)$, $T < \infty$, *is provided above.*

Since in case of equations with causal operators, say $\dot{x}(t) = f(t,x_t)$, where $x_t(s) = x(s)$ on $[0,t]$, the continuation of a solution is always possible, under standard conditions, for instance, as those mentioned before. It means that the global existence shown above for $\dot{x} = f(t,x(t))$, can be extended to equations of the form $\dot{x}(t) = f(t,x_t)$, $0 \leq t < T$, $T \leq \infty$.

We invite the readers to formulate the corresponding conditions in this case.

The criterion in Theorem 2.3 is called the criterion of Conti–Wintner for global existence by Burton [84]. Wintner criterion corresponds to $g(s)$ as mentioned earlier while $a(t) \equiv 1$.

In the above considerations, we have an example when the Liapunov function can be easily constructed based on data.

Another *example* we want to discuss is concerned with functional differential equations with retarded argument. We have proven, in Section 2.6, a global existence (and uniqueness) result, using the Bielecki variant of succesive approximations or contraction mapping method.

We refer to equation (2.83), keeping the same notations as in Section 2.6 and the same assumptions on data. In other words, the operator (of Niemytskii type), $(Fx)(t) = f(t,x(t))$ is defined on the space $C([-h,0],R^n)$ since we want to investigate only spaces of finite dimension for $x = x(t)$. The new method we sketch here is known as the *steps method* and reduces the existence problem for (2.83), under the initial condition (2.84) with $B = R^n$, to the existence for the recurrent differential equations

$$\dot{x}^{(k+1)}(t) = f(t, x^{(k)}(t)), \quad t_k = kh,\ h \geq 0,\ t \in [t_k, t_{k+1}), \tag{2.172}$$

where $x^{(0)}(t)$ is the initial function in $C([-h,0], R^n)$. Obviously, at each step we have to perform a quadrature (exact or approximative) of a vector function, as seen in (2.172). The initial condition, at each step, is determined by the value at the end of the interval $[t_{k-1}, t_k]$, of the function $x^{(k)}(t)$:

$$x^{(k+1)}(t) = x^{(k)}(t_k) + \int_{t_k}^{t} f(s, x^{(k)}(s))\, ds, \quad t \in [t_k, t_{k+1}]. \tag{2.173}$$

Let us point out that the functions $x^{(k)}(t)$, $k \geq 1$, are uniquely determined and the function

$$x(t) = \begin{cases} x^{(1)}(t), & t \in [0,h], \\ x^{(2)}(t), & t \in [h, 2h], \\ \dots & \dots \\ x^{(k)}(t), & t \in [(k-1)h, kh]. \end{cases}$$

is the solution of (2.83), $\dot{x}(t) = f(t, x_t)$, under initial functional condition $x_0 = x^{(0)}(t)$, $t \in [-h, 0]$. As mentioned earlier, we can have an exact solution or an approximate one. This depends on the way we estimate integrals.

The steps method is used, primarily, for the existence or approximation of the solution. The literature contains examples from stability theory and other global problems related to various kinds of functional equations (see Kalmar-Nagy [273]).

The *third* item we want to discuss in this section is related to the concept of invariance, presented in Section 2.12.

From our exposition of the concept and its characterization, one sees that some "boxes" are able to contain all trajectories (in the future) of all points belonging to them, can be naturally detected from the equation and its domain of investigation. The book by Prüss and Wilke [465] presents a criterion for the construction/detection of invariance domains to ordinary differential equations. We briefly present this criterion, which could lead to useful results concerning the problem of invariance in dynamical systems, particularly in their design.

Invariance criterion (Prüss, Wilke). Consider the system $\dot{x}(t) = f(t, x(t))$, with $x, f \in R^n$, f continuous on $R \times G$, $G \subset R^n$. By $x(t; t_0, x^0)$, $x^0 \in G$, one denotes the solution of the Cauchy problem $x(t_0) = x^0$ and assumes uniqueness of the solution for $t > t_0$ (to the right). Denote by

$\phi(x_1,x_2,\ldots,x_n)$ a continuously differentiable scalar function and assume $grad\phi(x_1,x_2,\ldots,x_n) \neq \theta$ for $x \in \phi^{-1}(a_1,a_2,\ldots,a_n)$, with $a = (a_1,a_2,\ldots,a_n)$ a regular point for ϕ (*i.e.*, $grad\,\phi(x) \neq \theta$, at $x = a$). Then, the following two conditions are equivalent for the considered system:

1) The set $D = \phi^{-1}((-\infty,a])$ is invariant (to the right).
2) $< f(t,x), grad\phi(x) > \leq 0$ for $t \in R$ and $x \in \phi^{-1}(a) = \partial D$.

The details of the proof are in Ref. [465].

2.14 BIBLIOGRAPHICAL NOTES

The existence of solutions of functional equations is the main problem in the investigation of such mathematical objects. At least, it is from the point of view of mathematicians. In the numerous applications of these equations, when the existence is not known, methods of constructing "approximate solutions" are in use, providing some help in approaching the real world or social phenomena.

The literature related to various classes of functional equations, including the classical ones (differential, integral, integro-differential) is of considerable dimensions. We shall dwell with those bibliographical items that are, by their nature, mathematical and besides the existence, we deal in this chapter with some related problems, such as uniqueness, global behavior, flow invariance. An application to optimal control theory is included, using only general results from Functional Analysis and differential equations.

Existence of solutions, of various classes of functional differential equations, involves usually, the equation itself and additional conditions: initial (Cauchy), boundary value, qualitative properties (like global existence), in general, to possess a certain property. All these features are emphasized in the vast existing literature, and partially illustrated in this chapter.

A good number of sources dedicated to the general theory of classical equations (ordinary differential or integral functional equations) is provided in Chapter 1. Also, reference is made to some contributions related to new types of functional equations, appeared and developed simultaneously with Functional Analysis. It is symptomatic that founders of Functional Analysis, like Volterra [528], Radon [467], and Riesz [470], also considered in their investigation some classes of functional equations. Tonelli [518] started the investigation of functional equations with abstract Volterra (causal) operators.

The paternity of the results, included in Chapter 2, is mentioned in each case.

We will now list, chosen from our section of references, several papers and authors whose results complete the picture of "general existence type," taking into account different function spaces in which the solution of the equation is sought.

In Burton's paper [82], the concept of a large contraction is defined, for a metric space, and then applied to functional equations of the form $V(t,x) = S(t, \int_0^t H(t,s,x(s))\,ds)$. The results obtained in this paper, also using an extension of Krasnoselskii's fixed point theorem and compactness criteria and implicit functions, are largely susceptible of generalization to other classes of functional equations. See also Burton's books [80, 81, 83, 84], which contain a wealth of results for functional equations, of different types, including existence and behavior.

In Ref. [52], Bedivan and O'Regan investigated the set of fixed points for abstract Volterra operators in Fréchet space.

Berezansky and Braverman [58] provided conditions for the existence of integrable solutions for impulsive delay differential equations.

Brauer and Sternberg [76] dealt with uniqueness problems, existence in the large of solutions and the convergence of the method of successive approximations (for ordinary differential equations only). Another source of extension to more general functional equations.

Campanini [86] investigated the existence in the large and the partial boundedness of systems of differential equations by the comparison method.

Cassago and Corduneanu [92] investigated the ultimate behavior for solutions of some nonlinear integro-differential equations.

Colombo et al. [109] studied the existence of continuous solutions of differential inclusions.

Corduneanu [117] proved existence and uniqueness for differential equations with Volterra operators. In Ref. [122] studied asymptotic behavior in systems with infinite delay. The paper [136] dealt with functional differential equations with Volterra operators (causal) and their control. In Ref. [142] he investigated existence in neutral functional equations, with abstract Volterra operators, proving existence of solutions. The same problem was dealt with, under different conditions, in Ref. [145]. The paper [146] was concerned with discrete systems of neutral type.

Corduneanu and Li [168] provided conditions assuring the global existence of solutions to general functional differential equations with causal operators. In Ref. [171], they proved the existence of solutions to certain neutral functional equations of convolution type.

Corduneanu and Mahdavi [175–177] proved existence results for neutral functional equations with causal operators. In Ref. [180], they established existence of solutions to neutral functional equations, on a half-axis.

Cushing [184, 185] obtained general admissibility results for equations with operators, in Banach space, building up a framework that covers many preceding cases, investigated in the literature.

Diagana and Nelson [194] obtained existence results for higher order evolution equations with operator coefficients.

Driver's papers [201,202] are seminal papers with regard to the basic results for functional differential equations. His book [203] is partly dedicated to fundamentals in the theory of functional differential equations with abstract Volterra operators.

Górniewicz's book [226] is concerned with fixed-point results for multivalued maps, which is a helpful procedure to prove existence of solutions set, to functional equations with multivalued unknown maps.

Hakl et al. [236] dealt with boundary value problems for scalar functional differential equations.

Opluštil [424] obtained existence of solutions, with periodic-type boundary value conditions, for a rather general class of functional differential equations.

Mawhin [389] applied topological degree methods in treating nonlinear boundary value problems.

The monograph [262] by Hino, Murakami, and Naito is dedicated to the theory of functional differential equations with infinite delay.

The choice of space of initial data, for infinite delay equations was considered by Hale and Kato [241] and Atkinson and Haddock [27], and a conclusion might be that there is no single answer to this problem.

Corduneanu [125] and Kwapisz [304, 305] applied Bielecki's method to prove global existence theorems for various classes of functional differential equations. In Ref. [305], Kwapisz advanced several extensions of the original method.

The monograph [311] by Lakshmikantham, Leela, Drici, and McRae is dedicated to the basic results in causal differential equations, including global existence, the comparison method with application to stability, and other global problems.

The paper by Lakshmikantham and Rama Mohana Rao [316] is a comprehensive source in the theory of integro-differential equations, with a list of references including the original contributors to the field.

Drakhlin and Litsyn [199] treated the problem of defining the Volterra operators in a general sense, while Litsyn [335, 336] investigated the linear Volterra operators in Banach space, and the nonlinear ones were considered in connection with infinite systems and normal forms of ordinary differential equations.

Lupulescu [341,342] treated basic problems for causal functional differential equations in Banach spaces, with functional analysis methods.

The paper [339] by Logemann and Ryan provides one of the best available sources in defining rigorously the concept of causal/Volterra differential equations, establishing all basic results: existence, uniqueness, and continuation of solutions.

Mahdavi [347, 348] discussed the basics of linear and neutral functional differential equations of causal type. In Ref. [349, 350] he investigated the asymptotic behavior, including second order equations in the above mentioned categories. Mahdavi and Li [352] presented the basic concepts for linear or quasilinear equations of causal type.

Marcelli and Salvadori [358] discussed the equivalence, of various definitions, for the concept of hereditary processes, in case of ordinary differential equations.

Mickens book [395] is devoted to difference equations, both theory and applications.

Mitropolskii [400] treated nonlinear differential equations with deviated argument. Mitropolskii et al. [401] investigated the dichotomy of linear systems of differential equations by means of Liapunov function method.

Myshkis and Eĺsgoĺtz, main contributions to the theory of (nonclassical) functional equations, presented in Ref. [412] the state of the theory of these classes of equations (to 1967).

Okonkwo [420] and Okonkwo et al. [421, 422] investigated existence, admissibility, and other properties for stochastic type of functional equations.

O'Regan and Precup [430] treated several problems related to functional equations, by the method of Leray–Schauder, including existence.

Pennequin [444] investigated the existence of solutions to discrete time systems in the Besicovitch space of almost periodic sequences, $B^2(Z,X)$.

Razumikhin [469] dealt with hereditary systems (mostly stability), and the presented his own details in regard to stability of this type of functional equations.

Rontó A. and Rontó M. [471], in an ample section of the *Handbook of Differential Equations* (Battelli and Fĕckan, Eds.), presented the successive approximations in the framework of nonlinear boundary value problems.

Rouche and Mawhin [475] offered a modern approach to the theory of ordinary differential equations, with main concern for stability and periodicity of solutions.

Rus [479], in his book on fixed points, included several existence results for various types of functional equations. In Ref. [481], he was concerned with Ulam kind of stability for ordinary differential equations.

Tabor [515] treated "Differential equations in metric spaces," a generalization of the classic theory.

Vath [527] investigated, in this book, basic problems for Volterra and integral vector functions.

Ważewski [533] laid, in this paper, the ground for further use of generalized differential inequalities, in the study of qualitative problems for functional differential equations.

O'Regan [425] provided conditions for the existence of solutions to the inclusion $x(t) \in (Nx)(t)$, $t \in [0,T]$, in a Banach space setting.

Aubin and Cellina, in their book [29], which is a classical reference work, provided the basic theory for differential inclusions in case of set-valued mappings. They also treated the viability theory.

Aizicovici and Staicu [14] investigated inclusions, with multi-valued Banach space maps, of the form $u'(t) \in -Au(t) + F(t,u(t))$, $t \in [0,T]$, $u(0) = g(u)$. Applications were given to obstacle problem (partial differential equations).

Corduneanu and Li [168] provided a general existence of global solutions for functional equations with causal operators.

We will now mention some contributions related to classes of functional equations we did not treat in this book. The categories we have in view are those for *fuzzy* functions, the *fractional* functional differential equations and time scale equations.

Diamond [195] presented elements of the general existence theory for fuzzy Volterra integral equations, of the form $y'(t) = G(t) + \int_0^t F(t,\sigma,y(\sigma))\,d\sigma$. The references include several recent contributions to the topic. Inclusions were also considered, as well as applications.

Seikkala [492] provided conditions for the solutions of the initial value problem, for fuzzy systems.

Sultana and A.F.M.K. Khan [510] constructed power series solutions to fuzzy differential equations. The method they used is based on the classical Picard one (successive approximations).

Puri and Ralescu [466] dealt with some basic concepts, related to fuzzy theory of functions, with particular concern for the differentiability.

Satco [490] dealt with a Cauchy-type problem in a *time scale* setting, providing some applications.

Kulik and Tisdell [300] also investigated Volterra equations on time scales, establishing some qualitative results and applying them to the solution of initial value problem on unbounded intervals.

Kilbas et al. [285] presented a book titled *Theory and Applications of Fractional Differential Equations*, Elsevier, 2006.

Tidke [516, 517] discussed the basic results for fractional evolution equations, with special attention on semilinear case.

Miller and Ross [397] treated the basic results and methods of the theory of fractional functional differential equations.

Abbas and Benchohra [1]: Fractional-order Riemann–Liouville integral inclusions with two independent variables and multiple time delays is dedicated to treat the fractional equations in case of several independent variables.

Abbas et al. [2] presented a book, with multiple topics, related to fractional differential equations.

Agarwal et al. [9] treated the existence problem for solutions of fractional integro-differential equations, with non-local three-point fractional boundary value conditions.

Benchohra et al. [54] were concerned with bounded solutions, on a half-line, to initial problems for fractional differential equations.

Cernea [95, 96] investigated the existence of continuous relations of solutions to fractional differential inclusions and discussed a result of Filippov in case of fractional equations.

Diethelm [196] provided in book form the analysis of fractional differential equations.

Lani-Wayda [321] investigated in depth the character of *erratic* solutions to simple delay equations. References are abundant.

Li, Zhao, and Liang were also concerned with the chaotic behavior in a class of delay differential equations [330], based on a long list of references.

Devaney [192] produced a book on the subject of chaotic dynamical systems and proposed a definition of "chaos."

Sofonea et al. [503] studied some dynamics of viscoplastic materials, by applying a new and interesting fixed point theorem.

Turinici [522] investigated functional (integral) inequalities, deriving the results from general theorems about iterated operators in ordered linear spaces.

Mawhin [389] investigated, in this monograph, a series of results related to various boundary value problem for several types of functional differential equations (one independent variable), providing also the necessary tools from nonlinear functional analysis, in particular, methods based on the topological degree. Periodic solutions and bifurcation theory were also treated in this volume.

Furumochi [215] proved the existence of periodic solutions for a special class of delay differential equations.

Muresan [409] studied the existence of solutions to a second-order functional differential equation.

Suarez [509] investigated functional differential equations with advanced argument, with applications in economics/business. This is related to Georgescu-Roegen's *Analytical Economics*, from Harvard University Press.

Petrusel et al. [450] is a presentation of fixed point theory, in generalized metric spaces (Luxemburg), with detailed proofs and a rich list of references.

Mustafa [410] studied the existence of solutions with prescribed asymptotic behavior, to second-order equations.

A. S. Muresan [408] applied a lemma due to Tonelli, to the study of certain global properties to differential equations.

Antosiewicz [25] used the comparison method to investigate the error when dealing with approximate solutions to ordinary differential equations.

Pavel [440] constructed approximate solutions for Cauchy problems with differential equations in Banach spaces.

Basit [48] relied on harmonic analysis to investigate the asymptotic behavior of solutions to abstract Cauchy problems.

Bernfeld et al. [60] investigated stability properties of invariant sets of functional differential equations.

Brauer and Sternberg [76] were concerned with local uniqueness, global existence, and convergence of successive approximations for differential equations.

Corduneanu and Lakshmikantham [167] surveyed the contributions to the theory of equations with infinite/unbounded delay, up to 1980. Corduneanu [144] surveyed over 100 contributions on abstract Volterra operators (i.e. causal), published from the inception of this concept [518] until the year 2000.

Azbelev [32] signaled new trends in the theory of nonlinear functional differential equations. Most of the contributions of this author, as well as those of several of his students/collaborators, dealt with functional differential equations of the form $(Ly)(t) = f(t)$, in the linear case, and $(Ly)(t) = (Ny)(t)$ in the nonlinear case.

Hokkanen and Morosanu [266] treated in this monograph several problems in the theory of partial differential equations. It is likely that some of the topics have place in case of functional differential equations with a single independent variable.

In concluding this list of references, related to the topics treated in our book, we will point out the fact that many more publications are dedicated by their authors to applied problems leading to various types of functional or functional differential equations.

We will mention again the book [291] by Kolmanovskii and Myshkis, which contains a large selection of applied problems that can be approached by the means of functional differential equations.

Gopalsamy [225] dedicated the whole book to applications in population dynamics.

Caputo [88], while treating applied problems in elasticity, also provided the methods of solution by introducing new classes of functional differential equations (Caputo's derivative).

3

STABILITY THEORY OF FUNCTIONAL DIFFERENTIAL EQUATIONS

As mentioned in Chapter 1, stability properties are one of the most encountered directions in investigating the properties of solutions to various classes of functional equations. Stability properties are particularly interesting for their applications in science and engineering.

Intuitively, the stability property of a solution to a functional equation means the continuous dependence of the solutions, in the neighborhood of the given one, with respect to the perturbations (initial or permanent). Unlike the case when the dependence is considered on a finite interval (which is a simpler problem), the *stability* involves the dependence on a whole semiaxis, which is a feature that is conducing to more complex problems. In this chapter, we examine the *stability* of solutions in its simplest formulation. In addition, we will examine some refinements of this concept, such as *uniform stability*, *asymptotic stability*, or *uniform asymptotic stability*. The *exponential asymptotic stability* is also frequently encountered in applications.

The *preservation* of a given kind of stability, when one perturbs the equations describing the dynamic process, is of interest and will be also dealt with in this chapter.

Functional Differential Equations: Advances and Applications, First Edition.
Constantin Corduneanu, Yizeng Li and Mehran Mahdavi

Our attention will be concentrated on the so-called comparison method in the theory of stability. This method has gained a wide range of applications and particular attention. It is still under development and consists of a systematic use of differential inequalities and one or more Liapunov functions/functionals, which are subject to such inequalities. With their help, the information obtained by means of the differential inequalities is transfered to the solutions of the equations under investigation. This method is one of the so-called reduction methods, aimed at reducing a problem to a simpler one (e.g., reducing the dimension of the space in which the problem is considered).

Other methods will be also illustrated in various cases of functional equations.

Due to the fact that the stability theory is of utmost significance for applications, the contributions are extensive and varied. In Section 3.12, we indicate references pertaining to particular fields of applied science where stability is the main concern.

The concept of *partial stability* is also encountered in many applications (e.g., the stability of an aircraft during a flight). While Liapunov's work contains some elements of the theory of partial stability for ordinary differential systems, the real progress was emphasized much later, particularly in the research conducted in Russia (former Soviet Union). This kind of stability will be also investigated in this chapter.

3.1 SOME PRELIMINARY CONSIDERATIONS AND DEFINITIONS

The stability theory, as it is presently understood, has developed over the course of more than two centuries. It is not our aim to provide a detailed account of this significant chapter of modern science (mathematics, mechanics, control theory, engineering sciences, stochastic analysis, population dynamics, etc.), which involves scientists like Lagrange, Poisson, Jacobi, Lejeune-Dirichlet, Poincaré, Zhukovskii, N., and Liapunov. We mention here a recent contribution by R.I. Leine [325], which contains a detailed discussion of the topic, illustrated by useful references. Both aspects, mathematical and physical, are emphasized and old Latin or current Russian appear in the text and references. Of course, this fact proves the extent of stability theory, in time and geographically.

Functional equations, in their varied realizations, constitute the most fertile ground for developments of the stability theory, a fact already proven by the progress made during their development and applications.

Nowadays, it would be a gigantic task to undertake the writing of a book on this subject, even when addressing only the main trends and aspects pertaining to it. We will focus primarily on the comparison method, adapted to various types of functional equations and illustrate other approaches by aiming at applications.

We want to stress the fact that the literature, in its vastity on this subject, contains a good deal of items related to its theoretical/mathematical aspects, as well as to the applications.

For instance, books entirely or partially dedicated to stability theory have been authored by Liapunov [332], Malkin [356], Cesari [97], Hahn [235], Yoshizawa [539], Bellman [53], Barbashin [43], Krasovskii [299], and Nemytskii and Stepanov [416].

The following references have entries relating to comparison method: Halanay [237], Sansone and Conti [489], Corduneanu [123], Lakshmikantham and Leela [309], Matrosov et al. [385], Matrosov and Voronov [387], Lakshmikantham et al. [311], Martynyuk [363,368], Rumiantsev and Oziraner [478], and Vorotnikov [531].

Before formulating some definitions in regard to the problem of stability in the theory of functional differential equations, let us recall that in Section 2.5, we presented the comparison method (sometimes called "comparison principle") in the case of ordinary differential systems/equations

$$\dot{x}(t) = f(t, x(t)),\ t \in [t_0, T),\ T \leq \infty,\ x, f \in R^n. \tag{3.1}$$

A basic hypothesis on the Liapunov's function $V = V(t,x)$, with real values, was its *differentiability* in $(t,x) \subset \Omega \subset R^{n+1}$, the domain Ω being usually a half space of R^{n+1} (say, $t \geq 0$), or a cylindrical shape domain, with symmetry about t-axis.

In the paper by Corduneanu [114], one considers more general assumptions on $V(t,x)$, namely, substituting the differentiability by a *local Lipschitz condition.* This approach leads to somewhat more general results, but it does not imply fundamentally different procedures in deriving Corollary 2.3. That is why we maintain our considerations, relying on Proposition 2.2, instead of dealing with differential inequalities in which the derivative is replaced by Dini's numbers. See Ważewski [533] for this general case.

In case of ordinary differential equations of the form (3.1), with $f(t, \theta) = \theta$, the solution function $x = x(t; t_0, x^0)$, $t \geq t_0 \geq 0$, $x^0 \in R^n$, which satisfies

$$x(t_0) = x(t_0; t_0, x^0) = x^0 \in R^n,\ |x^0| < a, \tag{3.2}$$

with $a > 0$, the following property is taken as definition for the *stability* of the zero solution of (3.1).

In the theory of stability, one deals only with the case $T = \infty$, which means all solutions exist for $t \geq 0$.

Definition 3.1 *The solution $x = \theta \in R^n$ of equation (3.1) is called stable, if and only if (iff) for each $\epsilon > 0$ ($\epsilon < a$), and $t_0 \geq 0$, one can find $\delta = \delta(\epsilon, t_0) > 0$, such that*

$$|x(t; t_0, x^0)| < \epsilon, \; t \geq t_0, \tag{3.3}$$

provided $|x^0| < \delta$.

Definition 3.2 *The solution $x = \theta$ of equation (3.1) is called uniformly stable, iff it is stable and $\delta(\epsilon, t_0)$ in Definition (3.1) is independent of t_0: $\delta(\epsilon, t_0) \equiv \delta(\epsilon) > 0$.*

Definition 3.3 *The solution $x = \theta$ of equation (3.1) is called asymptotically stable iff it is stable (Definition 3.1), and for $t_0 > 0$, there exists $\gamma(t_0) > 0$, such that*

$$\lim_{t \to \infty} |x(t; t_0, x^0)| = 0, \tag{3.4}$$

for all x^0 with $|x^0| < \gamma(t_0)$.

Definition 3.4 *The solution $x = \theta$ of (3.1) is called uniformly asymptotically stable, iff it is uniformly stable (Definition 3.2) and for $\epsilon > 0$, there exists $T(\epsilon) > 0$, such that*

$$|x(t; t_0, x^0)| < \epsilon \text{ for } t \geq t_0 + T(\epsilon), \tag{3.5}$$

for all x^0 with $|x^0| < \gamma_0$, $\gamma_0 = constant > 0$.

Definitions 3.1–3.4, formulated for systems of ordinary differential equations, will serve as model when dealing with other types of functional equations, when initial data are different from (3.2), or permanent perturbations are present in the system.

Besides the concepts of stability defined earlier, one encounters also the so-called exponential asymptotic stability. Its definition is as follows:

Definition 3.5 *The solution $x = \theta$ of equation (3.1) is called exponentially asymptotically stable iff there exist constants $k, \alpha > 0$, such that*

$$|x(t; t_0, x^0)| \leq k|x^0| e^{-\alpha(t - t_0)}, \; t \geq t_0, \tag{3.6}$$

for $|x^0| < \gamma_0$.

If $\gamma_0 = \infty$, then we say that we have global exponential asymptotic stability.

Definition 3.1 is the original Liapunov's definition. Definition 3.2 is due to Persidskii [447], while Definition 3.4 is due to Malkin [356].

It is clear that all definitions formulated in the text are concerned with the solution $x = \theta$, which means that one deals with the stability of the "equilibrium" state ($x = \theta$).

What happens if we are interested in the stability of an arbitrary solution $x = \bar{x}(t)$ of (3.1)?

The case can be reduced to the stability of the equilibrium state $x = \theta$, by means of the substitution $x = y + \bar{x}(t)$. Indeed, for y one obtains the equation

$$\dot{y}(t) = f(t, y(t) + \bar{x}(t)) - \dot{\bar{x}}(t),\ t \geq 0,$$

for which $y = \theta$ is obviously a solution. Clearly, this reduction method is rather theoretical than practical, but it is logically valid. It is applicable if $\bar{x}(t)$ is known from an existence result.

Besides the most frequently encountered case of equations/systems of the form (3.1), we will deal with more general types of functional differential equations/systems, such as the *causal functional differential equations* considered in Sections 2.1 and 2.2

$$\dot{x}(t) = (Vx)(t),\ t \in R_+, \tag{3.7}$$

under the initial condition

$$x(0) = x^0 \in R^n, \tag{3.8}$$

of the *first kind*, or, with $t_0 > 0$,

$$x(t) = \phi(t),\ t \in [0, t_0),\ x(t_0) = x^0 \in R^n, \tag{3.9}$$

the *second kind*, in which case one must assign $\phi(t)$, on $[0, t_0]$, from a certain function space $E = E([0, t_0], R^n)$.

Of course, here we are keeping the terms and notations we have used in Chapter 2.

The general form (3.7), for causal operator V, allows us to deal with various special choices, such as the integro-differential operators of Volterra type considered in Chapter 2.

The book by Corduneanu [149] contains results on the existence and stability for both *first* or *second* kind initial value problems. We rely on some results or methods from this reference, with regard to the stability theory.

In order to encompass the case of functional differential equations with finite delay, we will keep $t_0 > 0$ constant in (3.4), and formulate the definitions of stability, similar to Definitions 3.1–3.4. Assuming the uniqueness of the solution to (3.7) and (3.9), the following formulation is adequate.

Definition 3.6 *The solution of problem (3.7) and (3.9) is stable, iff for each* $\epsilon > 0$, *one can find* $\delta = \delta(\epsilon, t_1) > 0$, $t_1 > t_0$, *such that the solution corresponding to the initial condition*

$$y(t) = \psi(t), \ t \in [0, t_0], \ x(t_0) = x^1 \in R^n, \tag{3.10}$$

satisfies

$$|y(t) - x(t)| < \epsilon, \ t \geq t_1, \tag{3.11}$$

provided

$$|\psi(t) - \phi(t)|_E, \ |x^0 - x^1| < \delta, \tag{3.12}$$

where the index E to the norm denotes a function space $E([0, t_0], R^n)$, *chosen for initial data.*

Remark 3.1 *The space E can be chosen as dictated by the nature of the operator V in equation (3.1). If E consists of measurable functions, then the pointwise parts in (3.9) and (3.12) make sense. It does not, in case the space E is a space containing only continuous functions (in which case,* $E = C$ *or a subspace of C, and* $\phi(t_0) = x^0$, $\psi(t_0) = x^1$ *imply (3.11) due to the first part of this inequality).*

Definitions for other types of stability in case of problem (3.7) and (3.9), can be formulated following the same manner as in Definitions 3.1–3.5.

Let us conclude this section by making reference to the book by Corduneanu [149], in which the second type of initial conditions are considered in detail for linear V, and necessary and sufficient conditions are formulated for various types of stability. Also, in the papers by Corduneanu [138], and Li [328, 329], several results of stability for general functional differential equations with causal operators are given, as well as some open problems. Further results are available in the literature, and some of them will be featured in Section 3.12.

3.2 COMPARISON METHOD IN STABILITY THEORY OF ORDINARY DIFFERENTIAL EQUATIONS

When Liapunov [332] published his famous paper *Ustoichvosti Dvizhenya*, (*Stability of Motion*), it became clear that differential inequalities would play a significant role in the investigation and development of this theory. The French translation of this seminal paper (Annales de l'Université de Toulouse, 1905) has immensely contributed to the diffusion of this theory in the world's mathematical and engineering literature.

Later on, Chaplyguine published his work on approximation of solutions to ordinary differential equations, based on the theory of differential inequalities. His basic result is surprisingly simple; it states that the solution of a scalar differential inequality, like $y'(t) \leq f(t,y(t))$, $t \in [t_0,T)$, is dominated by the solution of the associated equation $x'(t) = f(t,x(t))$, if $y(t_0) \leq x(t_0)$.

Even though this elementary result has been used extensively by many authors to find estimates for solutions of ordinary differential equations, it is always considered in special cases (choices of f).

It was not until 1956 when R. Conti [110] and O. Boruwka [74] independently came up with results involving *general*-type Liapunov's functions and differential inequalities. Conti dealt with global existence of solutions to ordinary differential equations of the form (3.1), while Boruwka treated the uniqueness problem.

The following theorem of stability holds true for differential systems of the form (3.1):

Theorem 3.1 *Consider system (3.1) in Ω, where Ω denotes the half-cylinder*

$$\Omega = \{(t,x);\, t \in R_+,\, x \in R^n,\, |x| < a\}, \tag{3.13}$$

with $n \in N$, $a > 0$.

Assume that $x = \theta \in R^n$ is a solution of (3.1), while $f(t,x)$ is locally Lipschitz in Ω (hence, continuous there).

We also assume that the function $V = V(t,x) : \Omega \to R_+$ satisfies $V(t,\theta) = 0$, $t \in R_+$, and the differential inequality

$$V_t + f \cdot \mathrm{grad}_x V \leq \omega(t,V) \text{ in } \Omega. \tag{3.14}$$

The comparison function $\omega(t,y)$ is such that the scalar equation

$$\dot{y}(t) = \omega(t,y(t)) \tag{3.15}$$

possesses the properties of existence and uniqueness through each $(t_0,y_0) \in (R_+,R_+)$, while $\omega(t,0) = 0$, $t \in R_+$ (hence $y = 0$ is unique).

Then, denoting by $x(t;t_0,x^0)$ the unique solution of (3.1), whence

$$x(t_0;t_0,x^0) \equiv x^0, \ (t_0,x^0) \in \Omega. \tag{3.16}$$

The following statements are valid:

I. *If $V(t,x)$ also satisfies*

$$V(t,x) \geq \lambda(|x|), \ (t,x) \in \Omega, \tag{3.17}$$

where $\lambda(r)$, $r \in R_+$, is such that $\lambda(0) = 0$, $\lambda(r) > 0$ for $r \in (0,a)$ and increasing, the stability/asymptotic stability of the solution $y = 0$ of (3.15) implies the stability/asymptotic stability of the solution $x = \theta$ of (3.1).

II. *If $V(t,x)$ satisfies the properties in statement I above and also*

$$V(t,x) \leq \mu(|x|), \ (t,x) \in \Omega, \tag{3.18}$$

with $\mu : R_+ \to R_+$ having the same properties as λ in I, the uniform stability/uniform asymptotic stability of the solution $y = 0$ of equation (3.15) implies the uniform stability/uniform asymptotic stability of the solution $x = \theta$ of (3.1).

Proof. As seen in Section 2.4, if in inequality (3.14) we take $x = x(t;t_0,x^0)$, $t \geq t_0 \geq 0$, $|x^0| < a$, a solution which is defined in an interval $[t_0,T)$, $T \leq \infty$, and denote

$$v(t) = V(t,x(t;t_0,x^0)), \ t \geq t_0, \tag{3.19}$$

then we obtain

$$\dot{v}(t) \leq \omega(t,v(t)), \ t \in [t_0,T), \tag{3.20}$$

which constitutes a differential inequality associated to the comparison equation (3.15). Hence, on behalf of Proposition 2.2, one obtains

$$v(t) = V(t,x(t;t_0,x^0)) \leq y(t;t_0,y_0), \ t \in [t_0,T), \tag{3.21}$$

where $y(t_0) \geq V(t_0,x^0)$. But the solution $y = 0$ of (3.15) is stable, which means that for every $\epsilon > 0$, $t_0 > 0$, one can find $\delta(\epsilon,t_0) > 0$, such that

$$y(t;t_0,y_0) < \lambda(\epsilon), \ t \geq t_0. \tag{3.22}$$

Returning to (3.21), and relying on (3.17), one obtains from (3.22)

$$\lambda(|x(t;t_0,x^0)|) < \lambda(\epsilon),\ t \geq t_0, \tag{3.23}$$

if one chooses x^0 such that $V(t_0,x^0) \leq y_0$. But, from (3.18), one sees that this takes place when $\|x^0\| \leq \mu^{-1}(y_0)$. Therefore, (3.23) leads to

$$|x(t;t_0,x^0)| < \epsilon,\ t \geq t_0, \tag{3.24}$$

which implies the fact that $x(t;t_0,x^0)$ is defined on $t \geq t_0$, while $x = \theta$ is stable. The fact that $x(t;t_0,x^0)$, for $|x^0| < \mu^{-1}(y_0)$, is defined on $t \geq t_0$, follows from the elementary property stating that $x(t;t_0,x^0)$ cannot have "escape time" (see, e.g., Corduneanu [114]).

It remains to prove the asymptotic stability for the solution $x = \theta$ of (3.1) when this property is valid for the comparison equation (3.15). It suffices to consider again (3.21) and notice that

$$\lambda(|x(t;t_0,x^0)|) \leq y(t;t_0,y_0),\ t \geq t_0, \tag{3.25}$$

from which we derive

$$|x(t;t_0,x^0)| \leq \lambda^{-1}(y(t;t_0,x^0)),\ t \geq t_0. \tag{3.26}$$

Obviously, $|x(t;t_0,x^0)| \to 0$ as $t \to \infty$, because $y(t;t_0,x^0) \to 0$, according to our assumptions.

This ends the proof of part I of Theorem 3.1.

Moving to part II, in case of the uniform stability, we notice from the discussion that (3.22), according to our assumption (of uniform stability), the number $\delta = \delta(\epsilon,t_0)$ can be chosen independent of t_0, a property automatically transmitted to the system (3.1), if we rely on the inequality $V(t,x) \leq \mu(|x|)$.

For uniform asymptotic stability, the inequality (3.26) is conducing to

$$\lambda(|x(t;t_0,x^0)|) \leq y(t;t_0,y_0) < \lambda(\epsilon),\ t \geq t_0 + T(\epsilon), \tag{3.27}$$

for all $y_0 < \gamma_0$. Since for $t = t_0$ one obtains, from (3.27), $\lambda(|x^0|) \leq y_0$, there results $|x^0| < \lambda^{-1}(\gamma_0) = \mathit{constant}$, as required by Definition 3.4. One derives also from (3.27),

$$|x(t;t_0,x^0)| \leq \epsilon,\ t \geq t_0 + T(\epsilon), \tag{3.28}$$

which is required for uniform asymptotic stability of the solution $x = \theta$ of (3.1). This ends the proof of property II and of Theorem 3.1.

Remark 3.2 *The concepts of stability and asymptotic stability were introduced by Liapunov. The refinements of uniform stability and uniform asymptotic stability were introduced by Persidskii [447] and Malkin [356], respectively.*

We will conclude this section with a result of *instability* for the solution $x=\theta$ of system (3.1). Let us point out that instability is the contrary property of stability. In particular, if there are unbounded (on R_+) solutions of (3.1) with arbitrarily small (in norm) initial value $|x^0|$, then the property of instability is obviously assured.

Theorem 3.2 *Consider system (3.1), under the following assumptions:*

i) $f:\Omega\to R^n$, *with* Ω *defined by (3.14), is locally Lipschitz.*

ii) There exists an open set $G\subset\Omega$ *and a differentiable function* $V:G\to R$, *positive and bounded on* $\bar{G}$.

iii) $\partial G\cap R_+$ *contains a point* (T,θ), $T>0$.

iv) $V(t,x)=0$ *for* $(t,x)\in\bar{G}_\Omega\setminus G$, *where the index* Ω *stands for closure relative to* Ω.

v) The comparison inequality is valid in G.

$$V_t+f\cdot\operatorname{grad}_x V\geq\omega(t,V)>0.$$

vi) For $t_0>T$, *the solution* $y(t;t_0,y_0)$ *of equation (3.15), with arbitrarily small* y_0, *either does not exist on* $t\geq t_0$, *or it is unbounded there.*

Then, the solution $x=\theta$ *of system (3.1) is unstable.*

Proof. Let us notice that *iii*) implies the existence of points $(t_0,x^0)\in G$, $x^0\neq\theta$, as close as we want from (T,θ), while *ii*) and *v*) imply

$$V(t;t_0,x^0)\geq V(t_0,x^0)=y_0>0,\ t\geq t_0, \tag{3.29}$$

provided $(t,x(t;t_0,x^0))\in G$. But *ii*) implies that (3.29) takes place on the whole interval of existence of the solution $x(t;t_0,x^0)$. Hence, from *v*) we obtain that

$$\frac{d}{dt}V(t,x(t;t_0,x^0))\geq\omega(t,V(t;t_0,x^0)) \tag{3.30}$$

holds true on the whole interval of existence of $x(t;t_0,x^0)$. On behalf of the theory of differential inequalities, one obtains from (3.30) the relationship

$$V(t,x(t;t_0,x^0))\geq y(t;t_0,y_0),\ t\geq t_0, \tag{3.31}$$

on the interval of existence of the solution $x = x(t;t_0,x^0)$. But (3.31) leads to contradiction, due to our assumption *vi*). Indeed, if the interval of existence for $y(t;t_0,y_0)$ is finite, say $(t_0,t_1]$, then for $t \to t_1$ in (3.31) we have the contradiction because the left-hand side must remain bounded, while $y(t;t_0,y_0) \to \infty$ as $t \to t_1$. Similar argument can be used when the solution $y(t;t_0,y_0)$, of the comparison equation $y'(t) \geq \omega(t,y(t))$ exists on R_+, but is unbounded there.

This ends the proof of Theorem 3.2.

We shall point out that Theorem 3.2 has several corollaries, for instance Corduneanu [114], that include the so-called Liapunov second theorem on instability, in which case the comparison function is $\omega(t,y) = \lambda y$, $\lambda > 0$.

3.3 STABILITY UNDER PERMANENT PERTURBATIONS

Since stability is a property of continuous dependence of solutions with respect to initial data, on a semiaxis, it appears natural to ask whether it is possible to build up a stability theory in which one also takes into consideration the perturbations of the right-hand side of equations like (3.1), namely

$$\dot{x}(t) = f(t,x(t)) + r(t,x(t)),\ t \geq 0. \tag{3.32}$$

The term $r(t,x)$, which represents the perturbation of $f(t,x)$, is also defined in Ω and is usually subject to a condition of "smallness."

One distinguishes two different situations in regard to equation (3.32). The *first* type corresponds to the conditions to be imposed on the perturbing term, in order to assure a certain kind of stability (i.e. *integral stability* or stability *with respect to bounded permanent perturbations*). The *second* type of condition assures the stability of the zero solution for the perturbed system. Sometimes, by perturbing a system like (3.1), one obtains system (3.32). We are interested in same type of stability for (3.32) under adequate perturbations. Usually, one can get for (3.32) the same or weaker kind of stability as assumed for (3.1).

Both aspects of the problem will be discussed in this section.

In a rather general framework, we can consider a space $E = E(R_+,R)$ for the perturbations, and define the stability of the zero solution of system (3.1), which means that $f(t,\theta) = \theta$, $t \in R_+$, as follows:

Definition 3.7 *The solution $x = \theta$ of (3.1) is called stable with respect to E-perturbations iff for any $\epsilon > 0$ and $t_0 \geq 0$, there exists $\delta = \delta(\epsilon,t_0) > 0$, such that*

$$|x^0| < \delta \text{ and } \sup_{|x|\leq\epsilon} |r(t,x)|_E < \delta, \tag{3.33}$$

with $E = E([t_0,\infty),R^n)$, *imply*

$$|x(t;t_0,x^0)| < \epsilon,\ t \geq t_0, \tag{3.34}$$

where $x(t;t_0,x^0)$ *denotes the solution of the perturbed system (3.32), with* $x(t_0;t_0,x^0) = x^0$ *for* $t_0 \geq 0$.

Definition 3.7 is due to Vrkoč [532] in the case $E = L^1(R,R^n)$, and he has shown that, in fact, it suffices to discuss the simplest case, when the perturbed system has the form

$$\dot{x}(t) = f(t,x(t)) + r(t),\ t \in R_+, \tag{3.35}$$

with $r \in L^1(R_+,R^n)$, that is, only integrable perturbations on R_+ are accepted.

Consequently, Definition 3.7 should be rephrased as follows:

The solution $x = \theta$ of (3.1) is called *stable* with respect to E-perturbations, if for each $\epsilon > 0$, and $t_0 \geq 0$, there exists $\delta = \delta(\epsilon,t_0) > 0$, such that the solution $x(t;t_0,x^0)$ of system (3.35) satisfies $|x(t;t_0,x^0)| < \epsilon$, $t \geq t_0$, provided

$$|x^0| < \delta,\ |r(t)|_E < \delta, \tag{3.36}$$

The *first* possible choice for the space E, to be dealt with in this section, corresponds to the case investigated by Vrkoč [ ], is called *integral stability*. In this case, (3.36) becomes

$$|x^0| < \delta,\ \int_{t_0}^{\infty} |r(t)|\,dt < \delta. \tag{3.37}$$

The *second* choice for the space $E(R_+,R^n)$ is the space $BC(R_+,R^n)$, as defined in Chapter 1, for which the norm is the usual supremum norm of elements: $x \to |x|_{BC} = \sup\{|x(t)|;\, t \in R_+\}$.

Definition 3.7 can be adapted to this case, and we reformulate it as follows:

The solution $x = \theta$ of (3.1) is called *stable* with respect to *bounded permanent perturbations* if for each $\epsilon > 0$ and $t_0 \geq 0$, there exists $\delta = \delta(\epsilon,t_0) > 0$, such that the solution $x(t;t_0,x^0)$, of the perturbed system (3.35), satisfies the condition $|x(t;t_0,x^0)| < \epsilon$ for $t \geq t_0$, provided

$$|x^0| < \delta,\ \sup_{t \geq t_0} |r(t)| < \delta. \tag{3.38}$$

Similar definitions can be formulated for other choices of the space E of perturbations, such as $E = M$, which leads to the *permanent perturbations bounded in the mean*, which was investigated by Germaidze and Krasovskii [216].

In order to obtain results concerning the stabilities with respect to permanent perturbations mentioned before, we need some auxiliary results related to the *converse* theorems. These theorems are of the following type: Assuming that the solution $x = \theta$ of (3.1) possesses a certain kind of stability (as those in Theorem 3.1), construct a Liapunov function with the properties described in the direct theorem for that particular kind of stability.

For instance, in case of integral stability, such a converse result was given by Vrkoč [532]. In case of uniform stability, J. Kurzweil [303] obtained a converse result to statement II in Theorem 3.1. We will mention these results and use them in this section.

First, let us mention a more general result than Proposition 2.2, concerning differential inequalities. Namely,

Proposition 3.1 *Consider the scalar differential inequality*

$$D_{-}y(t) \leq \omega(t, y(t)),\ t \in [t_0, T) \tag{3.39}$$

with $\omega(t,y)$ continuous on a domain $\Delta \subset R^2$, containing the segment $t_0 \leq t < T$, or the semi-axis $t \geq t_0$ when $T = \infty$. $D_{-}y(t)$ is the Dini number

$$D_{-}y(t) = \varliminf_{h \to 0+} \frac{y(t+h) - y(t)}{h}. \tag{3.40}$$

If $z(t)$, $t \in [t_0, T)$ is the maximal solution of the comparison equation

$$\dot{z}(t) = \omega(t, z(t)),\ z(t_0) \leq y_0, \tag{3.41}$$

then

$$y(t) \leq z(t),\ t \in [t_0, T). \tag{3.42}$$

The *proof* is similar to that of Proposition 2.2 and it is omitted.

Let us notice that, for a Lipschitz-type function, the derivative may not exist at some points, while the Dini numbers do. And we will encounter such cases.

Now, let us return to system (3.1), under the assumption of continuity, or Carathéodory conditions for $f(t,x)$ (i.e., measurability in t for fixed x and continuity in x for almost all t), and considering a locally Lipschitz function $V(t,x)$ in Ω, one defines *the derivative of $V(t,x)$, with respect to system* (3.1), by the formula

$$V^{\cdot}(t,x) = \varliminf_{h \to 0+} \frac{V(t+h, x+hf(t,x)) - V(t,x)}{h}. \tag{3.43}$$

If in (3.43) we take $x = x(t)$ a solution of (3.1), then we obtain

$$V^{\cdot}(t,x(t)) = D_{-}V(t,x(t)),\ t > 0,\ x \in R^n. \tag{3.44}$$

In the next results, we will consider the comparison inequality

$$V^{\cdot}(t,x) \leq \omega(t, V(t,x)), \tag{3.45}$$

which is meant to be valid in the half-space ($a = \infty$),

$$\Omega = \{(t,x);\ t \geq 0,\ x \in R^n\}, \tag{3.46}$$

and the global Lipschitz condition

$$|V(t,x) - V(t,y)| \leq L|x-y|, \tag{3.47}$$

where $L > 0$ is a constant, for $(t,x),(t,y) \in \Omega$, defined by (3.46).

We can now formulate the next theorem, related to *integral stability* of the solution $x = \theta$ of (3.1).

Theorem 3.3 *Consider system (3.1), under the following conditions:*

1) *$f(t,x)$ is defined on Ω, as shown in (3.46), is locally Lipschitz there, and $f(t,\theta) = \theta$, $t \in R_+$, which means $x = \theta$, is the solution of (3.1), on the whole positive semiaxis.*
2) *$V(t,x)$ is a scalar function defined on Ω, locally Lipschitz, and satisfies condition (3.17) of positive definiteness.*
3) *$V(t,x)$ satisfies the global Lipschitz condition (3.47), and the comparison inequality*

$$V^{\cdot}(t,x) \leq \omega(t, V(t,x)),\ (t,x) \in \Omega, \tag{3.48}$$

with $V^{\cdot}(t,x)$ given by (3.43), and $\omega(t,y)$ as described in Theorem 3.1.

Then, the solution $x = \theta$ of system (3.1) is integrally stable, if the solution $y = 0$ of the comparison equation $\dot{y} = \omega(t,y)$ is integrally stable.

Proof. Before getting into details of the proof, let us point out that, in case $\omega(t,y) \equiv 0$, the comparison equation in the case of system (3.35) is $\dot{y} = L|r(t)|$, whose solution is $y(t;t_0,y_0) = y_0 + L\int_{t_0}^{t} r(s)\,ds$. But this implies

$$|y(t;t_0,y_0)| \leq |y_0| + L\int_{t_0}^{\infty} |r(s)|\,ds, \tag{3.49}$$

which guarantees the integral stability of the zero solution of the comparison equation $\dot{y} = L|r(t)|$, $t \geq t_0 \geq 0$.

The proof of Theorem 3.3 will be accomplished, when, relying on a converse theorem of Vrkoč [532], we shall be able to take $\omega(t,y) \equiv 0$ in (3.48).

Indeed, the converse theorem on integral stability of systems of the form (3.1), can be stated as follows:

Consider system (3.1), under aforementioned hypotheses (Theorem 3.3) on f. If the solution $x = \theta$ of (3.1) is integrally stable (Definition 3.6), then there exists a scalar function $V(t,x)$ defined in $\Omega = \{(t,x); t \geq 0, x \in R^n\}$, such that

1) $V(t,x)$ is positive definite, (3.17).
2) $V(t,x)$ satisfies the global Lipschitz condition (3.47).
3) $V^{\cdot}(t,x)$, as defined by (3.43), satisfies the inequality $V^{\cdot}(t,x) \leq 0$, since $V(t,x(t))$ is non-increasing for $t \geq t_0 \geq 0$, for any solution $x(t)$ of (3.1).

Conditions 1)–3) are exactly those used in the proof of Theorem 3.3, which ends the proof of this theorem.

Remark 3.3 *We omit the proof of the converse theorem of Vrkoč (for integral stability). It can be carried out based on concepts and methods pertaining to classical real analysis, and we send the readers to the source. It also contains additional results on asymptotic integral stability and the use of Liapunov's functions.*

Now we shall consider, in Definition 3.7, the case of *bounded permanent perturbations*, which are fully described by elements of the space $E = BC(R_+, R^n)$, as defined in Chapter 1. See also conditions (3.38) in this section for initial restrictions.

Theorem 3.4 *Consider system (3.1), under the following assumptions:*

1) *Same as in Theorem 3.3, with Ω defined by (3.13).*
2) *The perturbing term $r(t,x)$, in (3.32), satisfies conditions assuming the local existence and uniqueness for (3.32).*
3) *There exists a scalar function $V(t,x)$, positive definite on Ω, $V(t,\theta) = 0$ and satisfying the local Lipschitz condition there, as well as the global one (3.47).*
4) *$V^{\cdot}(t,x)$, as defined by (3.43), satisfies the comparison inequality (3.48), with $\omega(t,y)$ as described in Theorem 3.1, part II.*

Then, the zero solution of system (3.1) is stable under bounded permanent perturbations, if the solution $y = 0$ of the comparison equation $\dot{y} = \omega(t,y)$ is stable under bounded permanent perturbations.

The *proof* of Theorem 3.4 is based on a converse theorem of Massera [373], whose proof is omitted due to rather complex technicalities.

The converse theorem of Massera, regarding the uniform asymptotic stability of the solution $x = \theta$ of (3.1), is also obtained under the assumption that $f(t,x)$ verifies the global Lipschitz condition

$$|f(t,x) - f(t,y)| \leq L|x-y|, \tag{3.50}$$

for $(t,x),(t,y) \in \Omega$, with $L > 0$ a constant.

Namely, the converse theorem asserts the existence of a function $V(t,x)$, $(t,x) \in \Omega$, verifying the following conditions:

$$\lambda(|x|) \leq V(t,x) \leq |x|, \; V^{\cdot}(t,x) \leq -\mu(|x|), \tag{3.51}$$

with λ and μ possessing the properties stipulated in Theorem 3.1. The inequalities in (3.51) tell that V is positive definite, while $V^{\cdot}$ is negative definite, with V having sublinear growth, as $|x| \to \infty$. These are conditions implying the uniform asymptotic stability of the zero solution of system (3.1), as shown in Theorem 3.2.

In order to prove the result on stability with respect to bounded permanent perturbations (i.e., E from Definition 3.7 is taken as space BC) which is defined in Chapter 1, we need an auxiliary result in regard to the Liapunov function to be used in the proof. Namely, we will prove that under conditions (3.50) and (3.51) there exists a Liapunov function for (3.1), say $W(t,x)$, such that the following properties are true:

Proposition 3.2 *Assuming that for system (3.1), there exists a Liapunov function $V(t,x)$, defined on Ω, such that (3.50) and (3.51) are verified, then there exists a Liapunov function $W(t,x)$, with the properties*

$$\lambda_0(|x|) \leq W(t,x) \leq \mu_0(|x|), \; W^{\cdot}(t,x) \leq -W(t,x), \tag{3.52}$$

in Ω, where $\lambda_0(r)$ and $\mu_0(r)$ have the same properties as the functions $\lambda(r)$ and $\mu(r)$ from above. Moreover, $W(t,x)$ satisfies also a global Lipschitz condition in Ω, of the form

$$|W(t,x) - W(t,y)| \leq K\,|x-y|, \tag{3.53}$$

where $K > 0$ is a constant.

Proof. We shall construct W as $W = \rho(V)$, with $\rho(r)$ a differentiable function on $[0,a]$, also verifying $\rho(0) = 0$, and possessing the same properties as

$\lambda(r)$, $\mu(r)$ earlier. On behalf of the second inequality (3.51), the last inequality leads to $V^{\bullet}(t,x) \leq -\mu_0(V(t,x))$. Let us determine now $\rho(r)$, in such a way that the earlier inequality for $V(t,x)$ is transformed into a similar one for $W(t,x) = \rho(V(t,x))$. One obtains, assuming $\rho(r)$ to be differentiable on $[0,a)$, $\rho(0) = 0$ and nondecreasing

$$W^{\bullet}(t,x) = \left(\frac{d\rho}{dr}\bigg|_{r=V} \right) \cdot V^{\bullet}(t,x) = \rho'(V)\, V^{\bullet}(t,x), \tag{3.54}$$

where ρ' denotes the derivative of $\rho(r)$, with respect to r. If we want to achieve the last inequality in (3.52), we obtain from (3.51) and (3.54) the following inequality/equation

$$\rho'(V)\, V^{\bullet}(t,x) \leq -\rho'(V)\, \mu(V) = -\rho(V), \tag{3.55}$$

which shows that we can take

$$W(t,x) = \rho(V(t,x)) = e^{\int_a^V \frac{dr}{\mu(r)}}, \tag{3.56}$$

for $0 < r \leq a$, with $\rho(0) = 0$. Actually, this condition means $\int_a^r [\mu(r)]^{-1}\, dr \to -\infty$ as $r \to 0^+$, because (3.51) imply $V^{\bullet}(t,x) \leq -\mu(V(t,x))$, and since (3.51) also implies asymptotic stability (Theorem 3.1), there results—along any solution of (3.1)—that $V(t,x(t)) \to 0$ as $t \to \infty$. The comparison equation being $\dot{y} = -\mu(y)$, the condition $\int_a^r [\mu(r)]^{-1}\, dr \to -\infty$ as $r \to 0^+$ is obvious. This has as consequence the fact that $W(t,x)$ satisfies inequalities of the form (3.52), with $\lambda_0(r) = \rho(\lambda(r))$ and $\mu_0(r) = \rho(\mu(r))$, $0 \leq r \leq a$, as required for this comparison function.

It remains to prove that a Lipschitz condition of the form (3.53) holds for $W(t,x)$, as constructed before, in $\Omega = \{(t,x);\, t \geq 0,\, |x| \leq a\}$.

Without loss of generality, one can assume $\mu(r)$ to be continuously differentiable on $[0,a]$. If not, then in all previous considerations one can substitute $\mu(r)$, which is only assumed positive and continuous for $r \in (0,a]$, $\mu(0) = 0$, by the function

$$\bar{\mu}(r) = \int_0^r e^{-s}\, \mu(s)\, ds.$$

Hence, by mean value theorem for $W = \rho(V)$, we shall have for $(t,x), (t,y) \in \Omega$,

$$|W(t,x) - W(t,y)| = |\rho'(V^*)\, (x - y)|, \tag{3.57}$$

and because of the boundedness of $\rho'(r)$ on $[0,a]$, there results inequality (3.53). Now, we only need to deal with the comparison equation associated to our problem, namely,

$$y' = -y + r(t),\ t \in R_+, \tag{3.58}$$

for which we can write the solution in the form

$$y(t;t_0,y_0) = y_0 e^{-t} + \int_{t_0}^{t} e^{(-t+s)}\, r(s)\, ds,$$

which implies

$$|y(t;t_0,y_0)| \leq y_0 + \sup_{t \geq t_0} |r(t)|,\ t \in R_+. \tag{3.59}$$

Inequality (3.59) proves that the zero solution of (3.1) is stable with respect to the bounded permanent perturbations, which ends the proof of Theorem 3.4.

Remark 3.4 *Theorems 3.3 and 3.4 deal with the choices $E = L^1(R_+,R^n)$ or $E = BC(R_+,R^n)$.*

Another possible choice, with perturbation term $r(t,x)$ in (3.32), of a more general nature than L^1 or BC, is given by $E = E(R_+,R^n) = M(R_+,R^n)$, a case that has been discussed by Germaidze and Krasovskii [216]. Let us recall the fact that the norm of M is $x \to \sup_{t\in R_+} \int_t^{t+1} |x(s)|\, ds$, for those $x \in L^2_{\text{loc}}(R_+,R^n)$, for which the supremum is finite. There exists equivalent norm to $|x|_M$ defined already, as indicated in Chapter 1, such as

$$\sup_{t\in R_+} \left\{ \int_0^t e^{\alpha(t-s)}\, |x(s)|\, ds \right\},\ \alpha > 0 \text{ fixed.} \tag{3.60}$$

In order to treat this type of perturbations, we limit ourselves to the *linear case* for the unperturbed system, while, the perturbations will be chosen in the space $E = M(R_+,R^n)$, generally nonlinear. In other words, we will deal with perturbed systems of the form

$$\dot{x}(t) = A(t)x(t) + r(t,x(t)), \tag{3.61}$$

considered on the semiaxis R_+. It is well known (see, e.g., Sansone and Conti [489], Halanay [237], and Corduneanu [123]) that system (3.60) is equivalent, under rather general assumptions, with the integral equation

$$x(t) = X(t)X^{-1}(t_0)x^0 + \int_{t_0}^{t} X(t)X^{-1}(s)\, r(s,x(s))\, ds \tag{3.62}$$

with $X(t)$ defined by $\dot{X}(t) = A(t)X(t)$, $X(t_0) = I$ =the identity matrix. Of course, in (3.62), $x(t) \equiv x(t;t_0,x^0)$, $t \geq t_0 \geq 0$.

Like in preceding cases, it suffices to deal with the simpler case of the system

$$x(t) = X(t)X^{-1}(t_0)x^0 + \int_{t_0}^{t} X(t)X^{-1}(s)\,r(s)\,ds,$$

with $r \in M(R_+,R^n)$.

As shown in the earlier quoted books, the solution $x = \theta$ of the unperturbed system $\dot{x}(t) = A(t)x(t)$ is uniformly asymptotically stable, if and only if, there exist constants $k,\alpha > 0$, such that

$$\sup_{t\geq t_0\geq 0} |X(t)X^{-1}(t_0)| \leq k e^{-\alpha(t-t_0)}, \tag{3.63}$$

which shows that the zero solution of $\dot{x} = A(t)x$ is, in fact, asymptotically exponentially stable. Therefore, from the previous formula for $x(t)$, one can derive

$$|x(t;t_0,x^0)| \leq k|x^0|e^{-\alpha(t-t_0)} + k\int_{t_0}^{t} |r(s)|e^{-\alpha(t-s)}\,ds. \tag{3.64}$$

Further, we can write the inequality

$$|x(t;t_0,x^0)| \leq k\,|x^0| + k\sup_{t\geq t_0}\left\{\int_{t_0}^{t} |r(s)|e^{-\alpha(t-s)}\,ds\right\},$$

and taking into account (3.60), which provides a norm on M, equivalent to the original one, there results

$$|x(t;t_0,x^0)| \leq k\,|x^0| + k(\alpha)\,|r|_M, \tag{3.65}$$

$M = \{x;\ x \in L_{\text{loc}}([t_0,\infty),R^n)\}$.

Inequality (3.65) proves that the zero solution of $\dot{x}(t) = A(t)x(t)$ is stable under permanent perturbations from $M(R_+,R^n)$.

In concluding Section 3.3, we shall consider the problem of *preserving the property of stability* for the zero solution of (3.1), by perturbing this system, to obtain system (3.32),

$$\dot{x}(t) = f(t,x(t)) + r(t,x(t)),\ t \in R_+,$$

with $r(t,\theta) = \theta$, $t \in R_+$.

Consequently, the perturbed system (3.32) must display the same property as system (3.1), or even a weaker one, depending on the conditions imposed on f and r.

It turns out that both uniform stability and uniform asymptotic stability, for (3.1), can be preserved by convenient choices for $r(t,x)$.

Let us state the basic results in this regard and prove them, while also relying on comparison techniques.

Theorem 3.5 *Consider system (3.1), under the general assumption of existence and uniqueness of solutions for the initial Cauchy problem $x(t_0) = x^0$, $(t_0, x^0) \in \Omega$, as stipulated in Theorem 3.1.*

I. If the solution $x = \theta$ is uniformly stable, then there exists a function $d(r)$, $r \in [0,a)$, with the same properties like the functions λ and μ considered above, such that the perturbed system (3.32), with

$$|r(t,x)| \leq \phi(t) d(|x|), \; (t,x) \in \Omega, \tag{3.66}$$

has the zero solution uniformly stable, provided $\phi \in L^1(R_+, R_+)$.

II. If the solution $x = \theta$ is uniformly asymptotically stable, then there exists a function $d(r)$, $r \in [0,a)$, with the same properties as the "wedges" λ and μ, such that the zero solution of the perturbed system (3.32) is also uniformly asymptotically stable, provided ϕ satisfies

$$\int_{t_0}^{t} \phi(s)\, ds \leq k(t - t_0) + k_1, \; t \geq t_0 \geq 0, \tag{3.67}$$

a condition assured by any $\phi = \phi_1 + \phi_2$, such that $\phi_1 \in BC$, $\phi_2 \in L^1$, where k, k_1 are positive constants.

Proof. We shall deal, first, with the statement I in Theorem 3.5, starting from the fact that uniform stability implies the existence of a Liapunov function $V(t,x)$, defined in Ω and taking nonnegative values, while $V(t,\theta) = 0$ for $t \in R_+$. Moreover, the condition $V^{\boldsymbol{\cdot}}(t,x) \leq 0$ is verified, while $V(t,x)$ is positive definite in Ω (condition (3.17)) and also satisfies (3.18). Under the extra conditions, such as (3.50), the function $V(t,x)$ can also be constructed to satisfy (3.47).

If we calculate $V^{\boldsymbol{\cdot}}(t,x)$ with respect to the perturbed system (3.32), then we obtain

$$V^{\boldsymbol{\cdot}}(t,x) \leq 0 + L\phi(t) d(\|x\|), \tag{3.68}$$

taking into account (3.47) and (3.66). Inequality (3.68), which is obtained for $V(t,x)$ from the converse theorem on uniform stability, leads to the comparison inequality

$$V^{\cdot}(t,x) \leq L\phi(t)\, d(\lambda^{-1}(V(t,x))), \tag{3.69}$$

whose associated equation is

$$\dot{y}(t) = L\phi(t)\, d(\lambda^{-1}(y)),\ y \geq 0. \tag{3.70}$$

In order to assure the uniform stability of the zero solution of the comparison equation (3.70), one can take, for instance, $d(r) = \lambda(r)$, and obtain $\dot{y}(t) = L\phi(t)\, y(t)$. This equation, with separate variables, can be easily integrated and one obtains for $y_0 > 0$ ($y_0 < a$), the solution

$$y(t) = y_0\, e^{L\int_{t_0}^{t} \phi(s)\, ds},\ t \geq t_0. \tag{3.71}$$

On behalf of the assumption $\phi \in L^1(R_+, R_+)$, we obtain from (3.71) an inequality of the form

$$y(t; t_0, y_0) \leq k y_0,\ t \geq t_0, \tag{3.72}$$

with $k = e^{L\int_0^\infty \phi(s)\, ds}$, which proves that $y = 0$ is uniformly stable for $\dot{y} = L\phi(t)\, y(t)$.

Let us now move to the proof of statement II in Theorem 3.5. This time we will rely on Proposition 3.2, which states the existence of a Liapunov function $W(t,x)$ and satisfies conditions in (3.52). In other words, W is positive definite in Ω, has an upper wedge function $\mu(|x|)$ there, while its derivative with respect to system (3.1) verifies $W^{\cdot}(t,x) \leq -W(t,x)$. Proceeding as in part I of the proof, we obtain the following comparison equation

$$y' = -y + L\phi(t)\, d(\mu_0^{-1}(W)),$$

and if one chooses $d(r) = c^{-1}\mu_0(r)$, $c > 0$, one obtains the final comparison equation

$$\dot{y} = -y + c^{-1} L\phi(t)\, y,\ t \in R_+. \tag{3.73}$$

For each $y_0 > 0$, the solution of (3.73) is

$$y(t; t_0, y_0) = y_0\, e^{-(t-t_0) + c^{-1} L\int_{t_0}^{t} \phi(s)\, ds},\ t \geq t_0. \tag{3.74}$$

We now need to make a choice for the constant c, such that, relying on Condition (3.67), we obtain the uniform asymptotic stability of the solution $y = 0$ of (3.73).

Taking hypothesis (3.67) into account, we obtain from (3.74) the following estimate:

$$y(t; t_0, y_0) \leq y_0 e^{(c^{-1}Lk_1)} e^{(-1+c^{-1}Lk)(t-t_0)}, \tag{3.75}$$

which tells us that assuming $-1 + c^{-1}Lk < 0$, we will obtain exponential decay (hence exponential asymptotic stability) for the solutions of (3.73). This means that $c > 0$ has to be chosen, such that

$$c > Lk \tag{3.76}$$

with L from the Lipschitz condition for W, and k as shown in (3.67).

Denoting $-1 + c^{-1} Lk = -\alpha < 0$, and $e^{c^{-1}Lk_1} = N$, we obtain from (3.75)

$$y(t; t_0, y_0) \leq N y_0 e^{-\alpha(t-t_0)}, \; t \geq t_0 > 0, \tag{3.77}$$

which is what characterizes the concept of exponential asymptotic stability, stronger than uniform asymptotic stability.

We cannot conclude the exponential asymptotic stability for the solution $x = \theta$ of the perturbed system (3.32), under our assumptions, because $x(t; t_0, x^0)$ does not depend linearly (or sublinearly) with respect to $y(t; t_0, y_0)$, due to the wedges λ, μ, λ_0, and μ_0 intervening earlier in the proof. However, due to the properties of the wedges, one obtains the uniform asymptotic stability for the solution $x = \theta$ of (3.32).

This ends the proof of Theorem 3.5, which relates to the concept of preservation of stability. This property plays an important role in various applications, as each dynamical system represents a certain approximation of reality. In this way, we secure the stability of the equilibrium state for the whole class of perturbed systems associated with the basic system (3.1), using only knowledge of properties of system (3.1).

The next section will be concerned with functional equations, which are more general than the ordinary differential equations.

3.4 STABILITY FOR SOME FUNCTIONAL DIFFERENTIAL EQUATIONS

For general functional differential equations, unlike ordinary differential equations, we have considered in the preceding sections, we do not possess stability results with the same degree of generality. Nevertheless, there are many results available for general equations of the form $\dot{x}(t) = (Fx)(t)$, $t \geq 0$, where F denotes an *operator* acting on a certain function space $E(R_+, R^n)$, not necessarily representable as $(Fx) = f(t, x(t))$, which leads to the classical case of ordinary differential equations. This is true even in the case of Banach space valued functions; see, for instance, Daletskii and Krein [186]. We may say that the theory of stability, presented in the preceding sections, is concerned only with the very special case when the operator F, involved in $\dot{x}(t) = (Fx)(t)$, is a Niemytskii operator (see for this term, e.g., Krasnoselskii and Zabreiko [298]).

We will start with the functional differential equation

$$\dot{x}(t) = (Vx)(t),\ t \in R_+, \tag{3.78}$$

for which we have provided global existence in Section 2.2. Here we make a rather general assumption, related also to uniqueness of the initial value problem of the second kind, namely,

a) $V : L_{\text{loc}}(R_+, R^n) \to L_{\text{loc}}(R_+, R^n)$ satisfies the generalized Lipschitz condition

$$\int_0^t |(Vx)(s) - (Vy)(s)|\, ds \leq A(t) \int_0^t |x(s) - y(s)|\, ds,\ t \in R_+. \tag{3.79}$$

It is obvious that (3.79) implies the continuity of V on $L_{\text{loc}}(R_+, R^n)$, as well as its *causality* (generalized Volterra condition).

The initial condition could be the original Cauchy condition

$$x(0) = x^0 \in R^n, \tag{3.80}$$

or the second kind of initial condition (functional type)

$$x(t) = \begin{cases} \phi(t), & t \in [0, t_0)\ t_0 > 0, \\ x^0, & t = t_0. \end{cases} \tag{3.81}$$

The necessity of assigning the value of $x(t)$ at $t = t_0$ stems from the fact that, in case $\phi(t)$ is only measurable, a situation might arise in which $\phi(t)$ is known only almost everywhere on $[0, t_0]$. Hence, by assigning $x(t_0) = x^0$, we deal with precise data. The solution of (3.78), verifying (3.81), will be denoted by $x(t; t_0, \phi, x^0)$, and its existence and uniqueness is assured by the condition *a)* formulated above for the operator V in (3.78), under initial condition (3.81), with

$$\phi \in L^1([0, t_0], R^n). \tag{3.82}$$

It is useful to point out the fact that the second initial value problem is reducible to the first initial value problem, that is, (3.78) and (3.80). See, for instance, Corduneanu [149], where the linear case is investigated in detail. The original results are due to Y. Li [329].

We shall use the notations and definitions from Corduneanu [149], Ch. II and III], where stability theory is presented with details pertaining to linear systems.

For stability of the solution $x = \theta$ of system (3.78); hence, we assume $(V\theta)(t) = \theta$, $t \in R_+$, the following definitions will be used:

Stability. For each $\epsilon > 0$ and $t_0 > 0$, there exists $\delta = \delta(\epsilon, t_0) > 0$, such that

$$|x^0| < \delta \text{ and } \int_0^{t_0} |\phi(s)|\, ds < \delta, \tag{3.83}$$

imply

$$|x(t; t_0, \phi, x^0)| < \epsilon \text{ for } t \geq t_0. \tag{3.84}$$

Asymptotic Stability. This means stability as defined before and, for each $t_0 > 0$, there exists $\eta(t_0) > 0$, such that

$$|x(t; t_0, \phi, x^0)| \to 0 \text{ as } t \to \infty, \tag{3.85}$$

provided

$$|x^0| < \eta, \int_0^{t_0} |\phi(s)|\, ds < \eta. \tag{3.86}$$

Uniform Stability. Stability as before and the independence of $\delta(\epsilon, t_0)$ in $t_0 > 0$; that is, $\delta(\epsilon, t_0) \equiv \delta(\epsilon)$.

Uniform Asymptotic Stability. Uniform stability, as defined earlier, and existence of $\delta_0 > 0$ and $T(\epsilon) > 0$ for $\epsilon > 0$, such that

$$|x^0| < \delta_0 \text{ and } \int_0^{t_0} |\phi(s)|\, ds < \delta_0, \tag{3.87}$$

imply

$$|x(t; t_0, \phi, x^0)| < \epsilon \text{ for } t \geq t_0 + T(\epsilon). \tag{3.88}$$

Let us mention, without further preoccupation, that similar definitions can be formulated for other stability concepts, such as exponential asymptotic stability or stability under permanent perturbations (in such cases, of the operator V).

Before we formulate some results on stability for the solution $x(t) = \theta, t \geq 0$, of system (3.78), we need to define a property for functionals which will play the role of Liapunov's functions we have used in preceding sections. Namely, the property of positive definiteness could be obtained if the Liapunov's functional $W : L_{\text{loc}}(R_+, R^n) \to R_+$ satisfies the following property:

A. Let us consider the functional $W : L_{\text{loc}}(R_+, R^n) \to R_+$, with $W(t, \theta) = 0$, $t \in R$. We say that W has property A, if for each $\epsilon > 0$ and $t_0 > 0$, there exists $\delta = \delta(\epsilon, t_0) > 0$ such that

$$(Wx)(t) < \delta \text{ for } t \geq t_0, \tag{3.89}$$

implies

$$|x(t)| < \epsilon \text{ for } t \geq t_0. \tag{3.90}$$

We can now state and prove a stability result for system (3.78), using the comparison method.

Theorem 3.6 *Consider system (3.78), with initial data (3.81) and V satisfying a). Assume the following conditions are satisfied:*

(1) *There exists a Fréchet differentiable functional $W : L_{\text{loc}}(R_+, R^n) \to R_+$, such that $(W)(\theta) = 0$ on R_+.*
(2) *There exists a function $\alpha : R^2 \to R_+$, continuous and such that $\alpha(0,0) = 0$, and*

$$(Wx)(t_0) \leq k(t_0)\, \alpha(|x^0|, |\phi|_L), \tag{3.91}$$

with $k(t_0) > 0$, along the solution $x(t; t_0, \phi, x^0)$.
(3) *The comparison inequality*

$$(W'x)(t)\, [(Vx)(t)] \leq \omega(t, (Wx)(t)),\ t \in R_+ \tag{3.92}$$

or on the largest interval of existence of $x(t)$.
(4) *The comparison equation*

$$\dot{y}(t) = \omega(t, y(t)),\ t \geq 0, \tag{3.93}$$

enjoys the properties of existence and uniqueness through each point (t, y_0), $t_0 \geq 0$, $0 \leq y_0 < k_0$, $k_0 > 0$ being a constant (sufficiently large).

Then, the solution $x = \theta$ of (3.78) is stable (asymptotically stable) if the solution $y = 0$ of (3.93) is stable (asymptotically stable).

Proof. Let us first notice that in (3.92), the first term of the inequality represents the Fréchet derivative of Wx, at the moment t, taken in the direction of $(Vx)(t)$.

Another remark, necessary to conduct our discussion of the proof, is the fact that an estimate of the form (3.91) does exist in some cases, and we do not have a result stating that it always holds true under the sole assumption of Fréchet differentiability of the functional $W : L_{\text{loc}}(R_+, R^n) \to R_+$. At the end of the proof, we shall illustrate the fact that a relationship of the form (3.91) does exist for linear functional differential equations.

We also need to mention the formula

$$\frac{d}{dt}(Wx)(t) = (W'x)(t)\,[(Vx)(t)], \tag{3.94}$$

which is true anytime $x = x(t)$ is a solution of equation (3.78), which means, on behalf of condition (3) in Theorem 3.6, that along the solutions of (3.78), the inequality of comparison becomes

$$\frac{d}{dt}(Wx)(t) \le \omega(t, (Wx)(t)), \tag{3.95}$$

whose associated equation is (3.93).

Now, let us consider a solution $x = x(t; t_0, \phi, x^0)$ of (3.78), and substitute it in the comparison inequality (3.95). One obtains the estimate

$$(Wx)(t) \le y(t; t_0, y_0),\ t \ge t_0, \tag{3.96}$$

provided

$$(Wx)(t_0) \le y_0 = y(t_0; t_0, y_0). \tag{3.97}$$

Both terms in (3.96) are defined for $t \ge t_0$, and, relying on condition (2) in Theorem 3.6, we realize (3.97) in time, as soon as

$$k(t_0)\,\alpha(|x^0|, |\phi|_{L(0,t_0)}) \le y_0. \tag{3.98}$$

With reference to the property A of the functional W, let us choose now y_0, such that $y_0 < \delta$. This will take place according to (2), if we take

$$|x^0| < \eta(\epsilon, t_0) \text{ and } |\phi|_{L(0,t_0)} < \eta(\epsilon, t_0). \tag{3.99}$$

Taking (3.96) into account, we obtain

$$(Wx)(t) \leq \delta,\ t \geq t_0, \tag{3.100}$$

provided (3.99) are satisfied. According to property A, from (3.100) we derive

$$|x(t;t_0,\phi,x^0)| < \epsilon,\ t \geq t_0, \tag{3.101}$$

as soon as initial data satisfy (3.99). This statement is equivalent to the stability of the solution $x = \theta$ of (3.78).

To prove the asymptotic stability of the zero solution of (3.78), we have to rely on the comparison inequality (3.96). If the solution $y = 0$ of (3.93) is asymptotically stable, then one can obtain from A the inequality $|x(t;t_0,\phi,x^0)| < \epsilon$ for $t \geq t_0$, as soon as initial data satisfy (3.99). But from (3.96), under assumption of asymptotic stability for the zero solution of the equation (3.93), we also obtain $\lim |x(t;t_0,\phi,x^0)| = 0$ as $t \to \infty$.

This ends the proof of Theorem 3.6.

Remark 3.5 *In the case of linear equation (3.78), which means $(Vx)(t) = (Lx)(t)$, where L is linearly acting on $L^2_{\text{loc}}(R_+,R^n)$, assumed also continuous on this space and causal, the unique solution of $\dot{x}(t) = (Lx)(t)$, under initial data (3.81), is given by the formula*

$$x(t;t_0,\phi,x^0) = X(t,t_0)x^0 + \int_0^{t_0} \tilde{X}(t,s,t_0)\,\phi(s)\,ds,$$

where $X(t,t_0)$ is determined by $(\frac{d}{dt})X = L(X)$, $t \geq t_0$, $X(t_0,t_0) = I$ =the identity matrix. Based on the properties of the matrices $X(t,t_0)$ and $\tilde{X}(t,s,t_0)$, as shown in the book Corduneanu [149], one can easily derive from the formula given in the text, for $x(t;t_0,\phi,x^0)$, an estimate (sublinear) of the form (3.91).

Remark 3.6 *A similar result to Theorem 3.6 can be obtained in cases of uniform stability (uniform asymptotic stability). We invite the readers to formulate the result and carry out the proof based on the earlier definitions, by using the comparison method.*

Remark 3.7 *We also ask the readers to thoroughly examine the property A, which clearly substituted the classical property of positive definiteness. Indeed, if $(Wx)(t) \geq \lambda(|x(t)|)$, where $\lambda(r)$ is such that $\lambda(0) = 0$, $\lambda(r)$ increasing for $r > 0$ (as noticed, one can assume its continuity, without loss of generality), the property A takes place. Is property A equivalent to the usual definition of positive definiteness? Under what condition on W?*

Remark 3.8 *In the case of the linearity of operator V in (3.78), the various types of stability have been characterized in terms of the properties of the matrices $X(t,s)$ and $\tilde{X}(t,s,t_0)$. The matrix $\tilde{X}(t,s,t_0)$ has been obtained in the framework of L^2-theory (i.e., assuming $V = L$ to be continuous in the L^2-convergence). This has been done in Li [329], and included in Corduneanu [149]. In the paper by Mahdavi and Li [352], there is background for similar constructions in the case of any L^p, $p \geq 1$.*

Let us point out the fact that the stability investigated earlier can also be treated for nonlinear functional differential equations of the form (causal operators)

$$\dot{x}(t) = (Lx)(t) + (fx)(t), \tag{3.102}$$

in which the nonlinear term designates a nonlinear operator, and satisfies $(f\theta)(t) = \theta$ on R_+.

In concluding this section, we briefly present an example illustrating the concept of uniform asymptotic stability, due to Staffans [505]. Namely, one considers the integro-differential equation with infinite delay

$$\dot{x}(t) + A(t)x(t) = \int_{-\infty}^{t} C(t,s)x(s)\,ds, \; t \geq t_0, \tag{3.103}$$

under the initial condition

$$x(t) = \phi(t), \; t \in (-\infty, t_0], \; t_0 > 0, \tag{3.104}$$

with $\phi \in BC((-\infty,t_0],R^n)$. The following assumptions will be made:

a) $A(t) \in \mathcal{L}(R^n,R^n)$, for $t \geq t_0$, is continuous.

b) $C(t,s) \in \mathcal{L}(R^n,R^n)$, for $(t,s) \in \Delta$, with Δ the half plane $s \leq t$, is continuous.

c) There exists a positive definite matrix $B \subset \mathcal{L}(R^n,R^n)$, such that the matrix

$$R(t) = A^T(t)B + BA(t), \; t \in R_+$$

is positive definite.

d) If β denotes the largest eigenvalue of the matrix B, and

$$p(t) = \beta \inf \left\{ \frac{< x, Rx >}{< x, Bx >}, \; \theta \neq x \in R^n \right\},$$

then we can find constants $k > 0$, $\epsilon > 0$, with $\epsilon < (2\beta)^{-1}$, such that the inequality

$$2^{-1}k\int_{-\infty}^{t}|C(t,u)|_B\,du+(2k)^{-1}\int_{t}^{\infty}|C(t,v)|_B\,dv \leq (2\beta)^{-1}p(t)-\epsilon[1+p(t)],$$

for $t \in R$; $|.|_B$ stands for the matrix norm, when the usual scalar product in R^n is replaced by $< x,y >_B=< x,By >$.

e) The functions $C(t,s)$ and $p(t)$ satisfy

$$\sup\int_{t_0}^{\infty}[1+p(t)]^{-1}\left\{\int_{-\infty}^{t_0}|C(t,s)|\,ds\right\}^2 dt < +\infty,$$

the supremum being taken for $t_0 \in R$, and

$$\lim_{T\to\infty}\sup_{t_0\in R}\int_{t_0}^{\infty}[1+p(t)]^{-1}\left\{\int_{-\infty}^{t_0-T}|C(t,s)|\,ds\right\}^2 dt = 0.$$

Under conditions *a*)–*e*), the solution $x = \theta$ of system (3.103) is uniformly asymptotically stable.

The proof can be carried out using the surprisingly simple Liapunov function $W(x) = |x|^2$, $x \in R^n$. For details of the proof, the readers are invited to consult the original sources indicated before.

Let us point out the fact that the special case of system (3.103), corresponding to $C(t,s) = 0$ for $s \leq t$,

$$\dot{x}(t)+A(t)x(t) = \int_{0}^{t}C(t,s)x(s)\,ds,$$

has been investigated by different methods, by many authors. In book form, see Burton [84] and Lakshmikantham and Rao [316].

3.5 PARTIAL STABILITY

We shall start with the systems of ordinary differential equations, a priority which is warranted by the historical development of this concept. This will take us back to differential systems of the form (3.1), $\dot{x}(t) = f(t,x(t)), x,f \in R^n$.

The concept of partial stability has emerged in the work of Liapunov in the nineteenth century, but it received special attention only after the 1950s. See Rumiantsev [477], for instance, which was mostly motivated by the mechanical interpretations of stability. There are many engineering-oriented publications from which one can easily realize that, in investigating stability of such systems, there are only part of the parameters involved, presenting significance for the whole systems. Such a system is even an aircraft, so commonly used.

Let us proceed with the definition of the *partial stability*, also called by many authors—stability with respect to part of the variables.

We shall "split" the system (3.1), setting $x = \text{col}(y,z)$, $f = \text{col}(g,h)$, and writing it as

$$\dot{y} = g(t,y,z), \ \dot{z} = h(t,y,z), \tag{3.105}$$

with $y, g \in R^p$, and $z, h \in R^q$, ($p+q = n$-the dimension of x). Of course, the initial data will be of the form

$$y(t_0) = y^0 \in R^p, \ z(t_0) = z^0 \in R^q. \tag{3.106}$$

The conditions

$$g(t,0,0) = 0, \ h(t,0,0) = 0, \ t \geq 0, \tag{3.107}$$

show that system (3.106) has the zero solution. In (3.107), 0 means the null element in R^p, respectively R^q.

We shall assume in this section that for the initial value problem (3.105), (3.106), the existence and uniqueness of solution is assured, through each point $(t_0, y^0, z^0) \in \Omega = \{(t,x); t \geq 0, |x| = \sqrt{y^2+z^2} < a, a > 0\}$, as we have always considered in preceding sections.

The unique (local) solution of problem (3.105) and (3.106) will be denoted by

$$y(t) = y(t;t_0,y^0,z^0), \ z(t) = z(t;t_0,y^0,z^0). \tag{3.108}$$

At the points $t_0 \geq 0$, we will have in mind only the right solutions $(t > t_0)$.

Definitions of partial stability are formulated as follows (for the zero solution):

Partial Stability. The zero solution of system (3.105) is called partially stable with respect to y-variables (or y-stable), if for each $\epsilon > 0$ and $t_0 \geq 0$, there exists $\delta = \delta(\epsilon, t_0) > 0$, with the property that

$$|y(t;t_0,y_0^0,z^0)| < \epsilon, \ t \geq t_0, \tag{3.109}$$

provided $|y^0| < \delta$, $|z^0| < \delta$.

Let us point out the fact that (3.109) implies the existence of the function $y(t;t_0,y^0,z^0)$ on the semiaxis $t \geq t_0$.

The following question arises: What happens with the z-components of the solution, that is, with the function $z(t;t_0,y^0,z^0)$, for $t \geq t_0$? The earlier definition does not contain any information about it. Hence, there is the possibility that it is also defined for $t \geq t_0$, the graph remaining in Ω, or it ceases to exist at a certain moment $T > t_0$. Such a moment is known as the *escape time* of the solution $(y(t),z(t))$. Obviously, it represents escape time from Ω.

The first case of the alternative mentioned already is preferable for interpreting solutions that describe a motion.

Asymptotic Partial Stability. To the definition of stability, one has to add

$$\lim_{t\to\infty} |y(t;t_0,y^0,z^0)| = 0, \tag{3.110}$$

for all solutions with $|y^0|$, $|z^0| < \eta(t_0)$, $\eta > 0$.

Similar definitions can be formulated for the concepts of *uniform partial stability* and *uniform asymptotic partial stability*.

For various extensions and refinements of the concepts related to partial stability, we send the readers to the following book references: Halanay [237], Rumiantsev and Oziraner [478], and Vorotnikov [531]. The last book is a rather comprehensive source.

Let us find conditions, on system (3.105), assuring the stability/asymptotic stability of the zero solution $y = z = 0$. We will provide a rather general result, based on the comparison method.

Theorem 3.7 *Assume the following conditions are satisfied by the system (3.105):*

(1) *Local conditions, assuring local existence and uniqueness of the solution through each point $(t_0,y^0,z^0) \in \Omega$.*
(2) *There exists a Liapunov function $V = V(t,x) \equiv V(t,y,z)$, satisfying the comparison inequality*

$$V^{\boldsymbol{\cdot}}(t,x) \leq \omega(t,V(t,x)),\ (t,x) \in \Omega. \tag{3.111}$$

(3) *$V(t,y,z)$ is such that*

$$\lambda(|y|) \leq V(t,y,z),\ (t,y,z) \in \Omega, \tag{3.112}$$

with $\lambda(r)$ a lower wedge for V.
(4) *The comparison equation $\dot{u}(t) = \omega(t,u(t))$ is like in Theorem 3.1.*

Then, if the zero solution of the comparison equation is stable/asymptotically stable, the zero solution of system (3.105) is y-stable/y-asymptotically stable.

Proof. Proceeding as usual with the comparison method, one can write the following inequality, on the common interval of existence

$$V(t, y(t; t_0, y^0, z^0), z(t; t_0, y^0, z^0)) \leq u(t; t_0, u_0), \tag{3.113}$$

provided $V(t_0, y^0, z^0) \leq u_0$. Taking into account condition (3) of Theorem 3.7, we obtain that the last inequality is satisfied as soon as we choose $|y^0|, |z^0| < \delta(u_0, t_0)$.

If we now take $\epsilon > 0$ for u_0, then the earlier considerations lead to the conclusion that

$$V(t, y(t; t_0, y^0, z^0), z(t; t_0, y^0, z^0)) \leq \epsilon \tag{3.114}$$

for $t \geq t_0$, as soon as $|y^0|, |z^0| < \delta(\epsilon, t_0)$. Hence, on behalf of the assumption (3) in the statement of the theorem, we can write

$$|y(t; t_0, y^0, z^0)| < \lambda^{-1}(\epsilon),\ t \geq t_0, \tag{3.115}$$

as soon as $|y^0|, |z^0| < \delta(\epsilon, t_0)$. From (3.115), we derive the y-stability of the solution $x = \theta$ of system (3.105).

For the second statement of Theorem 3.7, we need to notice the inequality

$$\lambda(|y(t; t_0, y^0, z^0)|) \leq u(t; t_0, u_0),\ t \geq t_0, \tag{3.116}$$

derived from (3.113), and take again into account assumption (3). There results $|y(t; t_0, y^0, z^0)| \to 0$ as $t \to \infty$, for all y^0, z^0, such that $V(t_0, y^0, z^0) \leq u_0$, and since $V(t, 0, 0) = 0$, $t \in R_+$, we see that this is possible for $|y^0|, |z^0| < \eta(\epsilon, t_0)$, which proves the y-asymptotic stability.

Remark 3.9 *A completely similar result to Theorem 3.7 can be obtained for the concepts of uniform y-stability and for uniform y-asymptotic stability. We invite the readers to state and prove this result.*

In the remaining part of this section, we will emphasize a close connection between the concepts of stability and partial stability. The result we provide is due to Hatvani [251], and it shows, practically, that stability can be treated as a special case of partial stability.

The following result clarifies the connection between the two kinds of stability.

Theorem 3.8 *Consider system 3.1, $\dot{x}(t) = f(t, x(t))$, with $f(t, x)$ defined on $\Omega = \{(t, x); t \geq 0, |x| < a\}$ and satisfying the conditions assuring the (local)* *existence and uniqueness through any point* $(t_0, x^0) \in \Omega$. *Let there exist a function* $V = V(t,x)$ *and a comparison function* $\omega(t,x,u)$, *such that*

(1) $f(t,x,u)$ *is continuous on* $\Omega \times R_+$, *real-valued and* $f(t,0,0) = 0$ *on* R_+.
(2) $V(t,x)$ *is differentiable on* Ω, *and such that* $V(t,\theta) = 0$, $t \in R_+$; *moreover, it is positive definite.*
(3) *The comparison inequality*

$$V^{\cdot}(t,x) \leq \omega(t,x,V(t,x)), \text{ on } \Omega, \tag{3.117}$$

is satisfied.

Then, the following statements hold true.

(I) If the zero solution of the auxiliary system

$$\dot{x}(t) = f(t,x(t)), \; \dot{u} = \omega(t,x,u), \tag{3.118}$$

is u-stable (asymptotically u-stable), the zero solution of (3.1) is stable (asymptotically stable).

(II) If, besides the conditions in (I), *one adds the existence of an upper wedge for* $V(t,x)$, *say* $V(t,x) \leq \mu(|x|)$, $(t,x) \in \Omega$, *then from the uniform u-stability (uniform asymptotic u-stability) of the zero solution of (3.118), there follows the same properties for the zero solution of system (3.1).*

Proof. Let us notice, first, that the condition of positive definiteness, for $V(t,x)$, means that a condition of the form (3.17) takes place, that is, $V(t,x) \geq \lambda(|x|)$, with $\lambda(r)$ a wedge function.

Assume now, in view of proving statement (I), that the zero solution of (3.118) is u-stable. This implies that for $\epsilon > 0$ $(\epsilon < a)$ and $t_0 \geq 0$, there exists $\eta = \eta(\epsilon, t_0) > 0$, such that the solution $u = u(t; t_0, x^0, u_0)$ of (3.118) satisfies

$$u(t; t_0, x^0, u_0) < \lambda(\epsilon), \; t \geq t_0, \tag{3.119}$$

provided $|x^0|, u_0 < \eta$. Let us consider $\delta = \delta(\epsilon, t_0)$, defined by $\delta = \min(\epsilon, \eta)$, such that $V(t_0, x) < \eta$, for $|x| < \delta$.

Now, we claim that the assumption $|x^0| < \delta$ implies

$$|x(t; t_0, x^0)| < \epsilon, \; t \geq t_0, \tag{3.120}$$

which means that the zero solution of system (3.1) is stable.

Indeed, if we assume the contrary, we should be able to find $t_1 > t_0$ and $\bar{x}^0$ with $|\bar{x}^0| < \delta$, for which $|x(t_1;t_0,\bar{x}^0)| = \epsilon$. Therefore, according to the choice of δ and η, one must have

$$V(t_1,x(t_1;t_0,\bar{x}^0)) \geq \lambda(\epsilon). \tag{3.121}$$

Consider now the upper solution, say $m = m(t;t_0,u_0)$, of the problem

$$\dot{u}(t) = \omega(t,\bar{x}(t),u(t)),\ u(t_0) = u_0 = V(t_0,\bar{x}(t_0)),$$

where $\bar{x}(t) = x(t;t_0,\bar{x}^0)$. Of course, this upper solution exists under the hypothesis of continuity of the function $\omega(t,x,u)$ and has a maximal interval of existence $[t_0,T)$, $T \geq t_1$. See, for instance, Corduneanu [123].

Taking inequality (3.117) into account, that is, $V^{\boldsymbol{\cdot}}(t,\bar{x}(t)) \leq \omega(t,\bar{x}(t), V(t,\bar{x}(t))$, one obtains $V(t,\bar{x}(t)) \leq m(t)$, $t \in [t_0,t_1]$, with $|x^0| < \delta \leq \eta$. Consequently, based on the theory of differential inequalities, one obtains $V(t,\bar{x}(t_1)) \leq m(t_1) < \lambda(\epsilon)$, which contradicts (3.121). Therefore, inequality (3.120) is true, which proves the first part of statement in (I).

Concerning second part of (I), which starts from the assumption that the zero solution of system (3.118) is asymptotically u-stable, what we have to prove is the following assertion: for each $\epsilon > 0$ ($\epsilon < a$) and for some $\rho(t_0) > 0$, for the solution $x(t;t_0,x^0)$ of the system $\dot{x}(t) = f(t,x(t))$, with $|x^0| < \rho$, there exists $T = T(\epsilon,t_0,x^0)$, such that

$$|x(t;t_0,x^0)| < \epsilon \text{ for } t \geq t_0 + T. \tag{3.122}$$

This means that $|x(t;t_0,x^0)| \to 0$ as $t \to \infty$, for each solution of $\dot{x}(t) = f(t,x(t))$, as soon as $|x^0|$ is small enough, as required in asymptotic stability.

Now, let us fix the numbers ϵ and t_0. The zero solution of (3.118) being asymptotically u-stable, there exists $\sigma(t_0) > 0$, such that for each component $u(t;t_0,x^0,u_0)$ of the solution, with $|x^0| < \delta$ and $u_0 < \sigma$, one can find a number $S = S(\epsilon,t_0,x^0,u_0)$ with the property

$$u(t;t_0,x^0,u_0) < \lambda(\epsilon),\ t \geq t_0 + S. \tag{3.123}$$

Let $\rho = \rho(t_0) > 0$, such that $V(t_0,x) < \sigma(t_0)$ for $|x| < \rho$. As in the first part of the proof of (I), assuming the contrary, that is, that $x(t;t_0,x^0)$ does not satisfy property (3.122), we are led to contradiction, because with $\rho = \rho(t_0) > 0$ and $T = S(\epsilon,t_0,x^0,V(t_0,x^0))$, condition (3.122) is verified.

Regarding the statements about uniform stability and uniform asymptotic stability, it is sufficient to prove that, under assumption of existence of an upper wedge for $V(t,x)$, that is, $V(t,x) \leq \mu(|x|)$, and the uniform character of stability or asymptotic stability for the auxiliary system (3.118), the numbers δ

and ρ, which appear in the proof, can be chosen independently of t_0, while T is independent of t_0 and x^0, when η and σ do not depend on t_0, and while S is independent of t_0, x^0 and u_0.

Indeed, from the existence of the upper wedge μ, one derives $V(t,x) < s$ as soon as $|x| < \mu^{-1}(s)$, $s = \min\{(v,\mu(v)); 0 \leq v \leq a\}$. This means that δ, ρ and T can be defined by $\rho = \mu^{-1}(\sigma)$, $\delta = \mu^{-1}(\eta(\epsilon))$, $T = S(\epsilon)$, which proves the statement (II) in Theorem 3.8.

This ends the proof of Theorem 3.8.

Remark 3.10 *In case $\omega(t,x,u) = h(t,x)g(u)$, the following result is valid: Assume h and g to be continuous, with $g(0) = 0$ and $g(u) > 0$ for $u > 0$, while $\int_{0+}^{r} [g(u)]^{-1}\,du = \infty$ for $r > 0$, and $h(t,x)$ satisfies the condition*

$$\int_{t_0}^{t} h(s,x(s;t_0,x^0))\,ds \leq K(t_0),\ t \geq t_0, \tag{3.124}$$

with $K(t_0)$ finite, $t \geq t_0$, as soon as $|x^0| \leq \sigma(t_0)$, for some $\sigma(t_0) > 0$.

Then, the zero solution of system (3.118), with $\omega(t,x,u)$ as before is u-stable/asymptotically u-stable, the last situation occurring when

$$\int_{t_0}^{\infty} h(s,x(s;t_0,x^0))\,ds = -\infty. \tag{3.125}$$

The result, in Remark 3.10 follows from Theorem 3.8, and using also a result due to Rumiantsev [477].

3.6 STABILITY AND PARTIAL STABILITY OF FINITE DELAY SYSTEMS

In this section, we deal with some stability problems relating to the functional differential equations of the form

$$\dot{x}(t) = F(t,x_t),\ t \in [t_0,T), \tag{3.126}$$

with $T \leq \infty$, where $x_t(u) = x(t+u)$, $u \in [-h,0]$, $h > 0$, $x \in R^n$.

We assume that $F : R_+ \times C([-h,0],R^n) \to R^n$ is a continuous map and for each $t_0 \in R_+$, there exists a solution $x(t;t_0,\phi)$, $\phi \in C([-h,0],R^n)$, such that

$$x(t_0;t_0,\phi) = \phi, \tag{3.127}$$

with $x(t;t_0,\phi)$ defined on some interval $[t_0,t_0+\alpha)$, $\alpha > 0$ ($\alpha < T$).

One also assumes that $F(t,\phi)$ satisfies the condition $F(t;\theta) = \theta \subset C(R_+,R^n)$, θ designating the zero element, which implies the fact that $x = \theta$ is a solution.

In the space $C([-h,0],R^n)$ we will use several norms, namely $|\phi|_C = \sup\{|\phi(t)|; t \in [-h,0]\}$, or the L^2-norm: $|\phi|_2 = \{\int_{-h}^{0} |\phi(s)|^2 ds\}^{\frac{1}{2}}$. By $|.|$ we mean any norm in the space $C([-h,0],R^n)$

Definition 3.8 *The solution $x = \theta$ of (3.126) is said to be stable, if for any $\epsilon > 0$ and $t_0 \geq 0$, one can find $\delta = \delta(\epsilon,t_0) > 0$, with the property $|x(t;t_0,\phi)|_C < \epsilon$ for $t \geq t_0$, when $|\phi| < \delta$.*

The asymptotic stability of the solution $x = \theta$ of (3.126) means *stability* and the validity of the following property: for each $t_0 \geq 0$, one can find $\sigma = \sigma(t_0) > 0$, such that $|\phi|_C < \sigma$ implies $|x(t;t_0,\phi)| \to 0$ as $t \to \infty$.

Other types of stability are defined in the usual manner (uniform stability and uniform asymptotic stability), as we find in Section 3.1, in cases of ordinary differential equations.

We will need, for applying Liapunov's method, the concept of *derivative* of the scalar function $V(t,\phi)$, with respect to system (3.126). This concept is introduced by the following formula:

$$V^{\cdot}(t,\phi) = \lim_{h \to 0} \sup h^{-1}[V(t+h, x_{t+h}(t,\phi)) - V(t,\phi)].$$

It certainly makes sense, in case $V(t,\phi)$ is locally Lipschitz with respect to both variables.

In the case of ordinary differential equations, estimates must be imposed on both $V(t,\phi)$ and $V^{\cdot}(t,\phi)$ in order to obtain stability results. The comparison method relies on a differential inequality for $V(t,\phi)$. But the classical results of stability have been obtained without assuming such an inequality. Only adequate assumptions on $V(t,\phi)$ and $V^{\cdot}(t,\phi)$, separately, can lead to results.

We will now state a result concerning the asymptotic stability of the solution $x = \theta$ of system (3.126). Namely, one has

Theorem 3.9 *Let us consider system (3.126), under the following assumptions:*

1) *There exists a function $V(t,\phi)$, locally Lipschitz on $R_+ \times C([-h,0],R^n)$, and three wedges $a(r)$, $b(r)$, $c(r)$, such that*

$$a(|\phi(0)|) \leq V(t,\phi) \leq b(|\phi|_C), \ \phi \in C, \tag{3.128}$$

and

$$V^{\cdot}(t,x_t) \leq -\eta(t)\, c(|\phi|_C),\ \phi \in C. \tag{3.129}$$

2) $\eta \in L^1_{\mathrm{loc}}(R_+,R)$ *is nonnegative and satisfies*

$$\lim_{t\to\infty} \int_0^t \eta(s)\,ds = \int_0^\infty \eta(s)\,ds = -\infty. \tag{3.130}$$

Then, the zero solution $x = \theta$ of system (3.126) is asymptotically stable.

Proof. First, notice that our conditions imply the uniform stability of the zero solution of system (3.126). This means that for each $\epsilon > 0$, one can find $\delta = \delta(\epsilon) > 0$, such that $|\phi|_C < \delta$ implies $|x(t;t_0,\phi)| < \epsilon$ for $t \geq t_0$. Now, for an $\epsilon > 0$, choose $\delta > 0$ such that $a(\epsilon) = b(\delta)$, or $\delta = b^{-1}(a(\epsilon))$. From (3.128) and (3.129), for $|\phi|_C \leq \delta$, one finds $a(|x(t)|) \leq V(t,x_t) \leq V(t_0,\phi) \leq b(|\phi|_C) \leq b(\delta) = a(\epsilon)$. Hence, taking monotonicity of wedge functions into account, there results

$$|x(t;t_0,\phi)| \leq \epsilon,\ \text{if}\ |\phi|_C \leq \delta,\ t \geq t_0.$$

There remains to prove that

$$|x(t;t_0,\phi)| \to 0 \text{ for } t \to \infty, \tag{3.131}$$

provided $|\phi|_C \leq \bar{\delta}$, with $\bar{\delta} \leq \delta$ from above. Since $V(t,x(t;t_0,\phi))$ is non-increasing in t, due to (3.129), one obtains from (3.129) by integration on $(0,t)$,

$$V(t,x_t(t;t_0,\phi) \leq A + B \int_0^t \eta(s)\,ds,\ t \in R_+, \tag{3.132}$$

where $B \geq c(\bar{\delta}) > 0$. This inequality implies, as $t \to \infty$, $V(t,x_t(t;t_0,\phi)) \to -\infty$, because of (3.130). Therefore, if we accept that (3.131) does not hold, we obtain a contradiction and this ends the proof of Theorem 3.9.

Remark 3.11 *The conclusion of Theorem 3.9 can be improved with relative ease. Since the imposed conditions (3.128) and (3.129) imply the uniform stability of the zero solution, one should prove that (3.131) holds uniformly in t_0. In Hatvani [258], it is shown that this property is assured by the requirement*

$$\lim_{t\to\infty} \int_{t_0}^{t_0+t} \eta(s)\,ds = \infty,\ \text{uniformly for } t_0 \in R_+.$$

Remark 3.12 *The result in Theorem 3.9 has several variants, depending upon the choice of the norm used in (3.129). If this assumption is substituted for the following:*

$$V^{\cdot}(t,x_t) \leq -\eta(t)\, c(t, |x_t|_2),\ t \in R_+, \tag{3.133}$$

maintaining the rest, then asymptotic stability of the solution $x = \theta$ of system (3.129) is assured.

Before the Liapunov functionals were systematically used in stability problems for differential equations with delayed argument, rather elementary/classical procedures were applied to diverse kinds of functional differential equations with delayed argument. A good reference is the book by L. E. Elśgoĺts and S. B. Norkin [207]. This book has a good deal of literature for early Soviet contributions to the subject.

The book by N. N. Krasovskii [299] is the landmark in this regard (stability for finite delay equations). It extensively deals with the basic results, which are similar to those we have dealt with in Section 3.5. Also, the books by A. Halanay [237] and J. K. Hale [240] provide basic and advanced treatment of stability in the case of finite delay differential equations (and even in cases of infinite delays).

We now consider the problem of *partial stability* for linear finite delay systems, written in the form

$$\begin{aligned} \dot{x}(t) &= A(t,x_t) + B(t,y_t), \\ & \qquad\qquad t \in R_+, \\ \dot{y}(t) &= C(t,x_t) + D(t,y_t), \end{aligned} \tag{3.134}$$

with A, B, C, D linear and continuous operators on the spaces:

$A : R_+ \times C([-h,0],R^n) \to C(R_+,R^n)$;
$B : R_+ \times C([-h,0],R^m) \to C([-h,0],R^n)$;
$C : R_+ \times C([-h,0],R^n) \to C(R_+,R^m)$;
$D : R_+ \times C([-h,0],R^m) \to C(R_+,R^m)$.

To system (3.134) one attaches the initial conditions

$$x_0 = \phi \in C([-h,0],R^n),\ y_0 = \psi \in C([-h,0],R^m). \tag{3.135}$$

The existence and uniqueness of solution to problem (3.134) and (3.135) follows easily, considering the estimates for the operators A, B, C, and D. On behalf of our continuity assumptions and on their consequences, like

$$|A(t,\phi_1)-A(t,\phi_2)|\leq\lambda(t)|\phi_1-\phi_2|, \tag{3.136}$$

with $\lambda(t)$ a non-decreasing function on R_+, the method of successive approximations is uniformly convergent on any $[0,T]\subset R_+$.

Therefore, we have the solution to our problem, say

$$x=x(t;t_0,\phi,\psi),\ y=y(t;t_0,\phi,\psi), \tag{3.137}$$

for any $t_0\geq 0$ and $\phi\in C([-h,0],R^n)$, $\psi\in C([-h,0],R^m)$. These solutions are defined on the semiaxis $t\geq t_0$. Obviously, one has $x(t_0;t_0,\phi,\psi)=\phi$, $y(t_0;t_0,\phi,\psi)=\psi$.

The kind of stability we shall consider is the *partial exponential stability*. This concept is defined by the requirement

$$|x(t;t_0,\phi,\psi)|\leq M(|\phi|+|\psi|)e^{-\alpha(t-t_0)},\ t\geq t_0, \tag{3.138}$$

where M and α are some (fixed) positive numbers. We shall state and prove a result, characterizing the x-partial exponential stability, by means of a Liapunov function $V(t,\phi,\psi)$. Namely, the following statement will be proven.

Theorem 3.10 *The zero solution of system (3.134) is x-exponentially stable (or x-partially asymptotically stable, of exponential type) if, and only if, there exists a Liapunov functional $V(t,\phi,\psi)$, defined on $R_+\times C([-h,0],R^n)\times C([-h,0],R^m)$, satisfying the following conditions:*

$$|\phi|\leq V(t,\phi,\psi)\leq M(|\phi|+|\psi|), \tag{3.139}$$

$$|V(t,\phi_1,\psi_1)-V(t,\phi_2,\psi_2)|\leq M(|\phi_1-\phi_2|+|\psi_1-\psi_2|), \tag{3.140}$$

$$V^{\boldsymbol{\cdot}}(t,\phi,\psi)\leq-\alpha V(t,\phi,\psi), \tag{3.141}$$

where M and α are positive constants, for $t\in R_+$, $\phi,\phi_1,\phi_2\in C([-h,0],R^n)$ and $\psi,\psi_1,\psi_2\in C([-h,0],R^m)$.

Proof. Let us note that the derivative $V^{\boldsymbol{\cdot}}(t,\phi,\psi)$ is considered in the following sense:

$$V^{\boldsymbol{\cdot}}(t,\phi,\psi)=\limsup_{h\to 0^+} h^{-1}[V(t+h,x_{t+h},y_{t+h})-V(t,\phi,\psi)], \tag{3.142}$$

with $x_t=\phi$, $y_t=\psi$.

The sufficiency of conditions (3.139)–(3.141) can be easily obtained if we notice that (3.141) implies the exponential decay for V, as $t\to\infty$, which is transmitted to x_t by (3.139).

The necessity of conditions (3.139)–(3.141) is verified by constructing the Liapunov functional $V(t,\phi,\psi)$ with the formula

$$V(t,\phi,\psi) = \sup_{s\geq 0}\{|x_{t+s}(t,\phi,\psi)|e^{\alpha s}\}, \tag{3.143}$$

for any $t\in R_+$ and $\phi\in C([-h,0],R^n)$, $\psi\in C([-h,0],R^m)$, where $x_t(t;\phi,\psi)=\phi$. From (3.138) we see that the sup in (3.143) makes sense (is finite).

In order to derive (3.139), we notice that, for $s=0$, the quantity on the right-hand side of (3.143) reduces to $|\phi|$, appearing in (3.139). Further, relying again on (3.138), one sees that the right-hand side in (3.143) is dominated by $M(|\phi|+|\psi|)$.

We will now prove that formula (3.143) is defining a continuous function with respect to (t,ϕ,ψ). Indeed, the following sequence of inequalities is valid:

$$\begin{aligned}
&|V(t+h,\phi,\psi)-V(t,\phi_0,\psi_0)|\leq\\
&|V(t+h,\phi,\psi)-V(t+h,\phi_0,\psi_0)|\leq\\
&|V(t+h,\phi_0,\psi_0)-V(t+h,x_{t+h}(t,\phi_0,\psi_0),y_{t+h}(t,\phi_0,\psi_0)|\leq\\
&|V(t+h,x_{t+h}(t,\phi_0,\psi_0),y_{t+h}(t,\phi_0,\psi_0))-V(t,\phi_0,\psi_0)|.
\end{aligned} \tag{3.144}$$

Let us remark first, that $V(t,\phi,\psi)$, as defined by (3.143), is locally Lipschitz with respect to the last two arguments. Indeed, from (3.143) one obtains the inequality

$$|V(t,\phi,\psi)-V(t,\phi_0,\psi_0)|\leq \sup_{s\geq 0}\{|x_{t+s}(t,\phi,\psi)-x_{t+s}(t,\phi_0,\psi_0)|e^{\alpha s}\},$$

and, on behalf of the linearity of system (3.134), we can write

$$|V(t,\phi,\psi)-V(t,\phi_0,\psi_0)|\leq \sup_{s\geq 0}\{|x_{t+s}(t,\phi-\phi_0,\psi-\psi_0)|e^{\alpha s}\}.$$

The right-hand side of the last inequality, combined with formula (3.138), shows that $V(t,\phi,\psi)$ is (globally) Lipschitz with respect to (ϕ,ψ).

We now have to prove the continuity of the function $V(t,\phi,\psi)$. Since the Lipschitz condition implies continuity with respect to the arguments ϕ and ψ, we only need to show the continuity with respect to t ($h\to 0^+$). Relying on the definition of $V(t,\phi,\psi)$, given by (3.143), we obtain the inequality

$$|V(t+h,\phi,\psi)-V(t,\phi,\psi)|\leq \sup_{s\geq 0}\{|x_{t+s+h}(t,\phi,\psi)-x_{t+s}(t,\phi,\psi)|e^{\alpha s}\},$$

which leads to the conclusion, due to the continuous dependence of x_t with respect to t on each finite interval (see Krasovskii [299] and Hale [240]), and also due to the exponential decay at ∞ of x_t (under our assumptions).

This ends the proof of continuity of $V(t,\phi,\psi)$ on $R_+ \times C([-h,0],R^n) \times C([-h,0],R^m)$.

There remains to show that the differential inequality (3.141) is satisfied by $V(t,\phi,\psi)$ defined by (3.143). We will notice that the following identity takes place for the function $x_t(t,\phi,\psi)$:

$$x_{t+h+s}(t+h, x_{t+h}(t,\phi,\psi), y_{t+h}(t,\phi,\psi)) = x_{t+h+s}(t,\phi,\psi),$$

for $h \geq 0$ and $t,s \geq 0$. Indeed, both terms above are solutions of (3.126) and they coincide for $h = 0$. The uniqueness for system (3.126) applies, which leads to the equality when $t \geq 0$, $s \geq 0$. Now, based on (3.127), we can write

$$\begin{aligned}
&\limsup_{h\to 0+} h^{-1}[V(t+h, x_{t+h}(t,\phi,\psi), y_{t+h}(t,\phi,\psi)) - V(t,\phi,\psi)] \\
&\quad = \limsup_{h\to 0+} h^{-1}\{\sup_{s\geq 0} |x_{t+h}(t+h, x(t,\phi,\psi), y_{t+h}(t,\phi,\psi))| e^{\alpha s} \\
&\quad\quad - \sup_{s\geq 0} |x_{t+s}(t,\phi,\psi)| e^{\alpha s}\} \\
&\quad = \limsup_{h\to 0+} h^{-1}\{\sup_{s\geq h} |x_{t+s}(t,\phi,\psi)| e^{-\alpha h} - \sup_{s\geq 0} |x_{t+s}(t,\phi,\psi)| e^{\alpha s}\} \\
&\quad \leq \limsup_{h\to 0+} h^{-1}(e^{-\alpha h} - 1) \sup_{s\geq 0} |x_{t+s}(t,\phi,\psi)| e^{\alpha s} \\
&\quad = -\alpha V(t,\phi,\psi).
\end{aligned}$$

This ends the proof of Theorem 3.10.

Remark 3.13 *The proof of continuity of the function $V(t,\phi,\psi)$, with respect to t, can also be obtained from this observation: the existence of $V^{\boldsymbol{\cdot}}(t,\phi,\psi)$ implies, according to the definition,*

$$V(t+h, x_{t+h}(t,\phi_0,\psi_0), y_{t+h}(t,\phi_0,\psi_0)) - V(t,\phi_0,\psi_0) = O(h),\ h \to 0.$$

Taking into account the Lipschitz property for V, in (ϕ,ψ), one sees that the last term in (3.144) tends to zero, as $t \to h^+$. The proof in the text is requiring the continuous dependence of x_t, with respect to t, which we did not establish under our assumptions. Nevertheless, this is a consequence of an inequality of Gronwall's type and can be found in classical books(for instance, see Hale [240]).

Remark 3.14 *Let us derive from Theorem 3.10 a special case, which occurs when the y-variables are absent. In other words, we shall deal with exponential asymptotic stability for system (3.126). Of course, the definition of exponential asymptotic stability, derived from the case of partial asymptotic stability of exponential type, is obtained from the requirement*

$$|x(t;t_0,\phi)| \leq M|\phi|e^{-\alpha(t-t_0)}, \; t \geq t_0, \tag{3.145}$$

with M, α positive constants and $|\phi| = |\phi|_C$, and $x(t_0;t_0,\phi) = \phi$.

Then, the Liapunov function $V(t,\phi)$ will be defined by (taking $t_0 = 0$),

$$V(t,\phi) = \sup_{s\geq 0}\{|x_{t+s}(t,\phi)|_C \, e^{\alpha s}\}, \tag{3.146}$$

while its basic properties are (we omit the index for the norm):

$$|\phi| \leq V(t,\phi) \leq M|\phi|, \tag{3.147}$$

$$V^{\boldsymbol{\cdot}}(t,\phi) \leq -\alpha V(t,\phi). \tag{3.148}$$

The construction of $V(t,\phi)$ and its properties hold in case when system (3.126) is nonlinear. See, for instance, Hale [240].

The multiple applications to stability, boundedness and other asymptotic properties of the solutions to (3.126) make the earlier functions of type $V(t,\phi)$ a very handy tool. We illustrate here the case of boundedness for perturbed systems of the form

$$\dot{x}(t) = F(t,x_t) + R(t,x_t), \tag{3.149}$$

for $x : R_+ \to R^n$, $F : R_+ \times C([-h,0],R^n)$ and $R(t,x_t)$ like $F(t,x_t)$.

If we admit the exponential asymptotic stability for the zero solution of the non-perturbed system (3.126), then the existence of the Liapunov function $V(t,\phi)$, which satisfies (3.147) and (3.148), is guaranteed. If one admits, for instance, a global Lipschitz condition for $F(t,\phi)$ in ϕ, then $V(t,\phi)$ will also satisfy

$$|V(t,\phi_1) - V(t,\phi_2)| \leq L|\phi_1 - \phi_2|, \tag{3.150}$$

for $\phi_1, \phi_2 \in C([-h,0],R^n)$.

As seen in Section 3.3, the function $V(t,\phi)$ can be used to estimate its derivative $V'(t,\phi)$, with respect to the perturbed system (3.149). We ask the readers to investigate stability and boundedness of solutions to (3.149), following the procedure we have already used in Section 3.3 (see System (3.61).

In concluding this section, we will briefly indicate how various problems of asymptotic behavior for the solutions of system (3.149) can be dealt with by applying a method of reduction to equations with causal operators, as investigated in Chapter 2.

Indeed, when $F(t,x_t)$ is linear and continuous, as mentioned earlier, one can apply the formula of variation of parameters for functional differential equations with causal operates, as shown in Chapter 2. Formula (2.42) allows us to write, taking into account the linearity of F, the following equation in $x(t) = x(t;0,\phi)$,

$$x(t;0,\phi) = X(t,0)\,\phi(0) + \int_0^t X(t,s)\,R(s,x_s)\,ds, \tag{3.151}$$

with $x_0 = \phi$, where $X(t,s)$ is an $n \times n$ matrix, uniquely determined by $F(t,\phi)$, defined on $0 \le s \le t < \infty$.

The condition of exponential asymptotic stability, imposed on $\dot{x}(t) = F(t,x_t)$, implies

$$|X(t,s)| \le k e^{-\alpha(t-s)},\ t \ge s \ge 0, \tag{3.152}$$

where $k, \alpha > 0$ are fixed constants, which is a severe restriction on the kernel of the integral equation (3.151). In particular, in the autonomous case, when $X(t,s) = X(t-s)$, (3.152) implies $|X| \subset L^1(R_+,R)$. In general, Inequality (3.152) allows quite general type of $R(t,x_t)$, such as $|R(t,x_t)| \in M(R_+,R)$, with the space M defined in Chapter 1.

We invite the readers to investigate equation (3.149), which leads to results for equation (3.151), as one can see in the books by Corduneanu [123, 149]. For instance, a boundedness result for all solutions of (3.151) is assured when $|R| \in M$.

3.7 STABILITY OF INVARIANT SETS

In preceding sections, we have only discussed the concept of stability for a single solution of a functional differential equation. This section will be dedicated to the investigation of the stability concept for a *set*, which is not necessarily formed by solutions of a given equation/system. This type of problem makes, obviously, sense for several kinds of evolution equations. We consider here cases of finite delay equations.

Namely, the system can be written in the form

$$\dot{z}(t) = Z(t,z_t), \tag{3.153}$$

with $z_t(s) = z(t+s)$, $s \in [t-h,t]$, where $h > 0$ is fixed and $t \in R_+$. The initial condition is of the form

$$z_{t_0} = \phi \in C([t_0-h,t_0],R^n), \tag{3.154}$$

with $t_0 \in R_+, \phi : [t_0 - h, t_0] \to R^n$, usually a continuous map (since the existence of $\dot{z}(t)$, for $t \in R_+$, is suggesting the continuity property of ϕ, particularly when $t_0 = 0$). The initial space for (3.153) is thus $C([-h,0],R^n)$, with the supremum norm for its elements.

In view of presenting some results for the stability of a set, with respect to system (3.153), we shall need some definitions and notations.

First, we will rewrite system (3.153) in an equivalent form, by letting $z = \mathrm{col}(x,y) = \binom{x}{y}$, with $x \in R^m$, $m \leq n$, and $y \in R^p$, $p + m = n$, which allow to write (3.153) in the form

$$\begin{aligned} \dot{x}(t) &= X(t,x,y), \\ \dot{y}(t) &= Y(t,x.y), \end{aligned} \tag{3.155}$$

with X and Y determined from $Z = col(X,Y)$.

The initial condition (3.154) becomes

$$x_{t_0} = \psi, \; y_{t_0} = \chi, \tag{3.156}$$

where $\psi \in C([t_0 - h, t_0], R^m)$, $\chi \in C([t_0 - h, t_0], R^p)$. With the notation used in the preceding section, one has for $t \geq t_0$

$$x_t(t_0, \psi) = x(t; t_0, \psi), \; y_t(t_0, \chi) = y(t; t_0, \chi), \; t \in R.$$

We assume that all initial data are continuous and the norm, in their (Banach) space, is the supremum norm. These norms will be denoted by $\|\phi\|$, $\|\psi\|$, $\|\chi\|$, with $\|\phi\| = \max(\|\psi\|, \|\chi\|)$. One more notation is

$$C_\alpha = \{\phi; \text{ for } \|\psi\| \leq \alpha, \; \|\chi\| < +\infty\}, \; \alpha > 0. \tag{3.157}$$

The set for which we shall investigate the stability, with respect to system (3.153), is

$$M = \{\phi | \phi \in C([-h,0],R^n); \; \|\psi\| = 0, \; \|\chi\| < +\infty\}. \tag{3.158}$$

Definition 3.9 *A set $M \subset C([-h,0],R^n)$ is called positively invariant for system (3.153) if $t_0 \in R_+$ and $\phi \in M$ imply $z_t(t_0,\phi) \in M$ for any $t \geq t_0$.*

We can now formulate the definitions for various kinds of stability of the positively invariant set $M \subset C([-h,0],R^n)$, defined by (3.158).

Definition 3.10 *The set M is called stable for (3.153), if for any $\epsilon > 0$ and $t_0 \geq 0$ there exists $\delta = \delta(\epsilon, t_0) > 0$, with the property that $\phi \in C([-h,0],R^n)$, with $\|\psi\| < \delta$, imply $\|x_t(t_0,\phi)\| < \epsilon$ for $t \geq t_0$.*

The *uniform stability* takes place in case $\delta(\epsilon, t_0) = \delta(\epsilon)$, $t_0 \in R_+$.

The *asymptotic stability* of M means stability (as before), and in addition, the following property: for any $t_0 \geq 0$, one can find $\eta = \eta(t_0) > 0$, and for each $\epsilon > 0$, $\phi \in C([-h,0],R^n)$ with $\|\psi\| < \eta$, there exists $\sigma = \sigma(\epsilon, t_0, \phi) > 0$, such that $\|x_t(t_0, \phi)\| < \epsilon$ for any $t \geq t_0 + \sigma$.

The *uniform asymptotic stability* of the set $M \subset C([-h,0],R^n)$ means the asymptotic stability, and also the properties: $\eta(t_0)$ and $\sigma(\epsilon, t_0, \phi)$, from the definition of asymptotic stability, can be chosen independent of $t_0 \in R_+$, that is, $\eta(t_0) = \eta > 0$, $\sigma(\epsilon, t_0, \phi) = \sigma(\epsilon, \phi) > 0$.

All four cases, mentioned in Definition 3.10, are similar to those formulated in Chapter 2, for ordinary differential equations. They can be easily adapted to various types of functional differential equations. See, for instance, the case of functional differential equations with causal operators (Corduneanu [149]), with respect to the second kind initial value problem. Let us point out that, in this case, there are no available results in the literature (to the best of our knowledge).

In order to formulate a result related to the uniform asymptotic stability of the set M, we shall use the method of functionals of Liapunov–Krasovskii type [299], namely $V = V(t, \phi)$, $t \in R_+$, $\phi \in C([-h,0],R^n)$, assuming the continuity of V in a domain of the form C_α, as shown in (3.157), such that the derivative along the trajectories of (3.153) exists:

$$\dot{V}(t, x_t(t_0, \phi)) = \limsup_{h \to 0+} h^{-1}[V(t+h, z_{t+h}(t_0, \phi)) - V(t, z_t(t_0, \phi))]. \quad (3.159)$$

The lim sup in (3.159) makes sense when the local existence and uniqueness hold for (3.153). The functional $V(t, \phi)$ being differentiable or, at least, locally Lipschitz continuous in C_α.

It is possible to develop a full theory of stability with respect to an invariant set M, following the corresponding theory in the case of stability of a solution (see Section 3.2).

We are not getting into all of the details that pertain to the case of stability with respect to an invariant set, but we will state and use a result that provides a necessary and sufficient condition for uniform asymptotic stability of the set M, defined by (3.158).

Theorem 3.11 *A necessary and sufficient condition for the uniform asymptotic stability of the set M is the existence of a continuous, real-valued functional $V(t, \phi)$, defined on C_α defined by (3.157), such that the following conditions are satisfied:*

$$a(\|x_t\|) \leq V(t, x_t) \leq b(\|x_t\|), \quad (3.160)$$

$$\dot{V}(t, x_t) \leq -c(\|x_t\|), \quad (3.161)$$

where a, b, and c are wedge functions (i.e., from R_+ into itself, increasing and vanishing at 0*).*

Remark 3.15 *It is known that a wedge function can be assumed continuous. Indeed, for given $a(r)$ only increasing, one sees that $a_1(r) = \int_0^r e^{-t} a(t)\,dt = a_1(r) \leq a(T)$, while $\int_0^r e^t a(t)\,dt = a_2(t) \geq a(r)$. The term wedge was introduced by Burton [84]. Hahn [235] called such functions of class $\mathcal{K}$, and the continuity property was noticed by Massera [373].*

Proof. First, we will prove the sufficiency of conditions (3.160) and (3.161). In other words, we need to show that (3.160) and (3.161) imply the uniform asymptotic stability of the set M, with respect to system (3.153).

We will apply the comparison method to this special case. In this case, to do this, the comparison method we have used in Chapters 2 and 3 is not involved. But (3.160) and (3.161) lead easily to the comparison inequality

$$\dot{V}(t,z_t) \leq -c(b^{-1}(V(t,z_t))),\ t \in R_+, \tag{3.162}$$

and denoting $c(b^{-1}(V)) = \rho(V)$, we can rewrite the comparison equation associated to (3.162) in the form

$$\dot{y}(t) = -\rho(y(t)),\ t \in R_+, \tag{3.163}$$

and notice that this is an autonomous equation with $\rho : R_+ \to R_+$ a wedge function.

It is generally known that the stability or asymptotic stability for an autonomous equation/system is also uniform, (Halanay [237]). Hence, we need to show that under our assumptions, the zero solution of (3.163) is asymptotically stable. Indeed, excluding from consideration the trivial solution $y = 0$, any solution with $y(0) = y_0 > 0$ satisfies

$$\int_{y_0}^{y(t)} [\rho(y)]^{-1}\,dy = -t,\ t \in R_+. \tag{3.164}$$

Denoting by $\Omega(y)$ a primitive of $[\rho(y)]^{-1}$, on $y > 0$, (3.164) can be rewritten as

$$\Omega(y(t)) - \Omega(y_0) = -t,\ t \in R_+, \tag{3.165}$$

which yields, taking into account the invertibility of Ω,

$$y(t) = \Omega^{-1}(\Omega(y_0) - t),\ t \in R_+. \tag{3.166}$$

Any translation along the t-axis of the solution, namely

$$y(t;t_0) = \Omega^{-1}(\Omega(y_0) - t + t_0), \tag{3.167}$$

with $t_0 > 0$ and $y(t_0, t_0) = y_0 > 0$, is also a solution of (3.163) and all solutions can be obtained in this way. Therefore, if we consider all solutions (*i.e.*, all functions given by $y_0 > 0$ and $t_0 \in R_+$) arbitrary, these are all the solutions of the comparison equation (3.163).

Now it is easy to reach our conclusion about uniform asymptotic stability. Since each solution is monotonically decreasing on R_+, there results that $y(t;t_0) \leq y_0$ for $t \geq t_0$. This implies the uniform stability of the solution $y = 0$. Indeed, we know that $\lim_{t\to\infty} y(t, t_0) = y_\infty$ does exist (because of monotonicity and boundedness below by 0), which implies $\lim_{t\to\infty} y(t, t_0) = y_\infty = 0$. So far, we are sure about asymptotic stability of the solution $y = 0$ of (3.163). But the global uniqueness (which follows from Bompiani's uniqueness theorem) allows us to derive the uniform asymptotic stability of $y = 0$. Let us consider the solution $y(t; 0, y_0)$ of the comparison equation (3.163), for $t \in R_+$ and some $y_0 > 0$. Given $\epsilon > 0$, arbitrary, we can write on behalf of the above established asymptotic stability of $y = 0$,

$$y(t;0,y_0) < \epsilon, \ t \geq T(\epsilon), \tag{3.168}$$

for a fixed $T(\epsilon) > 0$. On the other hand, we can write $y(t;t_0,y_0) = y(t - t_0, 0, y_0)$ for $t \geq t_0$, which combined with (3.168) leads to

$$y(t,t_0,y_0) < \epsilon, \ t \geq t_0 + T(\epsilon). \tag{3.169}$$

This is the needed property, which combined with that of uniform stability, implies the uniform asymptotic stability of the zero solution of (3.163).

Hence, it is proven that the set M, defined by (3.158), is uniformly asymptotically stable. Equivalently, conditions (3.160) and (3.161) are *sufficient* to assure the uniform asymptotic stability of M, with respect to system (3.153).

The *necessity* part is somewhat lengthier, offering at the same time a *converse theorem on stability* (this time for a set M). We had a similar situation in Section 3.6, when the exponential asymptotic stability was discussed.

Therefore, under the basic assumption that the set M is uniformly asymptotically stable with respect to system (3.153), one must show the existence of the functional $V(t, \phi)$, such that (3.160) and (3.161) hold true.

In view of obtaining an extra important property for the functional $V(t, \phi)$, we will make one more assumption on system (3.153). Namely, we will assume that a Lipschitz-type condition is valid for $Z(t, \phi) = (X(t, \phi), Y(t, \phi))$:

$$|x(t,\phi_1) - x(t,\phi_2)| \leq L \|\psi_1 - \psi_2\|, \ L > 0, \tag{3.170}$$

where $\phi_1 = (\psi_1, \chi_1)$, $\phi_2 = (\psi_2, \chi_2)$, keeping the notations introduced before. On behalf of (3.170), one routinely obtains the inequality $(t \geq t_0)$

$$\begin{aligned} \|x_t(t_0,\phi_1) - x_t(t_0,\phi_2)\| &\leq \|\psi_1 - \psi_2\| \\ &+ L \int_{t_0}^{t} \|x_\tau(t_0,\phi_1) - x_\tau(t_0,\phi_2)\| \, d\tau, \end{aligned} \tag{3.171}$$

which is of Gronwall's type. Therefore, we get

$$\|x_t(t_0,\phi_1) - x_t(t_0,\phi_2)\| \leq \|\psi_1 - \psi_2\| \, e^{L(t-t_0)}, \; t \geq t_0, \tag{3.172}$$

which shows the *continuous dependence* of the X-part of the solution $z_t(t_0,\phi)$, with respect to the initial data, on each finite interval.

What we expect to get from improving the extra assumption (3.170) is the following type of Lipschitz condition for the functional $V(t,\phi)$, to be constructed:

$$|V(t,\phi_1) - V(t,\phi_2)| \leq \tilde{L} \, \|\psi_1 - \psi_2\|, \; \tilde{L} > 0, \tag{3.173}$$

besides conditions (3.160) and (3.161).

In the anticipation of constructing the Liapunov–Krasovskii functional $V(t,\phi)$, we shall write the formula of definition

$$V(t,\phi) = \int_t^\infty G(\|x_\tau(t,\phi)\|) \, d\tau + \sup_{\tau \in [t,\infty)} G(\|x_\tau(t,\phi)\|), \tag{3.174}$$

where $G(u)$ is a real valued function to be described in the following text.

We shall rely now on an auxiliary result, due to Malkin [356], Krasovskii [299], and Hahn [235] who dealt with the problem of uniform asymptotic stability in various stages.

First, let us notice that under assumption of uniform asymptotic stability of the set M, one can find a function $f : R_+ \to R_+$ continuous and decreasing, such that the following conditions are satisfied:

$$\|x_t(t_0,\phi)\| \leq f(t - t_0) \text{ for } t \geq t_0, \; \phi \in C_\alpha, \; f \in C_0(R_+,R_+). \tag{3.175}$$

Second, there exists a wedge function $G : R_+ \to R_+$, with G' also a wedge for which

$$\int_0^\infty G(f(\tau)) \, d\tau = N_1 < \infty, \quad \int_0^\infty G'(f(\tau)) \, e^{L\tau} \, d\tau = N_2 < \infty, \tag{3.176}$$

$$G'(f(t)) \, e^{Lt} \in BC(R_+,R_+), \tag{3.177}$$

the norm in $BC(R_+, R_+)$, for (3.177), will be denoted by $N_3 > 0$. The constant $L > 0$ is the one in (3.172).

We will now show that $V(t,\phi)$, given by (3.172), satisfies all the requirements for the case of uniform asymptotic stability of the invariant set M. Both integrals in (3.176) are convergent, as stipulated in (3.176), while the supremum in formula (3.172) is finite as it simply follows from (3.177).

Let us prove that conditions (3.160) and (3.161) are satisfied, which clears the case of uniform asymptotic stability.

In regard to conditions (3.160), we notice that $V(t,\phi) \geq \sup_{\tau\in[t,\infty)} [G(\|x_\tau(t,\phi)\|)]$, and since $x_\tau(t,\phi) = \psi$, there results $V(t,\phi) \geq G(\|\psi\|)$. Further, one has, according to (3.174),

$$V(t,\phi) < \int_0^\infty G(f(\tau))\,d\tau + G(f(0)) = N_1 + G(f(0)) = N_3 < \infty.$$

This shows the uniform boundedness of $V(t,\phi)$ on C_α. But $V(t,\theta) = 0, t \in R_+$, and if one denotes $b(r) = \sup\{V(t,\phi); t \in R_+, \|\phi\| \leq r, r > 0\}$, we obtain the right wedge from (3.160).

We shall postpone, one moment, the proof of property (3.161) for $V(t,\phi)$, proving first the Lipschitz-type condition (3.173). Indeed, one has from (3.174)

$$\begin{aligned}|V(t,\phi_1) - V(t,\phi_2)| \leq \int_t^\infty |G(\|x_\tau(t,\phi_1)\|) - G(\|x_\tau(t,\phi_2)\|)|\,d\tau \\ + \sup_{\tau\in[t,\infty)} |\|x_\tau(t,\phi_1) - x_\tau(t,\phi_2)\||.\end{aligned}$$

Taking (3.172) into account, we can write on behalf of the second mean value theorem,

$$\begin{aligned}&|V(t,\phi_1) - V(t,\phi_2)| \\ &\leq \|\psi_1 - \psi_2\| \left[\int_t^\infty G'(f(\tau - t))\,e^{L(t-\tau)}\,d\tau + \sup_{\tau\in[t,\infty)} G'(f(\tau - t))\,e^{L(\tau-t)}\right] \\ &\leq \tilde{L}\|\psi_1 - \psi_2\|,\end{aligned} \tag{3.178}$$

where $\tilde{L} \geq N_2 + N_3$, which proves (3.173). The Lipschitz condition, in C_α, is proven for $V(t,\phi)$. Actually, this is a partial Lipschitz condition with respect to only the x-coordinates ($x_0(t,\phi) = \psi$). But this type of property is needed to investigate the stability problem of the invariant set M, taking into account the permanent perturbations. In order to provide the proof of Theorem 3.11, we need to take care of condition (3.161) for the functional $V(t,\phi)$, given by (3.174), and the continuity of $V(t,\phi)$ in (t,ϕ).

Since (3.174) shows that the integral term on the right-hand side is differentiable, there remains to consider only the last term in formula (3.174), and estimate

$$\limsup_{h\to 0+} h^{-1}[\sup_{t_0+h\leq\tau} G(\|x_\tau(t_0+h, z_{t_0+h}(t_0,\phi))\|) - \sup_{t_0\leq\tau} G(\|x_\tau(t_0, z_{t_0}(t_0,\phi))\|)].$$

The first term in the brackets is smaller than the second, because

$$x_\tau(t, z_t(t_0,\phi)) = x_\tau(t_0,\phi), \text{ when } t_0 \leq t \leq \tau,$$

both terms being on the same x-trajectory, while they coincide for $\tau = t_0$. Therefore,

$$\begin{aligned}\dot{V}(t,\phi)\rfloor_{t=t_0} &\leq \left[\frac{d}{dt}\int_t^\infty G(\|x_\tau(t,\phi)\|)\,d\tau\right]_{t=t_0}\\ &\leq -G(\|x_{t_0}(t_0,\phi)\|)\\ &= -G(\|\psi\|). \end{aligned} \tag{3.179}$$

This is exactly condition (3.161).

It remains only to show the continuity of $V(t,\phi)$, with respect to t. Let us notice, from (3.179), that in accordance with the definition of Dini's numbers, the right upper number of V, along the solution $z_{t_0}(t,\phi)$, is finite. By a similar argument, one obtains the finiteness of the right lower Dini's number, which means that, for fixed $\phi \in C_\alpha$, for each $t > t_0$, the right Dini's number of $V(t, z_t(t_0,\phi))$ is finite. This implies the local Lipschitz-type condition at each $t \geq t_0$, from which we draw the conclusion of continuity at each $t \geq t_0$.

This ends the proof of Theorem 3.11.

To summarize the findings, we notice that conditions (3.160) and (3.161) are necessary and sufficient for the uniform asymptotic stability of the set M, with respect to the unperturbed system (3.153), if one also considers the Lipschitz condition (3.170), then one obtains for $V(t,\phi)$ the Lipschitz-type condition (3.173).

We shall be interested, in the last part of this section, in the uniform asymptotic stability of the invariant set M, defined by (3.158), for the perturbed system of system (3.153), namely

$$\dot{w}(t) = Z(t,w_t) + R(t,w_t), \tag{3.180}$$

with $w = \text{col}(u,v)$, $u \in R^m$, $v \in R^p$, $m+p=n$.

It will be admitted that $Z(t,\phi)$ and $R(t,\phi)$ are bounded on the set $R_+ \times C_{\alpha_1}$, with C_{α_1} defined by (3.157), $\alpha_1 > 0$, and satisfy there the global partial Lipschitz condition

$$|Z(t,\phi_1)-Z(t,\phi_2)| \leq L\|\psi_1-\psi_2\|, \qquad (3.181)$$
$$|R(t,\phi_1)-R(t,\phi_2)| \leq L\|\psi_1-\psi_2\|,$$

where $\phi_1 = col(\psi_1,\chi_1)$, $\phi_2 = col(\psi_2,\chi_2)$, and $L>0$ is a constant.

In order to secure the fact that (3.180) has the zero solution, we also assume

$$Z(t,\phi) = \theta_m,\ R(t,\phi) = \theta_p,\ t \in R_+,\ \phi = \theta \in R^n, \qquad (3.182)$$

where θ_m, θ_p denote the null elements of R^m, R^p $(m+p=n)$. These imply that system (3.180) has the solution $u=\theta_m$. We also assume that M is an invariant set for system (3.180).

The following theorem can be stated:

Theorem 3.12 *Consider the perturbed system (3.180), under the earlier formulated conditions, (3.181) and (3.182). Moreover, assume the invariant set M is uniformly asymptotically stable, with respect to the unperturbed system (3.153) whose domain of attraction contains C_{α_1}. Also, assume that $R(t,\phi)$ in (3.180) is such that*

$$|R(t,\phi)| \leq h(t),\ t \in R_+,\ \phi \in C_{\alpha_1}, \qquad (3.183)$$

with $h \in M_0(R_+,R_+)$, that is, $\lim_{t\to\infty}\int_t^{t+1} h(s)\,ds = 0$.

Then, M is uniformly asymptotically stable with respect to System (3.180), and there exists $\eta>0$, $\eta<\alpha_1$, such that C_η belongs to the domain of attraction of M.

The proof of Theorem 3.12 can be conducted on the same lines as that of Theorem 3.11, and we will send the readers to the papers of Bernfeld et al. [60] Corduneanu and Ignatyev [166] for details.

We emphasize the fact that in case M =the graph of a solution of systems (3.153) and (3.180), Theorems 3.11 and 3.12 provide conditions for the preservation of the uniform asymptotic stability of the null solution of each system (3.153) and (3.180).

3.8 ANOTHER TYPE OF STABILITY

This section will be dedicated to discussing and providing a few results that concern another type of stability for functional differential equations, namely the Ulam–Hyers–Rassias. These authors have considered this variant of stability, which is different from Liapunov's stability mentioned earlier in this section. Ulam [523] and Hyers et al. [268] contain all of the initial material and information. While both Liapunov and Ulam–Hyers–Rassias (U.H.R.)

stabilities constitute a kind of dependence of solutions with respect to data (initial or permanent), there are some differences in their nature. We proceed now to the investigation, directly treating the case of functional differential equations with causal operators. We have dealt with such equations in Section 3.4. Unlike the notations in that section, we will write the functional differential equation in the form

$$\dot{x}(t) = f(t, x(t), (Vx)(t)), \tag{3.184}$$

which in the very special case $f(t,x,u) \equiv u$, reduces to equation (3.78).

Equation (3.78) is only apparently less general than (3.184), in case we make the assumption that V is a causal operator (see also Chapter 2).

Indeed, the right-hand side of (3.184), which is a map from $R_+ \times R^n \times E$ into R^n, with E a function space, say $E = C(R_+, R^n)$, is an operator

$$(Ux)(t) = f(t, x(t), (Vx)(t)), \tag{3.185}$$

which is also causal. We will formulate in the sequel what specific conditions allow us to use the form (3.184) for the functional equations, instead of the form (3.78). The main reason for this change resides in the fact that hypotheses can be stated in a more convenient form, emphasizing the role of $x(t)$ in a direct manner.

Let us now formulate the definition of the new types of stability that we will investigate, in regard to the functional differential equation (3.184).

Definition 3.11 *Equation (3.184) is called Ulam–Hyers–Rassias stable, with respect to the function $\phi \in C(R_+, R_+)$, if there exists a constant $c_\phi > 0$, with the property that for each $y \in C^{(1)}(R_+, R^n)$, satisfying*

$$|\dot{y}(t) - f(t, y(t), (Vy)(t))| \leq \phi(t),\ t \in R_+, \tag{3.186}$$

one can find a solution $x(t) \in C^{(1)}(R_+, R^n)$ of equation (3.184), such that

$$|y(t) - x(t)| \leq c_\phi \phi(t),\ t \in R_+. \tag{3.187}$$

Remark 3.16 *Definition 3.11 can be easily reformulated, if one introduces the notation*

$$g_y(t) = \dot{y}(t) - f(t, y(t), (Vy)(t)),\ t \in R_+, \tag{3.188}$$

in which case condition (3.186) becomes $|g_y(t)| \leq \phi(t)$, $t \in R_+$.

In other words, a function $y(t) \in C^{(1)}(R_+, R^n)$ is a solution of (3.186), iff there exists a function $g_y(t) \in C(R_+, R^n)$, such that $|g_y(t)| \leq c_\phi \phi(t)$, $t \in R_+$, and $\dot{y}(t) = f(t, y(t), (Vy)(t)) + g_y(t)$, $t \in R_+$.

Remark 3.17 *If $y(t) \in C^{(1)}(R_+, R^n)$ is a solution of (3.186), then $y(t)$ is also a solution of the following integral inequality:*

$$\left| y(t) - y(0) - \int_0^t f(s, y(s), (Vy)(s))\, ds \right| \leq c_\phi \int_0^t \phi(s)\, ds, \qquad (3.189)$$

for all $t \in R_+$.

Inequality (3.189) is a direct consequence of Remark 3.16.

We shall state and prove a result which provides both existence and U.H.R. stability for equation (3.184).

Theorem 3.13 *Let us consider the equation (3.184), under the following assumptions:*

(*a*) *$f \in C(R_+ \times R^n \times R^n, R^n)$, V is an operator acting on the space $C(R_+, R^n)$.*

(*b*) *f satisfies a generalized Lipschitz condition, with respect to the last two arguments, namely*

$$|f(t, u_1, v_1) - f(t, u_2, v_2)| \leq \lambda(t)\, (|u_1 - u_2| + |v_1 - v_2|), \qquad (3.190)$$

for $t \in R_+$, $u_i, v_i \in R^n$, $i = 1, 2$, and $\lambda \in L^1(R_+, R_+)$.

(*c*) *There exists a constant $\mu > 0$, such that*

$$|(Vx)(t) - (Vy)(t)| \leq \mu\, |x(t) - y(t)|, \qquad (3.191)$$

for $t \in R_+$ and $x, y \in C(R_+, R^n)$.

(*d*) *The function $\phi \in C(R_+, R_+)$ is increasing, and satisfies the inequality*

$$\int_0^t \phi(s)\, ds \leq \gamma \phi(t),\ t \in R_+, \qquad (3.192)$$

for some $\gamma > 0$.

Then, equation (3.184) has a unique solution $x \in C^{(1)}(R_+, R^n)$ for each initial datum $x^0 \in R^n$ which is U.H.R. stable with respect to the function ϕ.

Proof. Let $y \in C^{(1)}(R_+, R^n)$ be a solution of inequality (3.186). Now, let us notice that equation (3.184) has a unique solution in $C^{(1)}(R_+, R^n)$, a fact which is a simple consequence of Theorem 2.2. Indeed, the Lipschitz-type condition (3.191) is implying the causal property on V. Therefore, the right-hand side of equation (3.184) represents a causal operator. We consider the unique solution $x \in C^{(1)}(R_+, R^n)$ of equation (3.184) satisfying the initial condition $x(0) = y(0)$. Hence, we can write for this solution

$$x(t) = y(0) + \int_0^t f(s, x(s), (Vx)(s))\, ds, \; t \in R_+. \tag{3.193}$$

If one takes into account (3.186), there results from (3.193)

$$\begin{aligned} |y(t) - x(t)| &\leq \left| y(t) - y(0) - \int_0^t f(s, y(s), (Vy)(s))\, ds \right| \\ &\quad + \int_0^t |f(s, y(s), (Vy)(s)) - f(s, x(s), (Vx)(s))|\, ds \\ &\leq \gamma\phi(t) + \int_0^t (1+\mu)\,\lambda(s)|y(s) - x(s)|\, ds. \end{aligned}$$

Retaining from these inequalities only the first and last terms, we obtain an integral inequality of Gronwall type, for $|y(t) - x(t)|$, which will lead to the final step. Indeed, the inequality looks (from above)

$$|y(t) - x(t)| \leq \gamma\phi(t) + \int_0^t (1+\mu)\,\lambda(s)|y(s) - x(s)|\, ds,$$

and this leads to the estimate (see Corduneanu [149, 69])

$$\begin{aligned} |y(t) - x(t)| &\leq \gamma\phi(t) + \int_0^t \gamma\phi(s)\,(1+\mu)\,\lambda(s)\, e^{\int_s^t (1+\mu)\,\lambda(u)\, du}\, ds \\ &\leq \gamma\phi(t)\left\{1 - \int_0^t e^{\int_s^t (1+\mu)\,\lambda(u)\, du}\, ds\right\} \\ &= \gamma\phi(t)\left[1 - \left[e^{\int_s^t (1+\mu)\,\lambda(u)\, du}\right]_0^t\right] \\ &= \gamma e^{(1+\mu)\int_0^t \lambda(u)\, du}\,\phi(t) \\ &\leq \gamma e^{(1+\mu)\int_0^\infty \lambda(u)\, du}\,\phi(t) = c_\phi\,\phi(t), \end{aligned}$$

with

$$c_\phi = \gamma e^{(1+\mu)\int_0^\infty \lambda(u)\, du}. \tag{3.194}$$

Hence,

$$|y(t) - x(t)| \leq c_\phi\,\phi(t), \; t \in R_+,$$

which is formula (3.187) in Definition 3.11.

In order to end the proof, one needs to show that inequality (3.192), in the statement of Theorem 3.13, has positive increasing solutions on R_+.

It suffices to substitute in (3.192) $\phi(t) = A\, e^{\alpha t}$, with $A, \alpha > 0$, to obtain that (3.192) holds true with $\alpha \geq \gamma^{-1}$.

Other solutions do exist for $\phi(t)$ and the readers are invited to solve this exercise.

The proof of Theorem 3.13 is ended, and we will now consider some applications. The rather general form of equation (3.184) allows several particular cases that can be easily treated.

First, we shall consider the usual ordinary differential equations of the first order, namely

$$\dot{x}(t) = f(t, x(t)),\ t \in R_+, \tag{3.195}$$

with $x, f \in R^n$. Condition (a) of Theorem 3.13 is verified if $f(t,x)$ is continuous on $R_+ \times R^n$. Condition (b) will be satisfied if we admit the generalized Lipschitz condition in x,

$$|f(t,x) - f(t,y)| \leq \lambda(t)\, |x - y|, \tag{3.196}$$

on the whole semi-space $R_+ \times R^n$, where $\lambda \in L^1(R_+, R_+)$. For condition (c) we need only to notice that $V = O =$ the zero operator. The condition (d) remains the same as in Theorem 3.13.

Therefore, equation (3.195) is stable with respect to the function $\phi(t)$, in the sense of U.H.R. Since there are infinitely many functions $\phi(t)$, as required in condition (d), there results that conditions (a) and (b) assure existence (on R_+) and uniqueness of solution, belonging to the space $C_g(R_+, R^n)$, with $g =$ any function satisfying (3.192). One may say that the U.H.R. stability is a property related to the spaces $C_g(R_+, R^n)$. These spaces have been introduced and utilized in case of Volterra–Hammerstein integral equations (see Corduneanu [135] for details). A more detailed analysis would be motivated by a desire to understand the connection between U.H.R. stability and the properties formulated in C_g spaces.

Second, let us consider functional equations of the form

$$\dot{x}(t) = f\left(t, x(t), \int_0^t k(t,s,x(s))\, ds\right), \tag{3.197}$$

where $x, f \in R^n$, $t \in R_+$, and $f(t,x,y)$ has its values in R^n; also, $x, y \in R^n$, while $k(t,s,x)$ is a vector valued function, with values in R^n, defined for $0 \leq s \leq t$ and $x \in R^n$. Such equations have been extensively investigated in the literature. See, for instance, Lakshmikantham and Leela [309], Lakshmikantham and Rao [316], Corduneanu [120, 135], and Gripenberg et al. [228]. A special case of (3.197), frequently encountered in the literature corresponds to the so-called Volterra–Hammerstein kernels, when

$$k(t,s,x) \equiv k(t,s)\,g(s,x), \tag{3.198}$$

which allows an efficient use of the results in the case of linear equations. See Corduneanu [120, 135] for certain results, including the convolution kernels $k(t,s) = k(t-s)$, where references are made to several sources in this field.

Third, let us briefly discuss the case of equations with causal operators of the form we considered in Chapter 2, namely

$$\dot{x}(t) = (Vx)(t),\ t \in R_+, \tag{3.199}$$

in which V is an operator acting on the space $BC(R_+,R^n)$ and verifying the global Lipschitz condition (3.191) in Theorem 3.13. This condition implies both causality and continuity of V. Rephrasing Definition 3.11 for (3.199), the readers are invited to formulate the exact conditions, besides (3.191), assuring the U.H.R. stability of equation (3.199).

Application of Theorem 3.13 is possible, but the simplicity of equation (3.199) suggests a straight approach.

In concluding this section, we mention as an open problem the searching of conditions for quasilinear functional equations of the form

$$\dot{x}(t) = (Lx)(t) + (Fx)(t),\ t \in R_+, \tag{3.200}$$

which can assure the U.H.R. stability. In (3.200), L stands for a linear continuous operator on a function space $E(R_+,R^n)$, consisting of continuous or measurable maps, while F is the nonlinear "perturbation." For background material, see Corduneanu [149, Ch. 4].

3.9 VECTOR AND MATRIX LIAPUNOV FUNCTIONS

In the preceding sections, we have relied almost exclusively on the method of investigation of the stability property, based on the use of an auxiliary function, generally known under the name of Liapunov function or functional. The development of the method of Liapunov has lead to many new concepts. Among these new concepts are those of *vector Liapunov function* and *matrix Liapunov function*. The mathematical literature is rich in contributions related to these extended Liapunov's methods, particularly due to Russian and former Soviet authors. For instance, the book by V. M. Matrosov, and A. A. Voronov [387] contains a rather comprehensive presentation of the *vector Liapunov functions* method, while the book by A. A. Martynyuk [366] does the same, for *matrix Liapunov functions*. The book by Matrosov and Voronov has a list of references with over 400 entries, which includes the pioneering work of R. Bellman, V. Lakshimikantham, V. M. Matrosov, and A. A. Martynyuk. Several papers have been published by Matrosov [376, 377], Matrosov and Anapolskii [384], and Matrosov et al. [385], and many other from the school of Matrosov. His seminal papers are from 1962 [376, 377], and deal with the use of more than a single auxiliary function. Also, his papers [375–383] contain a presentation of the method of vector Liapunov functions, based on vector differential inequalities in the sense of Ważewski [533]. The conditions imply a certain property of monotonicity different than the one commonly used in Hilbert space theory. For details regarding the aforementioned properties, we send the readers to the references indicated earlier.

If a vector Liapunov function is defined by means of a finite number of scalar type functions, say

$$\begin{gathered} v(t,x) \stackrel{\text{def}}{=} (v_1(t,x), v_2(t,x), \ldots, v_k(t,x)),\ k > 1, \\ (t,x) \in \Omega = \{(t,x); t \in R_+, x \in R^n,\ n \geq 1\}, \end{gathered}$$

then, there are also present in the literature, the matrix Liapunov functions,

$$V(t,x) = (V_{ij}(t,x)),\ 1 \leq i \leq m,\ 1 \leq j \leq n, \tag{3.201}$$

which in case $m = 1$ reduce to vector Liapunov functions. Such generalized auxiliary functions, playing the same role in stability theory, as the classical Liapunov function of scalar type, were introduced and thoroughly investigated by A. A. Martynyuk [361–368] and several of his followers (e.g., Djordjevic [197] and Lakshmikhantham et al. [312]).

We shall consider, in some detail, the case of matrix Liapunov functions, and present a result due to Martynyuk [362]. The adopted method will be emphasized due to its degree of generality.

We shall consider systems of ordinary differential equations of the usual form

$$\frac{dx}{dt} = f(t,x),\ t \geq 0,\ x \in R^n, \tag{3.202}$$

the case $x \in D \subset R^n$ being also possible (under adequate restrictions). D is always assumed to be a domain in R^n (open and connected); we shall deal with the case $D = R^n$.

Let us illustrate the matrix Liapunov method, by stating one result due to Martynyuk [362]:

Theorem 3.14 *Consider system (3.202), under conditions*

c_1. *$f : R_+ \times R^n \to R^n$ is a continuous function, enjoying also the property that the solution $x = x(t;t_0,x^0)$, $t_0 \geq 0$, $x^0 \in R^n$ is unique, while $f(t,\theta) = \theta \in R^n$ (null element).*

c_2. *There exists a continuous matrix-valued function $U : R_+ \times R^n \to R^m \times R^m$, and a vector $y \in R^m$, such that the function $v(t,x,y) = y^T U(t,x)y$ is locally Lipschitz in x for all $t \in R_+$.*

c_3. *There exist functions $\phi_{i1}, \phi_{i2}, \phi_{i3} \in KR$, $\tilde{\phi}_{i2} \in CKR$, $i = 1,2,\ldots,m$, and $m \times m$ matrices $B_j(y)$, $j = 1,2,3$, $\tilde{B}_2(y) \in CKR$, such that*

(a) $\phi_1^T(\|x\|)B_1(y)\phi_1(\|x\|) \leq v(t,x,y) \leq \tilde{\phi}_2^T(t,\|x\|)\tilde{B}_2(y)\tilde{\phi}_2(t,x)$, $\forall (t,x,y) \in R_+ \times R^n \times R^m$.

(b) $\phi_1^T(\|x\|)B_1(y)\phi_1(\|x\|) \leq v(t,x,y) \leq \tilde{\phi}_2(\|x\|)B_2(y)\phi_2(\|x\|)$, $\forall (t,x,y) \in R_+ \times R^n \times R^m$.

(c) $D^+v(t,x,y) \leq \phi_3^T(\|x\|)B_3(y)\phi_3(\|x\|)$, $\forall (t,x,y) \in R_+ \times R^n \times R^m$.

Then: if matrices $B_1(y)$, $B_2(y)$, and $\tilde{B}_2(y)$ are positive definite for $y \in R^m$, $y \neq \theta$, with $B_3(y)$ negative definite, under conditions (a) and (c), the solution $x = \theta$ of system (3.202) is globally stable; under conditions (b), (c), the solution $x = \theta$ of system (3.202) is uniformly globally stable.

The proof of Theorem 3.14 is included in the previously quoted work of Martynyuk, as well as in other publications of this author [362, 365]. Also, the explanations of the motivations and the definitions of involved concepts, while noticing the similarity with the case of the classical results (see Section 3.2). We shall also remark the complexity of this result.

Remark 3.18 *Similar results to Theorem 3.14 are given in aforementioned references for asymptotic stability or exponential asymptotic stability. Several applications of this class of results are illustrated in cases of large-scale systems, especially those occurring in solving stability problems for systems encountered in contemporary technology.*

Important and interesting problems for researchers in this field, are mentioned here:

1) How can we construct matrix Liapunov functions for a given dynamical system?
2) Which comparison functions should be used in cases of large-scale dynamical systems, to improve the quality of the results, in relation with the subsystems of a complex large-scale system?

3) How can we accurately estimate the attraction domain for the asymptotic stability of an equilibrium state?

These problems are still open, despite some progress achieved by several researchers.

To conclude this section, we will make the same remark, valid for many results in this book, that such results obtained for systems like (3.202) have not been yet extended, to more general classes of functional differential equations.

3.10 A FUNCTIONAL DIFFERENTIAL EQUATION

The preceding sections of this chapter have been dedicated to ordinary differential equations or to equations with finite delay. The first basic contributions have been brought during the period 1950–1990 by A. N. Krasovskii and J. K. Hale, who created research groups/schools, respectively, in Ekaterinburg (Russia) and Providence, R. I., and Atlanta, Georgia. Many other authors have contributed to the development of this relatively new branch, the interest being kept alive. Our list of references provides a portion of the qualitative results obtained in this field.

We consider, in this section on stability, functional differential equations of the form

$$\dot{x}(t) = (Lx)(t) + (Fx)(t),\ t \in R_+, \tag{3.203}$$

in which L designates a linear causal operator, on a convenient space, while F stands for a nonlinear operator, acting on the same space as L. Equation (3.203) represents a perturbed version of the linear equation

$$\dot{x}(t) = (Lx)(t),\ t \in R_+, \tag{3.204}$$

so that the results, we shall provide in this section, can be regarded as *stability* results *in the first approximation*. We shall choose as underlying space for our discussion, the space $L^2_{\text{loc}}(R_+, R^n)$, that is, $L, F : L^2_{\text{loc}}(R_+, R^n) \to L^2_{\text{loc}}(R_+, R^n)$. The results of this section are closely related to those included (on stability) in the book by Corduneanu [149]. They can also be regarded as results on the preservation of stability (of a certain kind), under perturbation.

In order to establish some results in this direction, we need a few auxiliary facts from the theory of linear associated equations of the form

$$\dot{x}(t) = (Lx)(t) + f(t), \tag{3.205}$$

with $f \in L_{\text{loc}}(R_+, R^n)$ under the following initial conditions:

$$x(t_0) = x^0 \in R^n,\ t_0 \geq 0;\ x(t) = \phi(t),\ t \in [0, t_0), \tag{3.206}$$

with $\phi \in L^2([0, t_0], R^n)$.

When L is a linear, causal, and continuous operator on $L^2_{\text{loc}}(R_+, R^n)$, the unique solution of the initial (Cauchy, second kind) value problem is given by the formula

$$x(t) = X(t, t_0)x^0 + \int_0^{t_0} \tilde{X}(t, s; t_0)\phi(s)\, ds + \int_{t_0}^{t} X(t, s) f(s)\, ds,\ t \geq t_0, \tag{3.207}$$

where $X(t, t_0)$, $t \geq t_0$ and $\tilde{X}(t, s; t_0)$, $s \in [0, t_0]$, are $n \times n$ matrices, completely determined by the data.

In particular, as shown in Corduneanu [149], for *uniform stability* of the solution $x = \theta$ of (3.204), it is the existence of $M > 0$, such that $|X(t, t_0)| \leq M$ for $t \geq t_0 \geq 0$; this is a necessary and sufficient condition.

For the *exponential asymptotic stability* of the solution $x = \theta$ of the homogeneous system (3.204), the following conditions are necessary and sufficient: there exist numbers $N > 0$, $\alpha > 0$, such that

$$|X(t, t_0)| \leq N e^{-\alpha(t - t_0)},\ \text{for } t \geq t_0, \tag{3.208}$$

$$\left[\int_0^{t_0} |\tilde{X}(t, s; t_0)|^2\, ds\right]^{\frac{1}{2}} \leq N[-\alpha(t - t_0)],\ t \geq t_0 \geq 0.$$

We shall use these estimates in establishing some results for system (3.203).

Theorem 3.15 *Consider the perturbed system (3.203), with $L : L^2_{\text{loc}}(R_+, R^n) \to L^2_{\text{loc}}(R_+, R^n)$ a linear, causal, and continuous operator. The operator $F : L^2_{\text{loc}}(R_+, R^n) \to L^2_{\text{loc}}(R_+, R^n)$ is assumed continuous, satisfying the growth condition*

$$|(Fx)(t)| \leq \lambda(t)\, |x(t)|,\ a.e.,\ t \in R_+, \tag{3.209}$$

or

$$|(Fx)(t)| \leq \mu\, |x(t)|,\ a.e.,\ t \in R_+. \tag{3.210}$$

Then, if the solution $x = \theta$ of (3.204) is uniformly stable and (3.209) holds with $\lambda \in L^1(R_+, R)$, the solution $x = \theta$ of (3.203) is also uniformly stable. If the solution $x = \theta$ of (3.204) is exponentially asymptotically stable, the solution $x = \theta$ of (3.203) holds the same property.

Proof. If we use formula (3.207) for the solution of (3.203), we obtain the following integral equation for the solution $x(t,t_0;x^0,\phi)$ of the perturbed equation (3.203):

$$x(t,t_0;x^0,\phi) = X(t,s)x^0 + \int_0^{t_0} \tilde{X}(t,s,t_0)\,\phi(s)\,ds + \int_{t_0}^{t} X(t,s)\,(Fx)(s)\,ds. \tag{3.211}$$

Let us now look at the uniform stability case. One derives from (3.211), the following inequality, taking into account (3.209):

$$|x(t;t_0,x^0,\phi)| \le M\,(|x^0| + |\phi|_{L^2}) + M\int_{t_0}^{t} \lambda(s)\,|x(s;t_0,x^0,\phi)|\,ds, \tag{3.212}$$

for $t \ge t_0$, as long as the solution exists. From (3.212), we obtain (Gronwall)

$$|x(t;t_0,x^0,\phi)| \le M\,(|x^0| + |\phi|_{L^2})\,e^{M\int_0^\infty \lambda(s)\,ds},\ t \ge t_0. \tag{3.213}$$

Since the right-hand side of (3.213) is a constant, independent of $t \in R_+$, the first assertion of Theorem 3.15 is proven. The second assertion follows from the inequality, similar to (3.212),

$$|x(t;t_0,x^0,\phi)| \le N\,(|x^0| + |\phi|_{L^2})\,e^{M(\mu-\alpha)(t-t_0)},\ t \ge t_0. \tag{3.214}$$

The positive number α is from (3.208), and it follows from (3.214) that one must have $\mu - \alpha < 0$, or $\mu < \alpha$, in order to secure the property of exponential asymptotic stability for the solution $x = \theta$ of (3.203).

This ends the proof of the second assertion of Theorem 3.15.

Remark 3.19 *If the condition (3.209) is replaced by the similar one*

$$|(Fx)(t) - (Fy)(t)| \le \mu\,|x(t) - y(t)|,\ t \in R_+, \tag{3.215}$$

then, following the same procedure as above, but estimating the difference $|x(t) - y(t)|$*, one obtains*

$$|x(t) - y(t)| \le M\,|x^0 - y^0| + |\phi - \psi|_{L^2}\,e^{-(\lambda-\mu)(t-t_0)}\ t \ge t_0,$$

under the assumption $\lambda > \mu$*. Such estimates for the difference of two solutions are used, for instance, when one deals with approximate data.*

Remark 3.20 *We shall consider here an example which shows a case when exponential asymptotic stability can occur for the null solution of functional equation of the form (3.204). We shall assume*

$$(Lx)(t) = Ax(t) + \int_0^t B(t-s)x(s)\,ds, \tag{3.216}$$

with A a constant $n \times n$ matrix and $B(t)$ a matrix function such that $|B| \in L^1(R_+,R)$. The Laplace transform of B is given by

$$\tilde{B}(s) = \int_0^\infty B(t)e^{-ts}\,dt,\ Re\, s \geq 0. \tag{3.217}$$

One knows that the condition (necessary and sufficient) for asymptotic stability is

$$\det[sI - A - \tilde{B}(s)] \neq 0 \text{ for } Re\, s \geq 0 \tag{3.218}$$

(see, e.g., Corduneanu [120]). Condition (3.218) does not imply necessarily the exponential asymptotic stability. The stronger condition is

$$\int_0^\infty |B(t)|e^{\lambda t}\,dt < \infty, \text{ for some } \lambda > 0$$

a fact showing that a property valid for ordinary differential equations ($B(t) = O$), may not be true for functional differential equations of a more general nature, even in case of some restrictive assumptions, like $|B(t)| \in L^1(R_+,R)$ and (3.218).

In concluding this section, we will embrace another approach for the investigation of the global behavior of the solutions of equation (3.203). Namely, we assume the following conditions hold for the operators involved in (3.203):

a) $L : L^2_{\text{loc}}(R_+,R^n) \to L^2_{\text{loc}}(R_+,R^n)$ is a linear negative definite operator, that is,

$$<(Lx)(t) - (Ly)(t), x(t) - y(t)> \leq -\alpha\,|x(t) - y(t)|^2,\ \alpha > 0,\ t \in R_+, \tag{3.219}$$

with $x,y \in L^2_{\text{loc}}(R_+,R^n)$. Let us notice that this condition on the operator L is in the sense of Euclidean scalar product (not in L^2_{loc} !). On the operator $F : L^2_{\text{loc}}(R_+,R^n) \to L^2_{\text{loc}}(R_+,R^n)$, we shall impose a weak positivity condition, namely

$$<(Fx)(t) - (Fy)(t), x(t) - y(t)> \leq \lambda(t)\,|x(t) - y(t)|^2,\ t \in R_+,$$
$$x,y \in L^2_{\text{loc}}(R_+,R^n). \tag{3.220}$$

Under these assumptions on the operators L and F, we obtain the following comparison inequality for the difference $x(t) - y(t)$:

$$\frac{1}{2}\frac{d}{dt}|x(t)-y(t)|^2 \leq -\alpha\,|x(t)-y(t)|^2 + \lambda(t)\,|x(t)-y(t)|^2,$$

or, denoting $|x(t)-y(t)|^2 = z(t)$, $t \in R_+$,

$$\frac{dz}{dt} \leq -2\alpha z + \lambda(t)\,z,\ t \in R_+, \tag{3.221}$$

which is a very simple first-order differential inequality. One obtains the following estimate for the difference $|x(t)-y(t)|$

$$|x(t)-y(t)| \leq |x^0 - y^0|\,e^{[-\alpha(t-t_0)+\frac{1}{2}\int_{t_0}^{t}\lambda(s)\,ds]}. \tag{3.222}$$

Inequality (3.222) provides an estimate of the difference of two solutions of equation (3.203). This allows us to derive several properties, of a qualitative nature, with regard to the solutions of equation (3.203), making different assumptions on the scalar function $\lambda(t)$, $t \in R_+$. For instance if we assume, for fixed $t_0 \geq 0$,

$$\lim_{t\to\infty}\left[-\alpha(t-t_0) + \frac{1}{2}\int_{t_0}^{t}\lambda(s)\,ds\right] = -\infty, \tag{3.223}$$

there results the global asymptotic stability of any solution of (3.203). Some elementary considerations will lead to this conclusion. Also, the boundedness of all solutions is assured, if we know there exists a single-bounded solution on R_+.

If one assumes only the boundedness above, instead of (3.223)

$$-\alpha\,(t-t_0) + \frac{1}{2}\int_{t_0}^{t}\lambda(s)\,ds \leq K,\ t \geq t_0 \geq 0, \tag{3.224}$$

obviously weaker than (3.223), then the conclusion is *the boundedness* of all solutions of system (3.203), on R_+. Another possible condition is

$$\lim_{t\to\infty}\left[-\alpha(t-t_0) + \frac{1}{2}\int_{t_0}^{t}\lambda(s)\,ds\right] = \ell \in R_-, \tag{3.225}$$

which is stronger than (3.224). Then, if there is a solution with finite limit at ∞, all solutions will have the same finite limit as $t \to \infty$. Such solutions are sometimes called *transient solutions*. This case implies stability.

We ask the readers to search other solutions (i.e., interplay between α and $\lambda(t)$) and find similar results to those sketched above.

3.11 BRIEF COMMENTS ON THE START AND EVOLUTION OF THE COMPARISON METHOD IN STABILITY

Since our exposition of stability theory is centered around the comparison method, we want to comment on the history of this method and indicate several sources that will hopefully contribute to further development. While this method has been used amply in cases of ordinary differential equations, we notice the fact that it is still in the infancy in cases of nonclassical equations.

First, let us mention the fact that the comparison method primarily relies on the theory of differential inequalities, a branch of investigation that appeared later in the theory of differential equations. One usually refers to Chaplyguine for inequalities of the form $\dot{x}(t) \leq \omega(t,x(t))$, which lead to the conclusion that a solution $x(t)$ of the inequality is smaller, at any moment $t > t_0$, than the solution of the associated equation, say $\bar{x}(t)$, if $x(t_0) \leq x_0 = \bar{x}(t_0)$. It is assumed that the equation $\dot{x}(t) = \omega(t,x(t))$ satisfies conditions assuring existence and uniqueness of the solution through (t_0,x_0).

The main idea that was used to construct the comparison method, first appeared to R. Conti [110] in 1956. He used the inequalities of the form $V^{\cdot}(t,x) \leq \omega(t,V(t,x))$, as described in Section 3.2 to obtain global existence results (instead of local ones). He obtained these from the application of traditional procedures (successive approximations, fixed point based on compactness). The last known criterion of that time was the Wintner criterion, a rather special case of the Conti result, which stated that the solutions of the equation $\dot{x}(t) = f(t,x(t))$ do exist on the same interval, on which the solution of the comparison equation $\dot{y}(t) = \omega(t,y(t))$ does.

Conti's result on global existence caught the attention of researchers. In (1958–1959) papers of F. Brauer [75], and F. Brauer and S. Sternberg [76] appeared. These papers discussed how to apply the comparison method to obtain uniqueness and other global properties of solutions.

In 1960, Corduneanu [114] published a paper dedicated to this new method in the theory of stability. All of these stability results concerning stability, asymptotic stability, uniform stability and uniform asymptotic stability, were obtained by comparison method. In 1961, the paper [115] appeared, showing the advantages of this method when obtaining results related to the concepts of stability under permanent *perturbations* or the *preservation* of stability, under adequate perturbations (including nonlinear ones).

The results mentioned above have been included in several monographs: Halanay [237], Lakshmikantham and Leela [309], Hahn [235], Corduneanu [123], Sansone and Conti [489], and Yoshizawa [539], and in monographs of other authors, particularly in the Soviet-Russian literature.

In 1962, Matrosov [377] published an interesting paper dedicated to stability theory, based on Liapunov vector functions and Ważewski's theory of differential inequalities [533] for systems (instead of scalar) of differential equations. Because of this paper, one can consider that the theory of Liapunov's vector functions has gained full recognition. Many authors have investigated stability by using this new tool (see, e.g., the book by Lakshmikantham et. al. [315], or Martynyuk [366–368]). New ideas relating to this relatively new approach in stability theory are due to a large number of researchers in the field. Our Bibliography, contains the works of many of these researchers. See Rouche and Mawhin [475], Rouche et al. [473], Rouche et al. [474], Kato et al. [281], Vorotnikov [530, 531], and Rumiantsev [477]. The papers (survey type) by Matrosov [378, 383] and Voronov [529] also provide results already achieved in stability theory. The book [385] by Matrosov, Anapolskii, and Vassyliev is entirely dedicated to the method of comparison in stability theory. Contributions related to this method are often encountered in the literature (see the books mentioned before for a wealth of information regarding the use of comparison method in stability theory).

Among the various uses for the method of comparison, important to note is its use in the theory of *dynamical systems* (see Matrosov, Voronov, and Martynyuk, in our Bibliography). During the second half of the past century, the authors Matrosov, Rumiantsev, and Martynyuk created schools with preoccupations in the theory of stability, which continue to operate today. These schools attracted researchers from countries like Hungary (where Hatvani led a group), who have made important contributions to the theory of stability, including contributions to the comparison method as well as other aspects of the theory of stability.

Similar results concerning stability or other properties (like boundedness) can be obtained by using *qualitative inequalities* for differential operators intervening in differential equations. See Corduneanu [128] for the continuous time case and Corduneanu [139] for the discrete time case. This last paper leads to results for discrete types of functional equations, a topic only briefly approached in our exposition.

More stability results and applications for neutral equations are provided in Chapter 5.

3.12 BIBLIOGRAPHICAL NOTES

The property of stability (of a motion) is one of the most important from an application point of view. As mentioned in Section 3.1, since our attention was concentrated on the comparison method, we will provide a rather limited account of the status of stability theory, as it appears nowadays in the literature (both mathematical and engineering). Also, most of the topics included are from the theory of ordinary differential equations, the field which displays the deepest available results, with emphasis on the applications of these equations. Besides results in the theory of ordinary differential equations, we are including the case of functional differential equations in the formulation $\dot{x}(t) = f(t, x_t)$, with $x_t(s) = x(t+s)$, $t \in R$, $s \in [-h, 0]$, where $h > 0$ represents the delay.

Landmarks in the theory of stability are the memoir of A. M. Lyapunov (Annales de l'Université de Toulouse, 1905, French translation of an earlier work, in Russian) and the Academic Press book *Stability of Motion*, 1966. Additionally, please refer to some early books mentioned in Section 3.1 and in Bibliography. Particular attention is due to Krasovskii's book and Hale's treatise [240]. The first generalized Liapunov's function method to the case of functional differential equations with bounded delay, while the second is characterized by an intensive use of Functional Analysis (linear and nonlinear) to the study of varied problems in the theory of functional differential equations, including stability, bifurcation, and other aspects. The case of linear functional differential equations was treated using semigroup theory [239], a feature present for the first time in this theory/publications.

The book of Myshkis [411] is the first in the literature; it is dedicated entirely to delay-differential equations and develops ideas originating from his Ph.D. thesis at Moscow University, under J. G. Petrovski. Also, the survey paper [412] by Myshkis and Eĺsgoĺtz analyzes the development of the theory of delay-differential equations until 1967. Equations with perturbed/deviated argument are also discussed.

The topics discussed in Sections 3.2–3.6 are taken (though, not "at litteram") from publications of Corduneanu, Hatvani, Rumiantsev, and Vorotnikov (especially partial stability). Some of them also appear in monographs/treatises by Sansone and Conti, Lakshmikantham and Leela, Halanay, Yoshizawa, Rouche et al., Hahn, Matrosov, and others. A journal paper that contains and provides applications of the comparison method to adaptive systems is Dzieliński [205].

A large number of researchers in former Soviet Union, mainly from the schools of Matrosov and Rumiantsev, have contributed. Our list of references contains many names of contributors: Hatvani (Hungary), Vorotnikov, Oziraner, Voronov, Anapolskii, Vasiliev, and others (see also Bibliography). In Ukraine, Martynyuk [361, 363, 368] developed the comparison method and introduced the matrix Liapunov functions, which were presented in Section 3.9. See also Y. A. Martynyuk-Chernienko [372], in which the Liapunov method was applied to the study of stability for uncertain dynamical systems.

The concept of partial stability (also called *stability* with respect to a part of the variables) appeared in Liapunov's work. It was revived by Rumiantsev [477] and its investigation was continued by Rumiantsev and Oziraner [478] and Hatvani [253], who established the right connection between stability and partial stability. The concept of partial stability was furthered by Corduneanu [116, 119, 121], who is concerned with functional differential equations with finite delay. The books by Vorotnikov [530, 531] contain a thorough account of the partial stability concept, with application to control problems. In [531], aside from ordinary differential equations, the author dealt with functional differential equations. Also, concepts from Game Theory were treated in the framework of control theory. Burton's books [81, 83, 84], as well as the paper [85] by Burton and Hatvani, constitute a rich source of stability results. Various classes of functional equations were investigated. For the stability of invariant sets, see Corduneanu and Ignatyev [166].

Regarding functional differential equations with causal operators, some basic results on stability were given in Corduneanu [149] and Li [329]. Primarily, linear, and quasilinear systems were considered.

A large number of results of stability have been written for functional differential equations with infinite/unbounded delay. We mention here those due to Liang and Xiao [331], Sedova [491], and Hamaya [246]. Chukwu [105] was concerned with stability and control for hereditary systems, with applications.

A. Corduneanu [112] studied a case of exponential stability for delay-differential equations. Stability (absolute) for feedback systems was treated in Corduneanu's book [120]. Many other contributors to the problem of absolute stability have written a large number of papers on this subject (both mathematical and engineering literature). See Corduneanu and Luca [173], Corduneanu [120, 148], Halanay [237], Hyers et al. [268], and Voronov [529]. Another early seminal contribution is in Driver [201].

In the case of partial stability, we mention here Akinyele [15–17], Corduneanu [119, 121], Hatvani [253–257], and Ignatyev [269].

The properties of boundedness and stability of solutions of various classes of functional differential equations are closely related (e.g., they are equivalent for linear systems $\dot{x}(t) = A(t)x(t)$). Many authors have established these connections. For instance, Aftabizadeh [4]; V. Barbu [[45], infinite dimension]; Benchohra et al. ([54], fractional-order differential equations); Campanini [86] treated partial boundedness for differential equations by comparison method; Cheban [99], Coleman and Mizel [107], and Curtain et al. [183] investigated stability problems for various classes of equations in infinite dimension; Daletskii and Krein [186] also investigated stability for systems in Banach spaces (one of the first books fully dedicated to this subject); Djordjevič [197] used the Liapunov matrix method in stability theory of nonlinear systems; Giĺ [219] considered the L^2-stability for nonlinear causal systems; Gopalsamy [225] treated stability problems for classes of delay-differential equations, with applications to population dynamics in his book; Hamaya [246] was concerned with stability properties for some equations with infinite delay;

Hatvani investigated instability in Refs. [254–256]; the whole book [263] by Hino et al. dealt with almost periodicity, including various types of stability; Hale and Lunel [242] investigated functional differential equations in R^n, and provided several examples of Liapunov functionals; Kolmanovskii and Nosov [293] investigated stability in connection with periodic regimes in control systems with delay; A. V. Kim [286] was concerned with stability by second Liapunov method in systems with delay; the book [292] by Kolmanovskii and Myshkis offers examples when stability is investigated; Ulam type of stability is treated in Otrocol [434]; Peiffer and Rouche [443] treated partial stability by Liapunov's second method; Popovici [459,460], at an early stage of the theory of functional equations, was searching classes of solutions; the book [451] by Pinney is among the first books entirely dedicated to difference-differential functional equations in this category; mostly dealing with stability problems is the paper [446] by Perruquetti et al., in relationship with control theory; Razumikhin [469] is the author of a stability monograph, dedicated to hereditary systems; Roseau [472] treated vibrations and stability; Salvadori [484] studied stability by means of a family (one parameter) of Liapunov functions; Shaikhet [493] used Liapunov functionals for difference equations with continuous time; Shestakov [494] generalized the Liapunov's classical method to systems with distributed parameters; Fedorenko [209] investigated stability properties of functional equations of the form $(\mathcal{L}x)(t) = (Fx)(t)$, with $\mathcal{L}$ linear operator and F nonlinear; Berezansky [56] related the properties of stability to those of the Cauchy matrix function (for systems of the form $(\mathcal{L}x)(t) = f(t)$, the Cauchy matrix function is the one by means of which one represents the solution in the form $x(t) = C(t,0)x^0 + \int_0^t C(t,s)f(s)\,ds$, $t \in R_+$); Ahmad and Sivasundaram [11] presented stability results for set valued functions/solutions; Hatvani [253] is a survey paper (Hungarian) providing an overview on stability and partial stability; Ashordia et al. [26] discussed the problem of necessary and sufficient conditions for the stability of systems of certain generalized differential equations; Azbelev et al. discussed, in the book [34], the stability of functional differential equations in detail, with many results; Barbashin [43] was primarily concerned with the finding of Liapunov functions for stability purposes; Basit and Günzler [49] provided results for differential equations in Banach space; Berezansky et al. [57] investigated boundedness and stability to impulsively perturbed equations in Banach spaces; in Ref. [60], Bernfeld et al. investigated the stability of invariant sets for functional differential equations with finite delay; Bhatia and Szegö dedicated a book [64] to stability theory of dynamical systems in the general setting defined by Birkhoff [66]; Burton [83] treated stability problems by fixed point method and in Ref. [84] showed the use of Liapunov type functionals for stability of integral equations (i.e., a class of functional equations without derivatives involved); Shaikhet [493] used Liapunov functions to investigate difference equations with continuous time; Cantarelli [87] studied the stability of the equilibrium state for certain mechanical systems; Castro and Guevara [93] dealt with stability of Volterra equations, in the sense of Hyers–Ulam–Rassias, in weighted spaces; the book by Cesari [97] is a classic source in stability and related problems; Corduneanu [138] was concerned with generalizing the comparison method to abstract Volterra functional differential equations; see also Corduneanu [144]; the paper by Corduneanu and Li [169] contains results on exponential asymptotic stability for equations with causal operators; Cushing [184, 185] provided a general framework for admissibility theory; Deimling [190, 191] exemplified abstract results for various classes of functional equations; Filippov [210, 211] constructed an axiomatic theory for spaces of solutions to ordinary differential equations and provided topological constructions of such spaces (something we are missing for other classes of functional equations!); Giĺ [220] investigated absolute and input–output stabilities for system with causal operators; Gripenberg et al., in their book [228], dedicated a special section to equations with causal operators; Jordan et al. [271] dealt with subspaces of solutions stable or unstable; the book [297] by Krasnoselskii et al. contains several results on stability for classes of functional equations; Kurbatov [301] studied functional differential equations under various aspects; Kwapisz [304] used Bielecki's method of weighted spaces to investigate functional differential equations of different types; Lakshmikantham et al. [309–311], provided results on stability of various classes of functional differential equations and surveyed existing literature; see also Ref. [316]; an examination of the stability concept, in its historical perspective, was given by Leine [325]; other contributions to stability of various classes of functional differential equations are due to MacCamy [345], Malygina [357], Martynyuk [361–368], Murakami [407], and Okonkwo et al. [421] in stochastic setting; Peiffer and Rouche [443] for partial stability; and Rus [481] for Ulam-type stability.

As seen in Section 3.3, it is useful to possess *converse* theorems/results on various types of stability, in terms of Liapunov functions. We will mention here the Kurzweil's result [303], for ordinary differential equations, regarding asymptotic stability. The result included in Section 3.6 regards functional differential equations with finite delay and is available in several monographs on this subject. Other contributions in this category are due to Hafstein [234], which contains several other sources on this matter; Vrkoč [532] studied the concept of integral stability in detail and proved the converse theorem for this type of stability; Kalmar-Nagy [273] investigated the stability of functional equations with finite delay, by the method of steps (which is usually encountered in existence and approximation of solutions) and Laplace transforms; Kappel [275] used Laplace transform for linear autonomous functional differential equations (see also, Kappel and Schappacher [276], a book concerned with evolution systems and their applications). Converse theorems were also obtained [511] by Sumenkov et al. [218] who used methods of linear programming (planar systems).

Many contributions in the area of functional differential equations are of the applicative type. We mention the books by Kolmanovskii and Myshkis [291, 292], both with a focus on functional differential equations encountered in applications to various fields of science, engineering, biology, and medicine. Both deterministic and stochastic models were emphasized.

Other contributions dealt with various types of population dynamics. For instance, Gopalsamy [225] investigated the stability properties of the models appearing in this field of applications, as well as the oscillatory solutions when natural elements are changing. Very interesting are the applications concerned with crystallography; see, for instance, Meyer [392] and the references therein. See additional entries under the names of contributors mentioned at the beginning of Section 3.1 or in our list of references (sometimes, the title of the paper provides a good indication about its nature). The paper by Curtain et al. [183] is a valuable contribution to the theory of stability of control systems based on frequency domain methods. Its list of references is very useful in establishing the history of such methods, starting with Popov's paper [456] in the 1960s. Another kind of result regarding the applications of stability or behavior of solutions for various types of functional differential equations are due to Halanay and Răsvan [238], who rely heavily on the use of Liapunov's functions/functionals.

4

OSCILLATORY MOTION, WITH SPECIAL REGARD TO THE ALMOST PERIODIC CASE

The oscillatory solutions (e.g., periodic and almost periodic) constitute a wide preoccupation of researchers, and several monographs have been dedicated to the subject. Recently, relatively new classes of almost periodic solutions have made their way into the literature (see Shubin [496, 497], Corduneanu [161], a.o.). For most of these classes, the series approach can be applied, even in nonlinear cases.

In this chapter, we define the AP_r-almost periodic functions and establish basic properties (the case of the function defined on R). Of course, we have in mind applications to various classes of functional equations, namely ordinary differential equations, integral equations, and convolution equations. The convolution extends from the classical cases, to functions in AP_r-almost periodic spaces. This chapter also provides several examples of functional differential or integro-differential equations, with regard to the existence of AP_r-almost periodic solutions, solutions in Besicovitch spaces of almost periodic functions, and also in the classical case (Bohr).

We would like to note that the *role of series* is considerably increased, in comparison to the frequency in the mathematical literature nowadays. That is

Functional Differential Equations: Advances and Applications, First Edition.
Constantin Corduneanu, Yizeng Li and Mehran Mahdavi

why one of the most significant aspects in connection with their applications to functional equations is the problem (to be solved!) of reconstructing the function when we know its series. This problem is much simpler in classical approaches.

4.1 TRIGONOMETRIC POLYNOMIALS AND AP_r-SPACES

The oscillatory motion is encountered often in various applied fields. One of the most generally known models is that of the motion of a simple/mathematical pendulum. This motion is fully described by the equation

$$x(t) = A \sin(\omega t + \delta), \tag{4.1}$$

where $A > 0$ is known as the amplitude of oscillation/motion, $\frac{\omega}{2\pi}$ represents the frequency, and δ describes the phase displacement.

Taking into account Euler's formula $\exp(i\lambda t) = \cos\lambda t + i\sin\lambda t$, $\lambda, t \in R$, the complex form of the equation (4.1) will be

$$x(t) = A_1 e^{i\lambda t} + A_2 e^{-i\lambda t}, \tag{4.2}$$

where A_1 and A_2 are complex numbers. It is obvious that the right-hand side of (4.2) represents a special case of a complex-valued function of the form

$$T(t) = A_1 e^{i\lambda_1 t} + \cdots + A_k e^{i\lambda_k t}, \tag{4.3}$$

with the $A_j \in \mathcal{C}$ and $\lambda_j \in R, j = 1, 2, \ldots, k$, with $\lambda_j \neq \lambda_k$ for $j \neq k$.

A function of the form (4.3), for any $k \in N$, is called a *trigonometric polynomial.*

The set of all trigonometric polynomials will be denoted by $\mathcal{T}$. It is easy to check that $\mathcal{T}$ is an algebra over the complex field $\mathcal{C}$.

Starting from $\mathcal{T}$ we will construct a family of Banach function spaces by using different norms on this linear space. We shall consider the following type of norms on $\mathcal{T}$:

$$\|T\|_r = \left(\sum_{j=1}^{k} |A_j|^r \right)^{\frac{1}{r}}, \ 1 \leq r \leq 2. \tag{4.4}$$

It can be easily shown that $\|T\|_r$ is a norm on the linear space $\mathcal{T}$. The normed space obtained by endowing $\mathcal{T}$ with the norm given by (4.4) will be denoted by $\mathcal{T}_r$. One can check that $\mathcal{T}_r$ is not a Banach space (normed and complete), which raises difficulties when we want to operate in such a space.

An instant idea coming to our attention is whether we can draw some advantages by completing these normed spaces in order to obtain Banach function spaces. That will be our approach in constructing spaces of *almost periodic functions*.

The Banach function space obtained by the completion of $\mathcal{T}_r$ will be denoted by $AP_r(R,\mathcal{C})$, $1 \leq r \leq 2$. The norm on $AP_r(R,\mathcal{C})$, generated by (4.4), will be defined by the following formula:

$$|f|_r = \left(\sum_{k=1}^{\infty} |A_k|^r \right)^{\frac{1}{r}}, \ 1 \leq r \leq 2. \tag{4.5}$$

The connection between f and its norm must be understood in the following sense: the formal series

$$A_1 e^{i\lambda_1 t} + A_2 e^{i\lambda_2 t} + \cdots + A_n e^{i\lambda_n t} + \cdots, \tag{4.6}$$

is such that its partial sums

$$T_n(t) = \sum_{k=1}^{n} A_k e^{i\lambda_k t}, \tag{4.7}$$

provides a sequence which converges to f in the norm $|.|_r$:

$$|f(t) - T_n(t)|_r \to 0 \ \text{as} \ n \to \infty. \tag{4.8}$$

Since (4.8) can be read as

$$\sum_{k=n+1}^{\infty} |A_k|^r \to 0 \ \text{as} \ n \to \infty, \tag{4.9}$$

we realize that the formal series (4.6), in general non-convergent in the usual sense (like pointwise), enjoys the property

$$\sum_{k=1}^{\infty} |A_k|^r \ \text{converges}. \tag{4.10}$$

It is useful to examine in some detail the case $r = 1$, which implies the fact that the numerical series

$$\sum_{k=1}^{\infty} |A_k| \ \text{converges}. \tag{4.11}$$

The condition (4.11) implies the absolute and uniform convergence, on R, of the complex trigonometric series (4.6). Indeed, from the equalities

$$|A_k e^{i\lambda_k t}| = |A_k|,\ k \in N,\ t \in R, \tag{4.12}$$

we obtain the equality

$$f(t) = \sum_{k=1}^{\infty} A_k e^{i\lambda_k t},\ t \in R, \tag{4.13}$$

which allows us to get the coefficients A_k in terms of the function $f(t)$. Indeed, if we multiply both sides of (4.13) by $\exp(-i\lambda_j t)$ and take into account the formula

$$\lim_{\ell\to\infty} (2\ell)^{-1} \int_{-\ell}^{\ell} e^{i\lambda t}\, dt = \begin{cases} 0 & \lambda \neq 0, \\ 1 & \lambda = 0, \end{cases} \tag{4.14}$$

then integrating both sides of the equality, we obtain the following:

$$\lim_{\ell\to\infty} (2\ell)^{-1} \int_{-\ell}^{\ell} f(t) e^{-i\lambda_j t}\, dt = A_j,\ j \in N. \tag{4.15}$$

Hence, the series in (4.13) is the Fourier (generalized!) series of $f(t) \in AP_r(R,\mathcal{C})$, which is a space of *almost periodic functions*, apparently used for the first time by H. Poincaré in his treatise *Nouvelles Methodes de la Mećanique Celeste.*

Unfortunately, this simple approach does not necessarily hold for $r \in (1,2]$ and the series appearing in (4.6) for which (4.10) holds true. It will surely hold for those series (4.6) that are uniformly convergent on R.

Let us now examine the other limit case for $r = 2$. We notice first, taking again into account (4.14), that

$$\lim_{\ell\to\infty} (2\ell)^{-1} \int_{-\ell}^{\ell} |T_n(t)|^2\, dt = \sum_{k=1}^{n} |A_k|^2, \tag{4.16}$$

because

$$|T_n(t)|^2 = T_n(t)\overline{T}_n(t) = \sum_{k=1}^{n} |A_k|^2 + \sum_{j\neq k}\sum A_k A_j e^{i(\lambda_k - \lambda_j)t}.$$

Since we assumed that in the representation (4.7) the $\lambda_k{}'s$ are distinct, we easily obtain (4.16).

But in the case $r = 2$, we have the property

$$\sum_{k=1}^{\infty} |A_k|^2 < \infty, \tag{4.17}$$

according to (4.10). We shall use (4.17) to attach to the series (4.6) a function $f(t)$ such that the series converges to $f(t)$ in the norm of the space $AP_2(R,\mathcal{C})$.

This association was very simple to be made in the case $r = 1$ and (4.13) provides the sum of the series in $AP_1(R,\mathcal{C})$. Moreover, from classical analysis, we easily obtained the continuity of $f(t)$ on R.

In the case $r = 2$, we will obtain for the function $f(t)$, that this "sum" of series (4.6) in the space $AP_2(R,\mathcal{C})$, a function which is locally square integrable in R, that is, belongs to the space $L^2_{\text{loc}}(R,\mathcal{C})$.

Let $n,p \in N$. Consider the section of the series in (4.6) from n to $n+p$. We will have

$$\sum_{k=n}^{n+p} A_k e^{i\lambda_k t} = A_n e^{i\lambda_n t} + \cdots + A_{n+p} e^{i\lambda_{n+p} t}, \tag{4.18}$$

which implies (compare with Formula (4.16)),

$$\lim_{\ell\to\infty} (2\ell)^{-1} \int_{-\ell}^{\ell} \left| \sum_{k=n}^{n+p} A_k e^{i\lambda_k t} \right|^2 dt = \sum_{k=n}^{n+p} |A_k|^2. \tag{4.19}$$

From (4.18) and (4.19), we derive the fact that the series in (4.6) is convergent in $AP_2(R,\mathcal{C})$. But what kind of elements does the space $AP_2(R,\mathcal{C})$ possess?

Let us denote

$$s_n(t) = \sum_{k=1}^{n} A_k e^{i\lambda_k t}, \tag{4.20}$$

and consider the limit

$$\lim_{\ell\to\infty} (2\ell)^{-1} \int_{-\ell}^{\ell} |s_{n+p}(t) - s_n(t)|^2 dt = \sum_{k=n+1}^{n+p} |A_k|^2. \tag{4.21}$$

Taking into account (4.17), (4.20), and (4.21), we can write

$$\lim_{\ell\to\infty} (2\ell)^{-1} \int_{-\ell}^{\ell} \left|A_{n+1} e^{i\lambda_{n+1} t} + \cdots + A_{n+p} e^{i\lambda_{n+p} t}\right|^2 dt < \epsilon, \tag{4.22}$$

provided $n \geq N(\epsilon)$, and $p = 1, 2, \ldots$. From (4.22), we derive the existence of a sequence $\{\ell_j\}$, with $\ell_j \to \infty$ as $j \to \infty$, such that

$$\lim_{j \to \infty} (2\ell_j)^{-1} \int_{-\ell_j}^{\ell_j} \left| A_{n+1} e^{i \lambda_{n+1} t} + \cdots + A_{n+p} e^{i \lambda_{n+p} t} \right|^2 dt \leq \epsilon, \tag{4.23}$$

which implies

$$\int_{-\ell_j}^{\ell_j} \left| A_{n+1} e^{i \lambda_{n+1} t} + \cdots + A_{n+p} e^{i \lambda_{n+p} t} \right|^2 dt \leq 2 \ell_j \epsilon, \tag{4.24}$$

for $j \geq N_j(\epsilon)$, and fixed j, while $p = 1, 2, \ldots$. Since $\epsilon > 0$ is arbitrarily small, from in (4.24) we obtain the convergence of the series in (4.6), in the L^2 – *norm*, on the interval $[-\ell_j, \ell_j]$. But $\ell_j \to \infty$ *as* $j \to \infty$, which implies the L^2 – *norm* convergence, of series in (4.6), on each compact interval of R.

Let us summarize the findings in the discussions carried out in the previous paragraph.

First we notice the fact that each space $AP_r(R, \mathcal{C})$, with $r \in (1, 2)$ belongs to $AP_2(R, \mathcal{C})$, which is also known as Besicovitch space of almost periodic functions, of index 2. This is a consequence of the property of Minkowski's norms (4.5): $|.|_p \geq |.|_q$, $1 \leq p \leq q \leq 2$.

Theorem 4.1 *For each complex trigonometric series of the form (4.6), one can associate a function which is in $L^2_{\text{loc}}(R, \mathcal{C})$ in the manner described earlier, provided condition (4.10) is satisfied with $r \in [1, 2]$. The spaces $AP_r(R, \mathcal{C})$ of almost periodic functions are obtained each by the completion of the normed space $\mathcal{T}$, endowed with Minkowski's norms, while their elements belong to the space $L^2_{\text{loc}}(R, \mathcal{C})$. Since any function in $AP_r(R, \mathcal{C})$ is the limit of a sequence of trigonometric polynomials, with respect to Minkowski's norms, we can consider these functions as a type of almost periodic functions. For any $r \in [1, 2]$, the functions in $AP_r(R, \mathcal{C})$ are almost periodic in the sense of Besicovitch (with index 2).*

Remark 4.1 *The discussions in this section, including Theorem 4.1, represent a summarized version of the definition of $AP_r(R, \mathcal{C})$ spaces, $r \in [1, 2]$. These spaces constitute a scale of spaces between $AP_1(R, \mathcal{C})$ = Poincaré space of almost periodic functions and $AP_2(R, \mathcal{C})$ = Besicovitch space $B_2(R, \mathcal{C})$. It is well known that $B_2(R, \mathcal{C})$ contains the space $AP(R, \mathcal{C})$ = Bohr space of almost periodic functions. The space $AP(R, \mathcal{C})$ is obtained by the completion of the space $\mathcal{T}$ in the supremum norm (on R). For further details, see Corduneanu [129, 156].*

Remark 4.2 *Let us discuss, in some detail, formula (4.16) and its consequences. This formula generates the so-called Parseval equality. Since the sequence of trigonometric polynomials $T_n(t)$, $n \in N$, converges with respect to the Besicovitch norm, to the function f constructed earlier, associated to the series in (4.6), formula (4.16) suggests*

$$\lim_{\ell \to \infty} (2\ell)^{-1} \int_{-\ell}^{\ell} |f(t)|^2 \, dt = \sum_{k=1}^{\infty} |A_k|^2. \tag{4.25}$$

This formula is known as Parseval's equality. The connection between the function f and the sequence $\{A_k\}$ is that in between the function and its Fourier coefficients. For details related to these concepts, we send the readers to one of the following references: Corduneanu [129, 156] Fink [213], Levitan and Zhikov [327], and Amerio and Prouse [21].

Remark 4.3 *In this section we have dealt with spaces of almost periodic functions in various senses, the space $AP_2(R,\mathcal{C}) = B_2(R,\mathcal{C})$ representing the richest one. We would now like to briefly remark about a space of almost periodic functions, introduced by Besicovitch, which is even richer than $AP_2(R,\mathcal{C})$.*

One can define this space denoted by $B(R,\mathcal{C})$ in the following manner: we start with the linear space $\mathcal{T}$, in which we consider the following norm:

$$\|T_n(t)\| = \lim_{\ell \to \infty} (2\ell)^{-1} \int_{-\ell}^{\ell} |T_n(t)| \, dt. \tag{4.26}$$

Because of the inequality

$$(2\ell)^{-1} \int_{-\ell}^{\ell} |f(t)| \, dt \leq \left[(2\ell)^{-1} \int_{-\ell}^{\ell} |f(t)|^2 \, dt \right]^{\frac{1}{2}}, \tag{4.27}$$

valid for each $f \in L^2_{\text{loc}}(R,\mathcal{C})$, one sees that the norm in B is dominated by the norm in B_2, which implies $B_2 \subset B$.

A basic property of almost periodic functions in $B(R,\mathcal{C})$ is the existence of the *mean value* (on the whole R!). Indeed, it was shown by Besicovitch [61] that

$$\lim_{\ell \to \infty} (2\ell)^{-1} \int_{-\ell}^{\ell} f(t) \, dt = M\{f\} \in \mathcal{C} \tag{4.28}$$

exists for each $f \in B(R,\mathcal{C})$. The map $f \to M\{f\}$ from B into $\mathcal{C}$ is a linear functional whose kernel/null space is, in general, different from the zero space $\{\theta\}$.

See Corduneanu [129, 156] for more properties related to the space $B(R,\mathcal{C})$, which can also be regarded as the completion of $AP(R,\mathcal{C})$ with respect to the norm $f \to M\{|f|\}$.

The main significance of $B(R,\mathcal{C})$, with the existence of the mean value for each of its elements, comes from the fact that a (generalized) Fourier series can be attached to any $f \in B(R,\mathcal{C})$.

Namely, one considers the functional

$$a(f,\lambda) = M\{f(t)e^{-i\lambda t}\}, f \in B \tag{4.29}$$

the left-hand side of (4.29) denotes a linear functional on B, depending also on the real parameter λ.

One can show (see, e.g., Corduneanu [156]) that there is only a countable set of $\lambda's$, say $\{\lambda_k; k \in N\} \subset R$, such that

$$a(f,\lambda_k) \neq 0, \ k \in N. \tag{4.30}$$

The numbers $\lambda_k, k \in N$, are called *Fourier exponents* of the function f.

The values of $a(f,\lambda_k)$ are nonzero, and one calls them *Fourier coefficients* of f:

$$A_k = a(f,\lambda_k), \ k \in N. \tag{4.31}$$

One writes

$$f \simeq \sum_{k=1}^{\infty} A_k e^{i\lambda_k t}, \tag{4.32}$$

to describe the fact that the series on the right-hand side of (4.32) is the *Fourier series* corresponding to the function $f \in B$.

Relationship (4.32) does not tell us when the series converges, when in case of "convergence" it has f as its sum. However, it is a remarkable fact that, at least in the case $f \in B_2 = AP_2(R,\mathcal{C})$, the Fourier series associated to the function allows the reconstruction of the function itself. For a proof of this property, see one of the references like Amerio and Prouse [21], Corduneanu [156], or Zhang [553].

An important fact is the lack of a one-to-one map from $B_2(R,\mathcal{C})$ into the set of Fourier series associated with elements of $B_2(R,\mathcal{C})$, that is, of the map

$$f \to \left\{ \sum_{k=1}^{\infty} a(f,\lambda_k) e^{i\lambda_k t}; f \in B_2 \right\}. \tag{4.33}$$

with all $a(f,\lambda_k) \neq 0$. It turns out that two distinct functions $f,g \in B_2(R,\mathcal{C})$, can generate the same Fourier series. It is known (see aforementioned references) that this situation occurs only in case $M\{|f-g|^2\} = 0$, which means that the difference of these functions belongs to the null space of the functional $f \to M\{|f|^2\}, f \in B_2$.

The situation described already leads to the following equivalence relation in $B_2(R,\mathcal{C})$: $f \simeq g$ if and only if $M\{|f-g|^2\} = 0$. This property is needed in organizing $B_2(R,\mathcal{C})$ as a factor space and obtaining a Banach space (with a norm, instead of the semi-norm),

$$f \to \left(M\{|f|^2|\}\right)^{\frac{1}{2}}. \tag{4.34}$$

These constructions are given, for instance, in Corduneanu [156].

Other properties related to the spaces of almost periodic functions $AP_r(R,\mathcal{C})$ will be provided when necessary.

4.2 SOME PROPERTIES OF THE SPACES $AP_r(R,\mathcal{C})$

As seen in Section 4.1, the spaces $AP_r(R,\mathcal{C})$, $r \in [1,2]$, are constructed starting from the complex trigonometric series (4.6), with condition (4.10) as the key in distinguishing various spaces in the scale.

In a concise form, we can say that the elements of $AP_r(R,\mathcal{C})$ are complex trigonometric series of the form (4.6), such that (4.10) is satisfied. It is always assumed that in formula (4.6), all of the exponents $\lambda_k, k \in N$, are distinct real numbers. The norm $|.|_r$ is given by (4.5), which is the norm in the ℓ^r-space of Minkowski. It is easy to check the properties of the norm for a normed space, as the completeness with respect to the norm $|.|_r$ is similar to that for ℓ^r-spaces.

We have seen in Section 4.1 how one can associate, in case $r = 2$, a function $f \in L^2_{\text{loc}}(R,\mathcal{C})$ to the series in (4.6), under condition (4.17). But since $AP_r(R,\mathcal{C}) \subset AP_2(R,\mathcal{C})$ for $r < 2$, the same fact is valid for any series in (4.6), which satisfies (4.17). Therefore, the correspondence between series defining $AP_r(R,\mathcal{C})$, and the functions in $L^2_{\text{loc}}(R,\mathcal{C})$ must be understood as explained at the end of Section 4.1.

Now, we have the choice to look at the elements of $AP_r(R,\mathcal{C})$, either as complex trigonometric series of the form (4.6) under condition (4.17), or as functions in $L^2_{\text{loc}}(R,\mathcal{C})$, for which Parseval's equality holds (because $AP_r \subset AP_2$ for $r < 2$).

In what follows, we will primarily use the series approach to construct almost periodic functions/solutions.

A remarkable property to note is the fact that in $AP_r(R,\mathcal{C})$, $1 \leq r \leq 2$, for each element of the form (4.6), the sequence of the partial sums—constituted by complex trigonometric polynomials—is convergent to that element. We may say, as we shall later clarify, that for each function in $AP_r(R,\mathcal{C})$, its Fourier series is convergent to that function. We need to be clear about the fact that nothing is said about the pointwise convergence. The problem of the connection between convergence in $AP_r(R,\mathcal{C})$, and classical types of convergences, will be investigated in the future.

Let us prove the assertion made earlier about the convergence in $AP_r(R,\mathcal{C})$.

Proposition 4.1 *Let us consider the element*

$$\sum_{k=1}^{\infty} A_k e^{i\lambda_k t} \subset AP_r(R,\mathcal{C}), \tag{4.35}$$

and the sequence of the partial sums in the series given in the text,

$$s_n(t) = \sum_{k=1}^{n} A_k e^{i\lambda_k t}, \ t \in R. \tag{4.36}$$

Then, denoting by f the element on the left-hand side of (4.35), one has

$$|f - s_n|_r = \left| \sum_{k=n+1}^{\infty} A_k e^{i\lambda_k t} \right|_r = \left(\sum_{k=n+1}^{\infty} |A_k|^r \right)^{\frac{1}{r}} < \epsilon, \text{for } n > N(\epsilon). \tag{4.37}$$

The proof is an immediate consequence of condition (4.17).

More elementary properties of functions/elements in $AP_r(R,\mathcal{C})$ are given in the following text.

Proposition 4.2 *Let $f \in AP_r(R,\mathcal{C})$, and denote f_h, $h \in R$, $f_h(t) = f(t+h)$. Then*

(1)

$$|f_h - f|_r \to 0 \text{ as } h \to 0, \tag{4.38}$$

which shows the uniform continuity on R of the map $h \to f_h$.

(2)

$$|f_h|_r = |f|, \ \forall h \in R. \tag{4.39}$$

(3) *The usual properties derived from the linear structure of $AP_r(R,\mathcal{C})$ hold true.*

Proof. In order to prove (1), we notice that

$$|f_h - f|_r = \left(\sum_{k=1}^{\infty} |e^{i\lambda_k h} - 1|^r |A_k|^r \right)^{\frac{1}{r}} \tag{4.40}$$
$$\leq \left(\sum_{k=1}^{N} |e^{i\lambda_k h} - 1|^r |A_k|^r + 2^r \sum_{k=N+1}^{\infty} |A_k|^r \right)^{\frac{1}{r}},$$

for any natural N. Let us choose $N = N(\epsilon)$ sufficiently large, such that

$$\sum_{k=N+1}^{\infty} |A_k|^r < \frac{1}{2} \left(\frac{\epsilon}{2} \right)^r, \tag{4.41}$$

and denote

$$\eta_k(h) = \left| e^{i\lambda_k h} - 1 \right|^r, \; k = 1, 2, \ldots, N. \tag{4.42}$$

Each $\eta_k(h)$, $k = 1, 2, \ldots, N$, is continuous in h, including $h = 0$, which implies that we shall have

$$\max_{1 \leq k \leq N} \eta_k(h) < \frac{\epsilon^r}{2} \left(\sum_{k=1}^{N} |A_k|^r \right)^{-1}, \tag{4.43}$$

provided $|h| < \delta(\epsilon)$. Taking (4.40)–(4.43) into account, one obtains

$$|f_h - f| < \left(\frac{\epsilon^r}{2} + \frac{\epsilon^r}{2} \right)^{\frac{1}{r}} = \epsilon, \tag{4.44}$$

which proves statement (1).

One can easily prove that (4.39) is equivalent to the following: for each $\epsilon > 0$, one can find $\delta(\epsilon) > 0$, such that for given $f \in AP_r(R,\mathcal{C})$ one has

$$|f_h - f_{h'}| < \epsilon \text{ for } |h - h'| < \delta.$$

Properties (2) and (3) can be checked easily.

Continuing the parallelism with the classical theory (Bohr), one can establish the following result.

Theorem 4.2 *Let $f \in AP_r(R,\mathcal{C})$, $1 \leq r \leq 2$. Then the set of its translates $\mathcal{F} = \{f_h; h \in R\}$ is a relatively compact set in $AP_r(R,\mathcal{C})$.*

Proof. What we have to prove is the following property: if $f \in AP_r(R, \mathcal{C})$, then from an arbitrary sequence $\{h_k; k \in N\}$, one can extract a subsequence for which f_h converges in $AP_r(R, \mathcal{C})$.

We will accomplish this in two steps. First, we will prove for the case when $f \in AP_r$ is a trigonometric polynomial, say

$$f(t) = T_n(t) = \sum_{k=1}^{n} a_k e^{i\lambda_k t}. \tag{4.45}$$

In case $n = 1, f(t) = ce^{i\lambda t}$, $c \in \mathcal{C}$, $\lambda \in R$. The sequence of translates is $\{ce^{i\lambda(t+h_k)}; k \in N\}$, and we can write

$$|f_{h_k} - f_{h_m}|_r = \left(|c|^r \left|e^{ih_k\lambda} - e^{ih_m\lambda}\right|^r\right)^{\frac{1}{r}} = |c| \left|e^{ih_k\lambda} - e^{ih_m\lambda}\right|. \tag{4.46}$$

Taking into account $|\exp(ih_k\lambda) - \exp(ih_m\lambda)| \leq 2$, and relying on Bolzano–Weierstrass theorem, we can extract a convergent subsequence $\{\exp(ih_k\lambda); k \in N\}$. Without loss of generality, we can assume that the sequence $\{\exp(ih_k\lambda); k \in N\}$ is convergent. Then, from (4.46) we derive

$$|f_{h_k} - f_{h_m}|_r \leq \epsilon, \text{ for } k, m \geq N(\epsilon). \tag{4.47}$$

This means Cauchy's criterion is satisfied for the sequence $\{f_{h_k}; k \in N\}$, and we have obtained the property in Theorem 4.2 in the case $f(t) = c\exp(i\lambda t)$, $c \in \mathcal{C}$, $\lambda \in R$.

In order to check the validity of the property for the case when $f(t)$ is a trigonometric polynomial, as shown in (4.45), we repeatedly apply this argument to the first term on the right-hand side of (4.45). From the subsequence obtained, we extract another subsequence which converges and provides convergence to the sequence of exponentials in $AP_r(R, \mathcal{C})$, and so on. After n repeated applications of the procedure applied for $f(t) = c\exp(i\lambda t)$, we obtain the validity in the case shown in (4.45).

The last step of the proof of Theorem 4.2 consists in the extension of the validity of the property to the general case, $f \in AP_r(R, \mathcal{C})$. Let $\mathcal{F} = \{f(t+h); h \in R\}$ be the family of translates, and take a sequence of trigonometric polynomials, say $\{T^{(n)}(t); n \geq 1\}$, such that $f = \lim T^n$, as $n \to \infty$. Then, given $\epsilon > 0$ arbitrary, we will have

$$|f - T^{(n)}|_r < \frac{\epsilon}{3}, \text{ for } n \geq N(\epsilon). \tag{4.48}$$

Let us point out that (4.48) is equivalent to the inequalities

$$|f_h - T_h{}^{(n)}|_r < \frac{\epsilon}{3}, \text{ for } h \in R. \tag{4.49}$$

Consider now an arbitrary sequence $\{h_m; m \in N\}$. Then, based on the fact that $T^{(n)}$, $n \geq 1$, are trigonometric polynomials from the sequence $\{T^{(1)}_{h_m}; m \geq 1\}$, we can extract a subsequence $\{T^{(1)}_{h_{1m}}; m \geq 1\}$, which is convergent in $AP_r(R,\mathcal{C})$, as $m \to \infty$. Further on, from the sequence $\{T^{(2)}_{h_{1m}}; m \geq 1\}$ we will extract a subsequence $\{T^{(2)}_{h_{2m}}; m \geq 1\}$ which converges in $AP_r(R,\mathcal{C})$. This procedure can be continued indefinitely, and after p steps we will have a sequence $\{T^{(p)}_{h_{pm}}; m \geq 1\}$, convergent in $AP_r(R,\mathcal{C})$. Each new sequence is obtained from the preceding one. As a result, each sequence $\{T^{(p)}_{h_{qm}}; m \geq 1, q \leq p\}$ is convergent in $AP_r(R,\mathcal{C})$. We now consider the diagonal sequence $\{h_{pp}; p \geq 1\}$, which represents a subsequence of each row of the infinite matrix, whose rows are the sequences h_{qm}, $q \geq 1$, $m \geq 1$. Consequently, each of trigonometric polynomials $\{T^{(n)}_{h_{mm}}; m \geq 1, n \geq 1\}$ converges in $AP_r(R,\mathcal{C})$, according to the procedure given in the text.

Let us now fix $n \geq N\left(\frac{\epsilon}{3}\right)$, according to (4.49), and we claim that we shall have

$$\left|T^{(n)}_{pp} - T^{(n)}_{qq}\right|_r < \frac{\epsilon}{3}, \tag{4.50}$$

provided $p,q \geq N_1(\epsilon)$. From the obvious inequality

$$|f_{h_{pp}} - f_{h_{qq}}|_r \leq \left|f_{h_{pp}} - T^{(n)}_{h_{pp}}\right|_r + \left|T^{(n)}_{h_{pp}} - T^{(n)}_{h_{qq}}\right|_r + \left|T^{(n)}_{h_{qq}} - f_{h_{qq}}\right|_r, \tag{4.51}$$

which implies

$$|f_{h_{pp}} - f_{h_{qq}}|_r < \epsilon, \text{ for } p,q \geq \max\left(N(\epsilon), N_1\left(\frac{\epsilon}{3}\right)\right), \tag{4.52}$$

we notice that (4.52) is the Cauchy condition for f_h, $h \in R$, and this ends the proof of Theorem 4.2, that is, the set $\mathcal{F} = \{f_h; h \in R\}$ is relatively compact.

The next result will show that Bohr's property, known in the case of the space $AP(R,\mathcal{C})$, is also valid in case of the spaces $AP_r(R,\mathcal{C})$, $1 \leq r \leq 2$. The result can be stated as follows:

Theorem 4.3 *Let $f \in AP_r(R,\mathcal{C})$, $1 \leq r \leq 2$. Then, for each $\epsilon > 0$, there exists a positive number $\ell = \ell(\epsilon)$, such that any interval $(a, a+\ell) \subset R$ contains a number τ, with the Bohr's property*

$$|f_\tau - f| < \epsilon. \tag{4.53}$$

Proof. Let us take $f \in AP_r(R,\mathcal{C})$ and $\epsilon > 0$ as arbitrary. Assume, on the contrary to (4.53), that there exists at least one $\epsilon > 0$, such that (4.53) is violated.

This means that for any $\ell > 0$, arbitrarily large, there exists at least one interval $(a, a+\ell) \subset R$, such that it does not contain any τ for which (4.53) holds true.

More precisely, if for some $\epsilon_0 > 0$ there is no length $\ell = \ell(\epsilon)$ as stipulated in (4.53), then we can find $h_1 \in R$, and an interval $(a_1, b_1) \in R$, with $b_1 - a_1 < 2|h_1|$, such that this interval does not include any number τ with property (4.53), corresponding to $\epsilon = \epsilon_0 > 0$. Then we shall denote $h_2 = \frac{1}{2}(a_1 + b_1)$ and notice that $h_2 - h_1 \in (a_1, b_1)$, which implies that $h_2 - h_1$ cannot be a τ for f, as in (4.53). Next, we choose an interval (a_2, b_2) with $b_2 - a_2 > 2(|h_1| + |h_2|)$, which does not contain a number τ, as it appears in (4.53). If $h_3 = \frac{1}{2}(a_2 + b_2)$, then one sees easily that $h_3 - h_2$ and $h_3 - h_1$ cannot be τ-numbers for f, as shown in (4.53). One continues this process indefinitely, obtaining a sequence $\{h_j; j \in N\}$ with the property that none of the differences $h_i - h_j$ can be taken as τ-numbers, as it appears in (4.53), for $\epsilon = \epsilon_0$.

Therefore, for any $i,j \in N$, one has

$$|f_{h_i} - f_{h_j}|_r = |f_{h_i - h_j} - f|_r \geq \epsilon_0 > 0. \tag{4.54}$$

The inequality (4.54) tells us that, from the sequence $\{h_i - h_j; i,j \in N\} \subset R$, it is not possible to extract a subsequence which converges in $AP_r(R, \mathcal{C})$. This means that our initial assumption is false and this implies the existence of the length ℓ, for any $\epsilon > 0$.

The proof of Theorem 4.3 is now complete.

In concluding this section, we will make a few considerations related to the topics discussed so far, in connection to the background needed for applications of the almost periodic function spaces $AP_r(R, \mathcal{C})$ to functional equations. These remarks will also contain the formulation of some problems appearing in relationship to these relatively un-investigated classes of almost periodic functions.

Remark 4.4 *Considering the definition of the spaces $AP_r(R, \mathcal{C})$ based on the approximation property by trigonometric polynomials in Minkowski's norms, it is appropriate to notice that Theorem 4.2 (Bochner characterization of Bohr's almost periodicity by the relative compactness of the family of translates) and Theorem 4.3 (Bohr's property taken as definition of classical Bohr's theory of almost periodicity) constitute equivalents of the same concept. This fact is also valid for the AP_r-spaces. Based on the theory of dynamical systems, David Cheban (private communications), the equivalence of various definitions can be easily established. We postpone this part of the theory until after a short presentation of the concept of dynamical systems (G. Birkhoff) in Section 4.7.*

Remark 4.5 *In addition to the spaces $AP_r(R,\mathcal{C})$ that we have defined and discussed in Sections 4.1 and 4.2, one can easily deal with the similar spaces $AP_r(R,R)$, $AP_r(R,R^n)$ or $AP_r(R,\mathcal{C}^n)$. Somewhat separate considerations are necessary in case of the spaces $AP_r(R,B)$, where B is an infinite-dimensional Banach space (or even a Hilbert space). We will also postpone further comments until we reach the applications, part of the discussion.*

Remark 4.6 *Besides the earlier indicated problem of constructing the theory of AP_r-spaces, when $\mathcal{C}$ is substituted by a Banach or Hilbert space, we shall mention here another open problem. In our opinion, if a solution is found, this problem will have some impact on the problems concerning the approximation of functions, or solutions of various functional equations admitting AP_r-solutions.*

For each function in the space $AP_r(R,\mathcal{C})$, or in a similar space of almost periodic functions, one can associate an almost periodic function in $AP_1(R,\mathcal{C})$, namely to

$$f \simeq \sum_{k=1}^{\infty} A_k e^{i\lambda_k t} \in AP_r(R,\mathcal{C}),$$

one associates the AP_1-function

$$f_1(t) = \sum_{k=1}^{\infty} |A_k|^r e^{i\lambda_k t}.$$

The map $f \to f_1$, from $AP_r(R,\mathcal{C})$ into $AP_1(R,\mathcal{C})$, which we denote by $A_r : f \to f_1$, is an operator "projecting" a more complex function space onto another more manageable function space (at least, its functions possessing absolutely and uniformly convergent series, in the supremum norm).

Another problem which we would like to mention here is related to the obvious fact that the theory of AP_r-spaces is strongly based on the theory of ℓ^p-spaces, $1 \leq p \leq 2$. Of course, the ℓ^p-spaces have been investigated (see Lindenstrauss and Tzafriri [334]), as well as more recent contributions on the Web, by many authors. It is to be expected that many known results, from the theory of ℓ^p-spaces, generate interesting results for almost periodicity.

Finally, new spaces whose elements are describing complex oscillatory motions, which are not in the category of almost periodicity, have been introduced recently by Zhang [553, 554]. Many types of rather complex oscillations occur in science and engineering.

4.3 AP_r-SOLUTIONS TO ORDINARY DIFFERENTIAL EQUATIONS

The preceding two sections of Chapter 4 were dedicated to defining and establishing basic properties of the $AP_r(R,\mathcal{C})$ spaces, $r \in [1,2]$.

In order to handle applications of these new spaces of almost periodic functions, it is necessary to present concepts generalizing those encountered in the classical space of Bohr, $AP(R,\mathcal{C})$. Certainly, it will be also necessary to deal with other spaces of the same nature as $AP_r(R,\mathcal{C})$, such as $AP_r(R,R^n)$, $AP_r(R,\mathcal{C}^n)$, $n \geq 1$, or even $AP_r(R,X)$, where X is an abstract space (for instance, a Banach or Hilbert space).

In parallel to the $AP(R,\mathcal{C})$ spaces, we may try to find out if a given functional equation has a solution in the $AP_r(R,\mathcal{C})$ space. One of the simplest problems, in this context, would be to provide conditions guaranteeing that the integral (primitive) of a function from $AP_r(R,\mathcal{C})$ belongs to the same space (or, maybe, another space in the same family).

So far, we have remained within the space $B_2 = AP_2(R,\mathcal{C})$ of Besicovitch. However, the Fourier analysis of almost periodic functions (in various senses) can be developed for the space $B_1 = B =$ the closure of $\mathcal{T}$ with respect to the semi-norm

$$|f|_B = \lim_{\ell \to \infty} (2\ell)^{-1} \int_{-\ell}^{\ell} |f(t)|\, dt = M\{|f|\}. \tag{4.55}$$

This space, which can also be built by completing $AP(R,\mathcal{C})$ with respect to the norm $f \to M\{|f|\}$, as well as AP_2, can be obtained by the completion of $AP(R,\mathcal{C})$ with respect to the norm $f \to M\{|f|^2\}^{\frac{1}{2}}$; we can place/embed the space $PAP(R,\mathcal{C})$ of Zhang [553], known as the space of pseudo almost periodic functions. The space $PAP(R,\mathcal{C})$, or another similar to it, when substituting $\mathcal{C}$ by R or R^n, has in applications the same significance as the space $AP(R,\mathcal{C})$, if not greater. It is obtained by perturbing the trigonometric polynomials by certain functions "small" in the sense of the combined norm $|.|_{AP} + |.|_B$. If one takes into account the fact that our description of phenomena involves obtaining "approximate" measures of the parameters associated to these phenomena, we understand that the description by such "perturbed" quantities, which still "mimic" the almost periodicity behavior, has proper interpretation in reality.

Let us point out the formula, valid for $f, f' \in B$,

$$a(\lambda, f') = i\lambda\, a(\lambda, f), \tag{4.56}$$

with $a(\lambda, f) = M\{f(t) \exp(-i\lambda t)\}$, $\lambda \in R$, as in (4.29). This is the so-called Bohr's transform. It is used in constructing the Fourier series (i.e., to define the Fourier exponents as those $\lambda \in R$ for which $a(\lambda, f) \neq 0$), as well as the

Fourier coefficients, $A_k = a(\lambda_k, f)$, where λ_k designates any value for which $a(\lambda, f) \neq 0$.

According to (4.56), when both its terms have a meaning, the derivative of a function, when it exists, will have the same Fourier exponents as the function. In addition, $\lambda = 0$; this means that without the exponential factor, its Fourier series has no constant term. Equivalently, $a(0, f') = M\{f'\} = 0$; this condition represents a necessary condition for a function to possess an integral (indefinite) in the space B.

It turns out that a result valid in the case of the space $AP(R, \mathcal{C})$, about the existence of the integral of a function $f \in AP(R, \mathcal{C})$, can be extended to the case of AP_r-spaces. In Corduneanu [129] it was proven in the case of $AP_1(R, \mathcal{C})$, and then in $AP(R, \mathcal{C})$. This result can be stated as follows.

Proposition 4.3 *Let $f \in AP_r(R, \mathcal{C})$, $1 \leq r \leq 2$, be such that*

$$f \simeq \sum_{k=1}^{\infty} A_k e^{i \lambda_k t}, \tag{4.57}$$

with Fourier exponents satisfying

$$|\lambda_k| \geq m > 0, \ k = 1, 2, \ldots. \tag{4.58}$$

Then f has an integral in AP_r, given by

$$F \simeq \sum_{k=1}^{\infty} A_k (i \lambda_k)^{-1} e^{i \lambda_k t}. \tag{4.59}$$

Proof. Basically, we have to prove that the series $\sum_{k=1}^{\infty} |A_k|^r |\lambda_k|^{-r}$ is convergent, which means that the series under discussion consists of terms forming a sequence in ℓ^1. Indeed, we have the inequality $|A_k|^r |\lambda_k|^{-r} \leq \frac{|A_k|^r}{m^r}$, $k = 1, 2, \ldots$, on behalf of (4.58), which means that

$$\sum_{k=1}^{\infty} |A_k|^r |\lambda_k|^{-r} < \infty. \tag{4.60}$$

Another proof can be based on the approximation by trigonometric polynomials and a lemma from Fink [213]. See Corduneanu [156] for the details.

Remark 4.7 *The relationship (4.55) is valid for $f \in PAP(R, \mathcal{C})$, that is, in cases of pseudo almost periodic functions. We are not aware of definitions or results for spaces similar to $AP_r(R, \mathcal{C})$, for example, cases that deal with the space $PAP(R, \mathcal{C})$ instead of $AP(R, \mathcal{C})$. But since $PAP(R, \mathcal{C})$ cannot be treated* *in the framework of the space $AP_2(R,\mathcal{C})$ because $PAP(R,\mathcal{C})$ implies B, it is worth mentioning that in case $r = 1$, a result like the one given in Proposition 4.3 is valid for the space $PAP(R,\mathcal{C})$. Condition (4.58) remains the same.*

It would be interesting to define the spaces analogous to $AP_r(R,\mathcal{C})$ in the more comprehensive framework of B instead of $B_2 = AP_2(R,\mathcal{C})$; in other words, to see what $PAP_r(R,\mathcal{C})$ really means, and proceed further with its properties.

Let us now consider the linear system of differential equations with constant coefficients,

$$x'(t) = Ax(t) + f(t), \ t \in R, \tag{4.61}$$

where $A = (a_{ij})$ is a constant matrix of type $n \times n$, while both x and f are maps from R into $\mathcal{C}^n$. We are interested in proving the existence of solutions in $AP_r(R,\mathcal{C}^n)$, for f belonging to the same space. We shall start with the case in which one also has uniqueness in $AP_r(R,\mathcal{C})$.

By complete analogy with the classical case, when solutions are sought in the space $AP(R,\mathcal{C}^n)$ of Bohr's almost periodic functions, we can prove the following result.

Proposition 4.4 *Assume the matrix A is such that*

$$det(A - i\omega I) \neq 0, \ \omega \in R. \tag{4.62}$$

Then, for each $f \in AP_r(R,\mathcal{C}^n)$, $1 \leq r \leq 2$, there exists a unique solution $x(t) \in AP_r(R,\mathcal{C}^n)$ of the equation (4.61).

Proof. Let us notice that, without loss of generality, we can assume the matrix A to be of the lower triangular form. It suffices to apply a transformation of the form $x = Ty$ in (4.61), with T a matrix of type $n \times n$, with $det\, T \neq 0$. Then, the transformed system will take the form

$$y' = By + \tilde{f}(t), \tag{4.63}$$

where $B = T^{-1}AT$, $\tilde{f} = T^{-1}f$, with B of lower triangular form, when T is conveniently chosen.

It is known that B and A have the same characteristic roots, and $\tilde{f}$ will be in $AP_r(R,\mathcal{C}^n)$ when $f \in AP_r(R,\mathcal{C}^n)$. Therefore, without loss of generality, we can deal with equation (4.61), under the (simplifying) hypothesis that A has the lower triangular form. Then, the main diagonal of A consists of all the characteristic roots of A, with repetition when we have multiple ones.

Since we regard the elements of $AP_r(R, \mathcal{C}^n)$ as Fourier series; that is,

$$f \simeq \sum_{k=1}^{\infty} f_k e^{i\lambda_k t}, \tag{4.64}$$

with $f_k \in \mathcal{C}^n$, $n \geq 1$, we shall seek x in the form

$$x \simeq \sum_{k=1}^{\infty} x_k e^{i\lambda_k t}, \tag{4.65}$$

where $x_k \in \mathcal{C}^n$, $k \geq 1$. What we must show is that

$$\{f_k : k \geq 1\} \in \ell^r, \tag{4.66}$$

implies the existence and uniqueness of $x \in AP_r(R, \mathcal{C}^n)$; that is,

$$\{x_k : k \geq 1\} \in \ell^r, \tag{4.67}$$

such that (4.61) is satisfied.

We need to keep in mind the fact that AP_r consists of classes of equivalent elements/functions, each element/function completely characterizing its class.

Since the derivative, in case it exists, has the form

$$x'(t) \simeq \sum_{k=1}^{\infty} (i\lambda_k) x_k e^{i\lambda_k t}, \; t \in R, \tag{4.68}$$

one obtains, by substituting in (4.61), the series defining f, x and x'

$$\sum_{k=1}^{\infty} (i\lambda_k) x_k e^{i\lambda_k t} \simeq \sum_{k=1}^{\infty} A x_k e^{i\lambda_k t} + \sum_{k=1}^{\infty} f_k e^{i\lambda_k t}, \tag{4.69}$$

which leads to the following linear system, of infinite order, for x_k, $k \geq 1$;

$$i\lambda_k x_k = A x_k + f_k, \; k \geq 1. \tag{4.70}$$

The system (4.70) can be also written in the form

$$(A - i\lambda_k I) x_k + f_k = \theta, \; k \geq 1, \tag{4.71}$$

$\theta \in \mathcal{C}^n$ being the null vector. Based on condition (4.62), we obtain from (4.71)

$$x_k = (i\lambda_k I - A)^{-1} f_k, \; k \geq 1. \tag{4.72}$$

Formula (4.72) shows that each x_k, $k \geq 1$, can be uniquely determined from the system (4.70). This fact does not mean that x, constructed with these data, belongs to the space $AP_r(R, \mathcal{C}^n)$. One should get an estimate of the type

$$|x_k|^r \leq M\,|f_k|^r, \; k \geq 1, \tag{4.73}$$

with $M > 0$ a constant independent of $k \geq 1$, which is our next step.

Let us go back now to equation (4.71) and rewrite its scalar form. We denote by μ_j, $1 \leq j \leq n$, the eigenvalues of A, with repetition of those which are multiple, and by x_{k1}, respectively f_{k1}, the coordinates of the vectors x_k, respectively f_k. Then (4.71), in scalar form, will be

$$\begin{gathered}(\mu_1 - i\lambda_k)x_{k1} + f_{k1} = 0,\\ a_{21}x_{k1} + (\mu_2 - i\lambda_k)x_{k2} + f_{k2} = 0,\\ \cdots\\ a_{n1}x_{k1} + a_{n2}x_{k2} + \cdots + (\mu_n - i\lambda_k)x_{kn} + f_{kn} = 0.\end{gathered} \tag{4.74}$$

Due to the fact that all the eigenvalues μ_j, $1 \leq j \leq n$, lie outside the imaginary axis, the existence of a positive number α, such that

$$|\mu_j - i\lambda_k| \geq \alpha, \tag{4.75}$$

follows for $j \geq 1$, $k \geq 1$. Actually, one can easily see that one can choose α such that

$$\alpha = \min|Re\,\mu_j|, \; j = 1, 2, \ldots, n. \tag{4.76}$$

From the first equation in (4.74), we obtain $|x_{k1}| \leq \alpha^{-1}|f_{k1}|$. Using this estimate, we obtain from the second equation in (4.74) the inequality $|x_{k2}| \leq \alpha^{-1}|f_{k2}| + \alpha^{-2}|f_{k1}|$. Continuing this process until the nth step, we obtain estimates for all $|x_{kj}|$, $1 \leq j \leq n$, in terms of $|f_{kj}|$, $1 \leq j \leq n$. The coefficients involved are dependent only on the matrix A, which leads immediately to an estimate of the form $|x_k| \leq N|f_k|$, $N > 0$, where $|x_k|$ and $|f_k|$ stand for any norm we want to choose in $\mathcal{C}^n$, with N independent of k, $k \geq 1$.

The discussion leads immediately to the estimate (4.73), with $M = N^r$. Therefore, the series in (4.65) defines a function $x \in AP_r(R, \mathcal{C}^n)$.

This ends the proof of Proposition 4.4.

Remark 4.8 *Condition (4.62) plays an important role in the validity of the conclusion of Proposition 4.4. What happens if the matrix A has purely imaginary eigenvalues? This is a singular case, and the simple example*

$x'(t) = ix(t)$, corresponding to $n = 1$, shows us that none of its solutions $x(t) = (t+c)\exp(it)$ can belong to $AP_r(R,\mathcal{C}^n)$, for $1 \leq r \leq 2$.

There remains an open problem, which is to find conditions replacing (4.62), such that Equation (4.61) will admit solutions in $AP_r(R,\mathcal{C}^n)$.

Remark 4.9 *Since equation (4.61) is linear, it appears natural to wonder whether nonlinear cases could be treated in a similar framework. For instance, if one considers the system*

$$x'(t) = Ax(t) + (fx)(t), \; t \in R, \tag{4.77}$$

in which f is an operator acting on the space $AP_r(R,\mathcal{C}^n)$, under what conditions on f is it possible to establish the existence of a unique solution in $AP_r(R,\mathcal{C}^n)$?

The answer is *positive*, if we admit an adequate condition on the operator f. Namely, we will assume that $x \to fx$ is Lipschitz continuous

$$|fx - fy|_r \leq \lambda |x-y|_r, \; x,y \in AP_r, \tag{4.78}$$

with small enough λ.

More precisely, considering equation (4.77) in the space $AP_r(R,\mathcal{C})$, with the matrix A like in Proposition 4.4, and assuming f satisfies (4.78), the contraction mapping principle can be applied to the auxiliary equation

$$x'(t) = Ax(t) + (fu)(t), \; t \in R. \tag{4.79}$$

Indeed, denoting $x = Tu$, on behalf of (4.79) where $x,u \in AP_r(R,\mathcal{C})$, we obtain the operator $T : AP_r \to AP_r$, for which we can establish the Lipschitz condition

$$|Tu_1 - Tu_2|_r \leq \lambda m |u_1 - u_2|_r, \tag{4.80}$$

valid on the whole (Banach) space $AP_r(R,\mathcal{C})$. The meaning of the constants λ and m is as follows:

$$|fx - fy|_r \leq \lambda |x-y|_r, \; x,y \in AP_r, \tag{4.81}$$

while m appears from the inequality (4.73), namely, $m = M^{\frac{1}{r}}$. Summing up the earlier discussion, (4.80) leads to the conclusion that T is a contraction mapping when $\lambda M^{\frac{1}{r}} \leq 1$. Hence, there is a unique fixed point of the operator T, and, letting $u = x$ in (4.79), we obtain equation (4.77) for x.

4.4 AP_r-SOLUTIONS TO CONVOLUTION EQUATIONS

Before dealing with convolution equations involving AP_r-spaces, we shall define a generalized *convolution product* as one that reduces to the usual one when the involved functions provide a meaning to the integral occurring in the classical convolution product of two functions.

In Corduneanu [120], the classical convolution product is considered, represented by the integral

$$(k*x)(t) = \int_R k(t-s)\,x(s)\,ds, \tag{4.82}$$

which makes sense when $k \in L^1(R,\mathcal{C})$ and $x \in E(R,\mathcal{C})$, where E stands for any of the spaces $AP_1(R,\mathcal{C})$, $AP(R,\mathcal{C})$, or $S(R,\mathcal{C})$ = the Stepanov space of almost periodic functions. Moreover, it is shown that each of the aforementioned spaces is invariant with respect to the convolution product, when the kernel belongs to $L^1(R,\mathcal{C})$. More precisely, one has the inequality

$$|k*x|_E \le |k|_{L^1}\,|x|_E, \tag{4.83}$$

for any E of the aforementioned spaces of almost periodic functions.

Let us notice that for $x \in AP_1(R,\mathcal{C})$ we can write

$$(k*x)(t) = \int_R k(t-s)\,x(s)\,ds = \int_R k(t-s)\left[\sum_{j=1}^{\infty} x_j e^{i\lambda_j s}\right] ds.$$

Because $x \in AP_1(R,\mathcal{C})$, and its Fourier series is absolutely and uniformly convergent on R, we can rewrite the equation as

$$\begin{aligned}(k*x)(t) &= \sum_{j=1}^{\infty}\int_R k(t-s)\,e^{i\lambda_j s}x_j\,ds = \sum_{j=1}^{\infty}\left[\int_R k(s)\,e^{-i\lambda_j s}x_j\,ds\right]e^{i\lambda_j t}\\ &= \sum_{j=1}^{\infty}\tilde{x}_j e^{i\lambda_j t},\end{aligned}$$

where

$$\tilde{x}_j = x_j \int_R k(s)\,e^{-i\lambda_j s}\,ds,\ j \ge 1. \tag{4.84}$$

These formulas suggest the following definition for the (generalized) *convolution product*:

$$(k*x)(t) = \sum_{j=1}^{\infty} \tilde{x}_j e^{i\lambda_j t}, \ t \in R, \tag{4.85}$$

for $k \in L^1(R,\mathcal{C})$ and $x \in AP_r(R,\mathcal{C})$, $1 \leq r \leq 2$; indeed, based on (4.84), we easily obtain the inequality

$$|k*x|_r \leq |k|_{L^1} \, |x|_r, \tag{4.86}$$

due to the inequalities, derived from (4.84),

$$|\tilde{x}_j| \leq |k|_{L^1} \, |x_j|, \ j \geq 1, \tag{4.87}$$

which also proves that $k*x \in AP_r(R,\mathcal{C})$, when $x \in AP_r(R,\mathcal{C})$.

We see from (4.84) that $\tilde{x}_j's$ can be easily obtained from the $x_j's$, if we know the Fourier transform of the kernel k, namely

$$\tilde{k}(s) = \int_R k(t)\,e^{-its}\,dt. \tag{4.88}$$

We can now state and prove the following existence result for the convolution equation

$$x(t) = (k*x)(t) + f(t), \ t \in R, \tag{4.89}$$

which is generalizing another well-known result. See, for instance, Corduneanu [161].

Proposition 4.5 *Let us consider equation (4.89), in which the convolution product is meant according to (4.81), with $k \in L^1(R,\mathcal{C})$ and $x \in AP_r(R,\mathcal{C})$, $1 \leq r \leq 2$.*

Assume that for the Fourier transform of the kernel k, the condition

$$\tilde{k}(s) \neq 1, \ s \in R, \tag{4.90}$$

is verified.

Then, for each $f \in AP_r(R,\mathcal{C})$, there exists a unique solution of (4.89), say $x(t)$, such that $x \in AP_r(R,\mathcal{C})$.

Proof. Let us search for the solution of (4.89) in $AP_r(R,\mathcal{C})$, in the forms used before

$$x(t) \simeq \sum_{j=1}^{\infty} x_j e^{i\lambda_j t}, \tag{4.91}$$

where $\{x_j; j \in N\} \in \ell^r$. Taking into account the definition of the convolution product, we are lead to the equation

$$\sum_{j=1}^{\infty} x_j e^{i\lambda_j t} \simeq \sum_{j=1}^{\infty} \tilde{x}_j e^{i\lambda_j t} + \sum_{j=1}^{\infty} f_j e^{i\lambda_j t},$$

which can take place only in case

$$x_j = \tilde{k}(\lambda_j) x_j + f_j, \ j \in N, \tag{4.92}$$

or

$$x_j = [1 - \tilde{k}(\lambda_j)]^{-1} f_j, \ j \in N. \tag{4.93}$$

But $|\tilde{k}(s)| \to 0$ as $|s| \to \infty$, which means, based on (4.90), that we can find $M > 0$ with the property $|1 - \tilde{k}(s)|^{-1} \leq M$. Therefore, we can write the inequalities

$$|x_j| \leq M |f_j|, \ j \in N, \tag{4.94}$$

which tells us that $\{f_j; j \in N\} \in \ell^r$ implies $\{x_j; j \in N\} \in \ell^r$, or $x \in AP_r(R,\mathcal{C})$.
This ends the proof of Proposition 4.5.

Remark 4.10 *In fact, we only need the condition $\tilde{k}(s) \neq 1$ for $s = \lambda_k$, $k \in N$. But since we want the statement to be valid for any $f \in AP_r(R,\mathcal{C})$, we need the assumption in the form (4.90). It is also worth mentioning that, as shown in the previously quoted reference, one can make E the space $AP(R,\mathcal{C})$, or $S(R,\mathcal{C})$, which keeps the validity of Proposition 4.5.*

We will now deal with the integro-differential equation

$$\dot{x}(t) = \int_R k(t-s) x(s)\, ds + f(t), \ t \in R, \tag{4.95}$$

where $k : R \to \mathcal{L}(R^n, R^n)$, and

$$|k| \in L^1(R,R), \tag{4.96}$$

while $x, f \in AP_r(R, \mathcal{C}^n)$, with x the unknown. Since the (integral) convolution product in (4.95) may not make sense for $r > 1$, we will rewrite (4.95) using the generalized convolution product as

$$\dot{x}(t) = (k * x)(t) + f(t), \ t \in R, \tag{4.97}$$

where

$$(k * x)(t) = \sum_{j=1}^{\infty} \left(\int_R k(s) e^{-i\lambda_j s} ds \right) x_j e^{i\lambda_j t}. \tag{4.98}$$

Substituting in (4.98) the series for x and f, one obtains

$$i\lambda_j x_j e^{i\lambda_j t} \simeq \sum_{j=1}^{\infty} \left(\int_R k(s) e^{i\lambda_j s} ds \right) x_j e^{i\lambda_j t} + \sum_{j=1}^{\infty} f_j e^{i\lambda_j t},$$

which implies

$$i\lambda_j x_j = \left(\int_R k(s) e^{-i\lambda_j s} ds \right) x_j + f_j, \ j \in N,$$

and finally

$$\left[i\lambda_j - \int_R k(s) e^{-i\lambda_j s} ds \right] x_j = f_j, \ j \in N. \tag{4.99}$$

In order to solve uniquely the infinite system for $\{x_j; j \in N\}$, we will impose the following condition on k:

$$det \left[i\omega I - \int_R k(s) e^{-i\omega s} ds \right] \neq 0, \tag{4.100}$$

for all $\omega \in R$. Then, each x_j, $j \in N$, can be uniquely determined from (4.99), namely

$$x_j = \left[i\lambda_j I - \tilde{k}(i\lambda_j) \right]^{-1} f_j, \ j \in N, \tag{4.101}$$

where $\tilde{k}(i\omega)$ is the Fourier transform of $k(t)$.

But (4.101), providing the sequence $\{x_j; j \in N\}$, does not provide yet the condition $\{x_j; j \in N\} \in \ell^r(N, \mathcal{C})$, sufficient for concluding $x \in AP_r(R, \mathcal{C})$.

We will now prove that (4.100) is equivalent with the existence of some $m > 0$, such that

$$\left| det[i\omega I - \tilde{k}(i\omega)] \right| \geq m > 0, \ \omega \in R. \tag{4.102}$$

Indeed, since $|\tilde{k}(i\omega)| \to 0$ as $|\omega| \to \infty$, a condition implying the boundedness of $\tilde{k}(i\omega)$ on R, there results that the dominant term in $det\,[i\omega I - \tilde{k}(i\omega)]$ is $|\omega|^n$. This allows us to write the inequalities

$$\left| det\,[i\omega I - \tilde{k}(i\omega)] \right| > m_1 \text{ for } |\omega| > M_1, \tag{4.103}$$

with m_1 and M_1 both positive. Furthermore, for $|\omega| \leq M_1$ we will find $M_2 > 0$, such that

$$\left| det\,[i\omega I - \tilde{k}(i\omega)] \right| \geq m_2 > 0, \; |\omega| \leq M_1. \tag{4.104}$$

Therefore, we can write (4.102), on behalf of (4.103) and (4.104), with $m = \max\{m_1, m_2\} > 0$. From (4.101) and (4.102), we derive the inequalities

$$|x_j| \leq m^{-1} |f_j|, \; j \in N. \tag{4.105}$$

The inequalities (4.105) show that $f \in AP_r(R, \mathcal{C}^n)$ implies $x \in AP_r(R, \mathcal{C}^n)$.

The conclusion of the discussion carried previously, with regard to equation (4.95), can be formulated as follows.

Proposition 4.6 *Assume the following conditions hold for equation (4.95):*

(1) $k : R \to \mathcal{L}(R^n, R^n)$ *satisfies (4.100), and* $|k| \in L^1(R, R)$;
(2) $f \in AP_r(R, \mathcal{C}^n)$, *for fixed* $r \in [1, 2]$.

Then, there exists a unique solution of (4.95) in $AP_r(R, \mathcal{C}^n)$, *for each* $f \in AP_r(R, \mathcal{C}^n)$.

As a direct application of Proposition 4.6, we shall deal with the nonlinear, or rather semi-linear equation

$$\dot{x}(t) = (k * x)(t) + (fx)(t), \; t \in R. \tag{4.106}$$

The following result can be obtained, using the same procedure as in case of equation (4.77), in order to use the fixed point principle for contraction mappings.

Proposition 4.7 *The existence and uniqueness of a solution* $x \in AP_r(R, \mathcal{C}^n)$, *to equation (4.106), are guaranteed by the following assumptions:*

(1) *The same as in Proposition 4.6;*
(2) $f: AP_r(R,\mathcal{C}^n) \to AP_r(R,\mathcal{C}^n)$ *satisfies the Lipschitz condition*

$$|fx - fy|_r \leq \lambda |x-y|_r, \ x,y \in AP_r(R,\mathcal{C}^n),$$

with sufficiently small λ.

Proof. Based on Proposition 4.6, we define the operator $T : AP_r(R,\mathcal{C}^n) \to AP_r(R,\mathcal{C}^n)$ by means of the equation

$$\dot{x}(t) = (k*x)(t) + (fu)(t), \ t \in R, \tag{4.107}$$

with $u \in AP_r(R,\mathcal{C}^n)$, and x the unique solution of the linear equation (in x !), (4.107): $x = Tu$. From (4.105), we obtain for the solution of the linear equation (4.95), the estimate

$$|x|_r \leq M |f|_r, \tag{4.108}$$

with $M > 0$ independent of f, which combined with condition (2) in the statement of Proposition 4.7 and (4.107), provides

$$|Tu - Tv|_r \leq \lambda M |u-v|_r, \ u,v \in AP_r(R,\mathcal{C}^n) \tag{4.109}$$

which justifies the assertion of Proposition 4.7.

This ends the proof of Proposition 4.7.

Remark 4.11 *Instead of the Lipschitz condition (2) in the statement of Proposition 4.7, it is desirable to have another type of growth condition on the nonlinear term fx, and use the Schauder fixed-point result for existence of solutions. This approach, which is rather common in the theory of differential equations, requires a compactness criterion in the underlying space* $AP_r(R,\mathcal{C}^n)$.

Since $AP_r(R,\mathcal{C}^n)$ is constructed by means of the space $\ell^r(N,\mathcal{C}^n)$, it is natural to think using something similar. But each ℓ^r-space, with $1 \leq r < \infty$, belongs to the space c_0 of the sequences (in $\mathcal{C}^n$) convergent to the null element of the linear space $\mathcal{C}^n$. In this space, the criterion can be stated as follows: "A set $\mathcal{M} \in c_0(N,\mathcal{C}^n)$ is compact in the topology of uniform convergence on N if the following two conditions are satisfied:

1) The sequences belonging to $\mathcal{M}$ are uniformly bounded on N, that is, there exists a number $M > 0$, such that $\sup\{|x_k|; x \in \mathcal{M}\} \leq M$, $x = (x_1, x_2, \ldots, x_k, \ldots)$.
2) $\lim_{k\to\infty} x_k = 0$, uniformly with respect to $x \in \mathcal{M}$."

Based on this criterion, a similar one can be formulated for the function spaces $AP_r(R,\mathcal{C}^n)$ and then used to prove the existence result by means of Schauder's fixed point theorem.

We leave readers the task of formulating and proving the existence result. A hint: one should keep in mind that the elements of AP_r can be regarded as formal trigonometric series whose coefficients form a sequence in ℓ^r.

In concluding this section, we will provide another example of a mixed differential-convolution equation that can be treated in the same way as the previous one. Namely, let us consider the equation/system

$$\dot{x}(t) = Ax(t) + (k * x)(t) + f(t), \ t \in R, \tag{4.110}$$

where the data have the same significance as in Propositions 4.4 and 4.6. If we look for solutions in $AP_r(R,\mathcal{C}^n)$, in the form

$$x(t) \simeq \sum_{j=1}^{\infty} x_j e^{i\lambda_j t}, \ x_j \in R^n, \ j \in N, \tag{4.111}$$

then the equations determining the coefficients $x_j \in R^n$, will be

$$i\lambda_j x_j = Ax_j + \left(\int_R k(s) e^{-i\lambda_j s} ds \right) x_j + f_j, \ j \in N. \tag{4.112}$$

In order to assure the existence and the uniqueness of each x_j, $j \in N$, one has to assume that

$$det[i\lambda_j I - A - \tilde{k}(i\lambda_j)] \neq 0, \ j \in N. \tag{4.113}$$

If we want to deal with any $f \in AP_r(R,\mathcal{C}^n)$, then we have to require the much stronger condition

$$det[i\omega I - A - \tilde{k}(i\omega)] \neq 0, \ \omega \in R. \tag{4.114}$$

This condition can be handled the same manner as Propositions 4.4 and 4.6, in order to obtain estimates for the coefficients x_j.

We propose to the readers to carry out the details of this procedure.

4.5 OSCILLATORY SOLUTIONS INVOLVING THE SPACE *B*

The spaces $AP_r(R,\mathcal{C})$, $1 \leq r \leq 2$, are covering the whole spectrum of almost periodicity, starting with the space $AP_1(R,\mathcal{C})$, which is likely the most important in the classical framework, and ending with the Besicovitch B_2. Any almost periodic function in Bohr's sense or Stepanov's sense (with index 2) is in B_2.

Still, there are spaces of generalized almost periodic functions which contain elements outside B_2. One example is the space S (Stepanov, index 1) which contains the space of periodic functions, of the same period, such that on each interval, equal to the period, are in L^1-space but not L^2-space. Also, the space $PAP(R,\mathcal{C})$ of pseudo almost periodic functions (see Zhang [553]) requires, in its definition, the use of the norm of B.

A problem arising in this context is that of defining/finding spaces of almost periodic functions which lie in between B_2 and B.

Let us remember that B_2 is the completion of $AP(R,\mathcal{C})$ with respect to the norm $f \to [M\{|f|^2\}]^{\frac{1}{2}}$, while B is the completion of $AP(R,\mathcal{C})$ with respect to the norm $f \to M\{|f|\}$. As we know, $B_2 \subset B$, and we can ask ourselves, what is in between?

It is known (see Andres et al. [24]) that the Weyl's space of almost periodic functions $APW(R,\mathcal{C})$ is richer than B, and this implies that each function in $B \setminus B_2$ belongs to the Weyl's space. This fact did not lead to the use of Weyl's norm (somewhat more complicated than the norms of other traditional spaces of almost periodic functions), in building other types of almost periodicity.

Zhang [553] had the idea of simultaneously using, two norms from the arsenal of almost periodicity in order to get new almost periodic function spaces. Indeed, the space of pseudo almost periodic functions makes use of the norm in Bohr's space, as well as the Besicovitch norm of B. The representation of each function $f \in PAP(R,\mathcal{C})$ is

$$f = g + \phi, \tag{4.115}$$

with $g \in AP(R,\mathcal{C})$ and $\phi \in BC(R,\mathcal{C}) \cap B_\theta$, where BC stands for bounded and continuous with supremum norm and B_θ for the null space of B:

$$B_\theta = \left\{ \phi : \phi \in B,\ |\phi|_B = \lim_{\ell \to \infty} (2\ell)^{-1} \int_{-\ell}^{\ell} |\phi(t)|\, dt = 0 \right\}. \tag{4.116}$$

We can say that the space $PAP(R,\mathcal{C})$ is the direct sum of $AP(R,\mathcal{C})$ and of the zero manifold of $B(R,\mathcal{C})$:

$$PAP(R,\mathcal{C}) = AP(R,\mathcal{C}) \oplus [BC(R,\mathcal{C}) \cap B_\theta]. \tag{4.117}$$

As we can see from (4.115) to (4.117), two norms are engaged in defining $PAP(R,\mathcal{C})$: the supremum norm in $AP(R,\mathcal{C})$ and $BC(R,\mathcal{C})$, as well as the semi-norm of B.

The success of the construction stems from the fact that $PAP(R,\mathcal{C})$ is a closed subspace of $BC(R,\mathcal{C})$, and in such a case, the metric properties of the space B are also involved in defining $PAP(R,\mathcal{C})$.

So far, it seems that the only viable construction, except the Weyl's almost periodic functions for which applications are missing, is represented by the $PAP(R,\mathcal{C})$ functions.

If one tries to extend this construction, involving the space $AP_r(R,\mathcal{C})$, instead of $AP(R,\mathcal{C})$, we should deal with functions that can be represented as $f = g + \phi$, with $g \in AP_r(R,\mathcal{C})$, while ϕ must be in $B_\theta(R,\mathcal{C})$, with B_θ defined by (4.116), and also belonging to another function space. By comparison with the case of $PAP(R,\mathcal{C})$, one could think of choosing this space to be $AP_r(R,\mathcal{C})$ itself, or $B_2 \supset AP_r(R,\mathcal{C})$, $r \in [1,2]$, or even another space. This hypothetical construction is an open problem.

Since the space B is a rather rich space, and its topology is given by a semi-norm

$$|f|_B = \lim_{\ell\to\infty} (2\ell)^{-1} \int_{-\ell}^{\ell} |f(t)|\, dt = M\{|f|\}, \tag{4.118}$$

its structure is less efficient than in cases with a more sophisticated one. Even in the case of the space B_2, we have the Parseval equality, which shows a very useful connection between the semi-norm/norm and the Fourier coefficients (see formula (4.25)).

For the space B, however, we count on scarce information about the Fourier coefficients of a given function. Nevertheless, a general property of the coefficients should be emphasized. If $f \in B$ is such that

$$f \simeq \sum_{k=1}^{\infty} f_k e^{i\lambda_k t}, \tag{4.119}$$

we know that f_k is given by

$$f_k = \lim_{\ell\to\infty} (2\ell)^{-1} \int_{-\ell}^{\ell} f(t)\, e^{-i\lambda_k t},$$

which implies for each $k \geq 1$

$$|f_k| \leq \lim_{\ell\to\infty} (2\ell)^{-1} \int_{-\ell}^{\ell} |f(t)|\, dt = M\{|f|\}. \tag{4.120}$$

In other words, the sequence of Fourier coefficients of a function in B is in the sequence space $\ell^\infty(N,\mathcal{C})$.

On the other hand, when trying to extend properties encountered in the case of periodic functions to almost periodic ones, one has to deal with a much more complex and less organized structure.

This fact suggests that restricting our considerations to a subset of the space B, say a subspace—not necessarily closed, we may obtain similar results to those valid in the periodic case.

Accordingly, we shall limit our interest to the subspace of $B = B(R, \mathcal{C})$, defined as follows: we choose a sequence of frequencies (Fourier exponents), say $\{\lambda_k; k \geq 1\} \subset R$, and consider all functions in $B(R, \mathcal{C})$ having this sequence as their Fourier exponents. These functions are related to their Fourier series as shown in (4.119). It is obvious that these functions form a linear manifold (over $\mathcal{C}$) in the space B. Further along in our exposition, we shall impose some conditions on the set of Fourier exponents.

Let us point out the fact that in Zhang's construction of the space $PAP(R, \mathcal{C})$, he used only the null manifold of B.

We want to treat the existence problem for the differential equation

$$x'(t) = \mu x(t) + f(t), \ t \in R, \tag{4.121}$$

with $\mu \in \mathcal{C}$, and f in the linear manifold specified earlier. We will denote this manifold by $B_\lambda(R, \mathcal{C})$, $\lambda = \{\lambda_k; k \geq 1\}$, the $\lambda_k{}'s$ being those that appear in the representation of its elements.

Regarding equation (4.121), a restriction we shall impose on μ is

$$\mu \in \mathcal{C} \setminus \{i\omega : \omega \in R\}. \tag{4.122}$$

We notice that (4.122) implies the existence of a number $\rho > 0$, such that

$$|\mu - i\lambda_k| \geq \rho, \ k \geq 1, \tag{4.123}$$

having always in mind that the $\lambda_k{}'s$ are those involved in the definition of the manifold B_λ.

Proceeding as in the preceding section, we shall look for the solution of (4.121) in the form of its Fourier series, namely

$$x \simeq \sum_{k=1}^{\infty} x_k e^{i\lambda_k t}, \tag{4.124}$$

with $x_k \in \mathcal{C}$. The function $f \in B_\lambda$ will be represented by

$$f \simeq \sum_{k=1}^{\infty} f_k e^{i\lambda_k t}, \tag{4.125}$$

where $|f_k| \leq M\{|f|\} = |f|_B$, as noticed earlier in this section. That is, the only information we have about the $f_k's$.

Substituting (4.124) and (4.125) in (4.121), we obtain for $x_k's$ the following system of simple equations:

$$i\lambda_k x_k = \mu x_k + f_k, \ k \geq 1. \tag{4.126}$$

Each equation in (4.126) is uniquely solvable because of (4.122), (4.123), even in case one of the $\lambda_k's$ is zero.

One finds the following estimate for the coefficients x_k, $k \geq 1$, of the solution we are searching for

$$|x_k| \leq \rho^{-1}|f_k| \leq \rho^{-1}|f|_B, \ k \geq 1. \tag{4.127}$$

Unfortunately (4.127) is not telling us, for instance, that $x \in B$. Therefore, we shall process in a different manner the equation (4.126). Namely, we find for $\mu = \alpha + i\beta$,

$$x_k = \frac{-1}{\mu - i\lambda_k} f_k, \ k \geq 1, \tag{4.128}$$

or

$$|x_k| = [\alpha^2 + (\lambda_k - \beta)^2]^{-\frac{1}{2}} |f_k|, \ k \geq 1. \tag{4.129}$$

Taking (4.129) into account, as well as $\{f_k; k \geq 1\} \in \ell^\infty$, we can state that the series (4.125) will be the Fourier series of a function in $B_2(R, \mathcal{C})$, provided

$$\sum_{k=1}^{\infty} [\alpha^2 + (\lambda_k - \beta)^2]^{-1} < \infty,$$

which is the same as

$$\sum_{k=1}^{\infty} \lambda_k^{-2} < \infty, \tag{4.130}$$

because $|f_k| \leq M\{|f|\}, \ k \geq 1$.

Condition (4.130) is the one we impose on $\lambda = \{\lambda_k; k \geq 1\}$ to give meaning to the series solution of equation (4.121). More precisely, the above constructed solution is a Besicovitch almost periodic function that belongs to the space $B_2 = AP_2(R, \mathcal{C})$.

Let us notice that the periodic case, with any period $T > 0$, is covered by condition (4.130). Also, in application, one encounters $\lambda = \{\lambda_k; k \geq 1\}$, which form the eigenvalues of a Sturm–Liouville boundary value problem. As we know, conditions such as (4.130) are often satisfied in these cases.

We encourage readers to find conditions similar to (4.130), such that the solution x constructed before will belong to the $AP_r(R,\mathcal{C})$ space, $1 \leq r \leq 2$.

To summarize the discussion regarding equation (4.121), we can say this equation has a unique solution in $B_2(R,\mathcal{C})$, provided: $f \in B_\lambda(R,\mathcal{C})$, with $\lambda = \{\lambda_k; k \geq 1\}$ satisfying (4.130) and μ satisfying (4.122).

A discussion that parallels the one in Section 4.3, concerning the system (4.61), leads us to the following result.

Proposition 4.8 *Consider the differential system (4.61), under condition (4.62) for the matrix* $A = (a_{ij})_{n\times n}$, $a_{ij} \in \mathcal{C}$, $i,j = 1,2,\ldots,n$. *Let* $f \in B_\lambda(R,\mathcal{C}^n)$, *with* λ *satisfying (4.130).*

Then, there exists a unique solution $x \in B_2(R,\mathcal{C}^n)$ *of (4.61).*

The Fourier series of this solution can be constructed as shown in the case $n = 1$, and, in general, in Proposition 4.4.

Remark 4.12 *Proposition 4.8 is an illustration of the involvement of the richest Besicovitch space of almost periodic functions,* $B(R,\mathcal{C}^n)$, *in the theory of differential equations.*

4.6 OSCILLATORY MOTIONS DESCRIBED BY CLASSICAL ALMOST PERIODIC FUNCTIONS

In this section, we are concerned with ordinary differential equations in Banach spaces. The equation to be investigated regards the existence of almost periodic solutions and has the form

$$\dot{x}(t) = A(t,x(t)) + f(t), \ t \in R, \tag{4.131}$$

with f and x taking values in a real Banach space X, and with the norm denoted by $\|.\|$. A stands for a continuous operator from $R \times X$ into X satisfying the condition $A(t,\theta) = \theta \in X$, where θ is the null element of X, $\forall t \in R$.

In order to formulate a hypothesis, we need to introduce the following functional from $X \times X$ into R, by the formula

$$[x,y] = \lim_{h\to 0+} \frac{(\|x+hy\| - \|x\|)}{h}. \tag{4.132}$$

We see from (4.132) that $[x,y]$ represents the right derivative of $\|x\|$, in the y-direction. For the basic properties of the functional $[x,y]$, we direct the reader to Martin [1]. However, we would like to mention the following: (i) $[x,y+z]$

$\leq [x,y]+[x,z]$; (ii) $[x,y] \leq \|y\|$; (iii) if $u: I \to X$ is strongly differentiable at the interior point $t_0 \in I$, then the right derivative of $\|u(t)\|$ exists at t_0 and $D_+ \|u(t)\| = [u(t_0), u'(t_0)]$, at $t = t_0$.

The following result can be found in Martin [359] when it concerns the existence of a bounded solution to (4.131) on the whole real axis. With regard to the almost periodicity, it can be found in Kato and Sekiya [282].

Theorem 4.4 *Consider the differential equation (4.131), under the aforementioned assumptions on the operator A, and*

$$[x-y, A(t,x)-A(t,y)] \leq -p\|x-y\|, \; p > 0, \tag{4.133}$$

for all $(t,x),(t,y) \in R \times B_r(\theta)$, with $B_r(\theta) = \{x \in X; \|x\| \leq r, r > 0\}$. Assume further that $f: R \to X$ is continuous and bounded, that is, there is $M > 0$, such that $\|f(t)\| \leq M$, for all $t \in R$.

Then, if $M < pr$, there exists a unique solution of (4.131), such that

$$\|x(t)\| \leq Mp^{-1} < r, \; \forall t \in R. \tag{4.134}$$

If $f(t)$ is (Bohr) almost periodic and $A(t,x)$ is almost periodic in t, uniformly with respect to $x \in B_r(\theta)$, then the bounded solution $x(t)$ of (4.131) is also almost periodic.

Since the proof of existence of a bounded solution can be found in Martin [359], we provide a simple proof of the statement concerning almost periodicity, based on qualitative differential inequalities (see Corduneanu [128]).

Let $x(t)$ be the (unique) bounded solution of (4.131); then, for any real T, one has on behalf of (4.133)

$$\begin{aligned} D_+ \|x(t+T)-x(t)\| \leq -p\|x(t+T)-x(t)\| + \|f(t+T)-f(t)\| \\ + \|A(t+T,x(t)) - A(t,x(t))\|. \end{aligned} \tag{4.135}$$

If we make T a common $(\frac{\epsilon}{2})$-almost period for $f(t)$ and $A(t,x)$, then (4.135) immediately leads to the qualitative inequality

$$D_+ \|x(t+T)-x(t)\| \leq -p\|x(t+T)-x(t)\| + \epsilon,$$

from which we derive, after taking into account the boundedness of $x(t+T) - x(t)$ on the real axis, and applying the qualitative result in Corduneanu [128], the following inequality

$$\|x(t+T)-x(t)\| < p^{-1}\epsilon, \; t \in R,$$

which implies the almost periodicity of $x(t)$ in the sense of Bohr.

What follows is a variant of Theorem 4.4 that somewhat strengthens condition (4.133) but weakens the almost periodicity hypothesis. Namely, instead of Bohr type almost periodicity, we will use Stepanov almost periodicity.

Theorem 4.5 *Consider the differential equation (4.131) under the assumptions of Theorem 4.4, with the difference that (4.133) holds in $R \times X$, while $f(t)$ is Stepanov almost periodic.*

Then, there exists a unique solution $x(t)$ to (4.131) bounded on R and Bohr almost periodic.

Proof. Because Stepanov almost periodic functions are not continuous, in general, the solutions of (4.131) will be meant in Carathéodory sense.

For given $f \in S(R,X) =$ the Stepanov space of almost periodic functions, we can find a sequence of Bohr almost periodic functions such that the sequence $\{f_k; k \in N\}$ converges to f in $S(R,X)$. In particular, one can choose the terms of this sequence to be trigonometric polynomials over X.

Let us now consider the auxiliary differential equations

$$\dot{x}(t) = A(t,x(t)) + f_k(t), \quad k \geq 1. \tag{4.136}$$

Each equation (4.136) has a unique almost periodic (Bohr) solution according to Theorem 4.5. Let us denote by $x_k(t)$, $k \in N$, that solution of (4.136). From (4.136), we derive

$$\dot{x}_k(t) - \dot{x}_m(t) = A(t,x_k(t)) - A(t,x_m(t)) + f_k(t) - f_m(t),$$

for $k,m \in N$, and all $t \in R$. From the above equality, we obtain the qualitative inequality

$$D_+ \|x_k(t) - x_m(t)\| \leq -p\|x_k(t) - x_m(t)\| + \|f_k(t) - f_m(t)\|,$$

which implies, on behalf of qualitative inequalities,

$$\sup_{t \in R} \|x_k(t) - x_m(t)\| \leq K \sup_{t \in R} \int_t^{t+1} \|f_k(s) - f_m(s)\| \, ds, \tag{4.137}$$

where $K > 0$ is a fixed number.

From (4.137), we obtain the convergence of the sequence $\{x_k(t); k \in N\}$ in $AP(R,X)$ (i.e., uniformly on R). Denoting $x(t) = \lim x_k(t)$ as $k \to \infty$, we derive from

$$\dot{x}_k(t) = A(t,x_k(t)) + f_k(t), \quad k \in N,$$

that $x(t)$, the limit function, verifies equation (4.131) on R.

Theorem 4.5 is thus proven.

We will now provide a generalization of previous results, substituting the linear inequality (4.133) with a nonlinear counterpart,

$$[x-y, A(t,x)-A(t,y)] \leq -p(\|x-y\|), \tag{4.138}$$

where $p(r)$, $r \in R_+$, is continuous, strictly increasing on R_+, $p(0)=0$ and $p(r) \to \infty$ as $r \to \infty$. Moreover,

$$\int_{0+} [p(r)]^{-1} dr = +\infty,$$

while the function

$$\frac{d}{dr} \exp\left\{\int^{r} [p(u)]^{-1} du\right\}$$

is nondecreasing on R_+.

we can now state the following result of Bohr–Neugebauer type (i.e., on assuming the existence of a solution) to prove its almost periodicity.

Theorem 4.6 *Consider equation (4.131) under the previously specified assumptions, including Carathéodory property (i.e., measurable in t for fixed x and continuous in x for almost all $t \in R$). Moreover, A will be Stepanov almost periodic in t, uniformly with respect to x in bounded subsets of X. Also, $f(t)$ is assumed to be Stepanov almost periodic.*

Then, if $x(t)$ is a bounded solution on R, of (4.131), $x(t)$ is Bohr almost periodic.

Proof. The proof will be carried out in the same manner as the preceding cases, relying this time on a nonlinear qualitative inequality. In order to simplify the formulas, we will denote

$$g(t) = f(t+T) - f(t) + A(t+T, x(t)) - A(t, x(t)), \ t \in R, \tag{4.139}$$

for T to be specified later. From (4.131), (4.138), and (4.139), one derives the inequality

$$D_+ \|x(t+T) - x(t)\| \leq -p\left(\|x(t+T) - x(t)\|\right) + \|g(t)\|. \tag{4.140}$$

The inequality (4.140) is of the type

$$\dot{y}(t) \leq -p\left(y(t)\right) + f(t), \ t \in R, \tag{4.141}$$

which is the nonlinear qualitative inequality, implying, for a bounded on R, function $y(t)$

$$\sup_{t\in R} y(t) \le p^{-1}\left(K\,\|f\|_S\right). \tag{4.142}$$

Applying inequality (4.142) to (4.140), we obtain the following estimate

$$\sup_{t\in R} \|x(t+T)-x(t)\| < p^{-1}\left(K\,\|g\|_S\right). \tag{4.143}$$

If one chooses T to be an $(\frac{\epsilon}{2})$-almost period, common to f and A, then, (4.143) provides

$$\sup_{t\in R} \|x(t+T)-x(t)\| < \epsilon.$$

This last inequality concludes the proof of Theorem 4.6.

We will now consider functional differential equations of the form

$$\dot{x}(t) = (Lx)(t) + (fx)(t),\ \ t\in R, \tag{4.144}$$

where L denotes a linear operator on function space $E(R,R^n)$ or $E(R,C^n)$, with E standing for the space of Bohr almost periodic functions or other function spaces related to AP-spaces.

The space E will stand for the following spaces:

1) $E = BC$, consisting of all bounded and continuous functions on R, with the usual supremum norm

$$|x|_{BC} = \sup_{t\in R} |x(t)|;$$

2) $E = BUC$, consisting of all bounded and uniformly continuous functions on R, with values in R^n or C^n. The norm is the same as in case of BC;
3) $E = AP$, representing the space of almost periodic functions in the Bohr sense, also with supremum norm;
4) $E =$ any subspace of the space BC, defined earlier.

The following inclusions follow from the earlier definitions of function spaces $BC \supset BUC \supset AP$. These spaces are interrelated; the last two in the sequence are subspaces (hence, closed) of the Banach space BC. Another subspace of BUC, and also of BC, is the space C_ℓ, which contains all functions that have a limit at $\pm\infty$. Usually, one deals with spaces like $C_\ell(R_+,R^n)$ or $C_\ell(R_+,C^n)$.

In addition to equation (4.144), we will consider the linear counterpart

$$\dot{x}(t) = (Lx)(t) + f(t), \ t \in R, \tag{4.145}$$

and we must always consider the connection between the linear and nonlinear cases.

Another feature that we would like to emphasize is the construction of a space $E^{(1)}(R,R^n)$, associated with the space $E(R,R^n)$ as follows:

$$E^{(1)}(R,R^n) = \{x; x, \dot{x} \in E(R,R^n)\}, \tag{4.146}$$

with the norm

$$|x|_1 = \sup\{|x(t)| + |\dot{x}(t)|; t \in R\}. \tag{4.147}$$

It is easy to check that $E^{(1)}$ is also a Banach space.

We will now establish a rather general theorem of the type "Theorem in the first approximation," involving both linear and nonlinear equations, (4.144) and (4.145).

Theorem 4.7 *Consider equation (4.144) under the following assumptions:*

1. *$L : E(R,R^n) \to E(R,R^n)$ is a linear continuous operator, where E is a subspace of $BC(R,R^n)$.*
2. *Equation (4.145) has a unique solution $x \in E(R,R^n)$ for each $f \in E(R,R^n)$.*
3. *The map $f : E(R,R^n) \to E(R,R^n)$ is satisfying the global (i.e., on E) Lipschitz condition*

$$|fx - fy|_E \leq L_0 |x - y|_E, \tag{4.148}$$

with L_0 sufficiently small.

Then, equation (4.144) has a unique solution x in $E(R,R^n)$.

Proof. Let us consider the linear operator T, from $E^{(1)}$ into E defined by

$$(Tx)(t) = \dot{x}(t) - (Lx)(t), \tag{4.149}$$

and notice the fact that $T^{-1} : E \to E^{(1)}$, $x = T^{-1}f$, is well defined on E, according to our assumption 2. From (4.149), we obtain the following estimate

$$|Tx|_{BC} \leq |\dot{x}|_{BC} + |Lx|_{BC}, \ x \in E^{(1)},$$

which combined with

$$|Lx|_{BC} \leq K\,|x|_{BC},\ K > 0,$$

implied by the continuity of L, leads to

$$|Tx|_{BC} \leq K_1\,(|\dot{x}|_{BC} + |x|_{BC}),$$

where $K_1 = \max\{1, K\}$. One can rewrite the preceding formula as

$$|Tx|_{BC} \leq K_1\,|x|_1,\ x \in E^{(1)}.$$

we have thus proved the continuity of the operator T on $E^{(1)}$.

According to a well known theorem of Banach, regarding the continuity of the inverse operator, we can write for the operator $x = T^{-1}f$ the continuity condition $|x|_1 \leq M\,|f|_E$, for some $M > 0$, which leads on behalf of (4.147) to

$$|x|_E \leq M\,|f|_E. \tag{4.150}$$

We are now able to prove the existence and uniqueness of the solution of the equation (4.144). The method of proof relies on the application of the Banach fixed point principle to the operator U, defined as follows $x = Uy$ if x is the unique solution (on behalf of condition 2 in the statement of Theorem 4.7) of the auxiliary equation

$$\dot{x}(t) - (Lx)(t) = (fy)(t),\ t \in R. \tag{4.151}$$

Now let $x_1 = Uy_1$ and $x_2 = Uy_2$. From (4.151) we derive the equation

$$(x_1 - x_2)^{\cdot} - L(x_1 - x_2) = fy_1 - fy_2,\ y_1, y_2 \in E,$$

which can be rewritten as

$$T(x_1 - x_2) = fy_1 - fy_2,$$

or

$$x_1 - x_2 = T^{-1}(fy_1 - fy_2).$$

On behalf of (4.148) and (4.150), we obtain

$$|Ux_1 - Ux_2|_E \leq M\,L_0\,|y_1 - y_2|_E. \tag{4.152}$$

The inequality (4.152) shows that the operator U is a contraction on E, provided $ML_0 < 1$, or

$$L_0 < M^{-1}. \tag{4.153}$$

Hence, the Banach contraction theorem applies, and from (4.151) we see that the unique fixed point is the solution of the semi-linear equation (4.144).

This ends the proof of the existence and uniqueness of the solution of equation (4.144).

Remark 4.13 *Particularizing the subspace E, one obtains existence and uniqueness in different classes of functions from R into R^n or C^n (bounded solutions, Lipschitz continuous solutions, or almost periodic solutions).*

Notice that Theorem 4.7 remains valid when we consider function spaces whose elements take value in a Banach space.

There is another important special case, provided by the space $PAP(R,R^n)$ of pesudo almost periodic functions, that was introduced and investigated by Zhang [553]. The case of Banach space valued functions has been investigated by T. Diagana [193].

Remark 4.14 *A variant of Theorem 4.7 involves cases where the existence (but not uniqueness) is obtained by means of Schauder's fixed point theorem. This approach will require compactness criteria in the function spaces involved, a feature that is encountered in the spaces $AP(R,C^n)$ and $C_\ell(R,C^n)$, for which the compactness condition is of a simpler nature.*

To illustrate this situation, we will provide a result from Corduneanu [154].

Let us consider the functional differential equation (4.144) under the following hypotheses:

1) $L: AP(R,C^n) \to AP(R,C^n)$ is a linear continuous operator.
2) Equation (4.145) has a unique solution in $AP(R,C^n)$ for each $f \in AP(R,C^n)$.
3) $f: AP(R,C^n) \to AP(R,C^n)$ is a continuous operator, taking bounded sets into relatively compact sets (i.e., compact);
4) If $\lambda(r) = \sup\{|(fx)|,\ |x| \le r\}$, then

$$\lim_{r\to\infty} \sup \left[\frac{\lambda(r)}{r}\right] = \alpha,$$

with α small enough.

Then, there exists a solution of equation (4.144) in $AP(R,\mathcal{C}^n)$, such that $|x|_{AP} \leq r$, with r satisfying the inequality $M\lambda(r) \leq r$, where M denotes the norm of the operator T, defined in the proof of Theorem 4.7.

A similar result, related to the space $C_\ell(R,\mathcal{C}^n)$ is valid for equation (4.144), and we invite the readers to formulate and prove it. First, the compactness criteria in C_ℓ should be formulated, a task that can be easily accomplished relying on the classical Ascoli–Arzelà criterion of compactness.

In the last part of this section, we will investigate functional equations of the form

$$x(t) = (Lx)(t) + (fx)(t), \ t \in R, \tag{4.154}$$

which can be viewed as nonlinear perturbation of the associated linear equation

$$x(t) = (Lx)(t) + f(t), \ t \in R. \tag{4.155}$$

Generally speaking, in this section L and f will have the same meaning as before. Of course, we could deal with these equations taking various underlying spaces. However, we will limit our focus to function spaces whose elements are defined on R and take their values in $\mathcal{C}^n$. The space $AP(R,\mathcal{C}^n)$ will constitute our main objective.

The following result, similar to that of Theorem 4.7, can be obtained using similar arguments.

Theorem 4.8 *Consider equation (4.154), under the following assumptions:*

1) *The operator $L : AP(R,\mathcal{C}^n) \to AP(R,\mathcal{C}^n)$ is a linear continuous operator.*
2) *The linear equation (4.155) is uniquely solvable for each $f(t) \in AP(R,\mathcal{C}^n)$.*
3) *The map $f : AP(R,\mathcal{C}^n) \to AP(R,\mathcal{C}^n)$, generally nonlinear, is Lipschitz continuous with sufficiently small L_0,*

$$|fx - fy|_{AP} \leq L_0 \, |x - y|_{AP}. \tag{4.156}$$

Then, equation (4.154) has a unique solution $x \in AP(R,\mathcal{C}^n)$.

Proof. From equation (4.155) and condition 2) of the theorem, we obtain that the operator $f \longmapsto x$ is one-to-one on $AP(R,\mathcal{C}^n)$. Since the equation (4.155) can be written as $(I - L)x = f$, there results $x = (I - L)^{-1}f, f \in AP(R,\mathcal{C}^n)$. The continuity of the operator $(I - L)^{-1}$ is the consequence of the same theorem of Banach, concerning the continuity of the inverse operators, because $I - L$ is continuous and one-to-one on $AP(R,\mathcal{C}^n)$.

Based on these considerations, we can rewrite equation (4.154) in the operator notations,

$$x = (I - L)^{-1} f x. \tag{4.157}$$

Equation (4.157) will provide existence and uniqueness of solution if we assure the condition of contractibility of the operator $T = (I - L)^{-1} f$. But $(I - L)^{-1}$ is a linear continuous operator, hence it has a finite norm, say M. Taking into account the Lipschitz continuity for f, we find that T is a contraction as soon as

$$ML_0 < 1 \text{ or } L_0 < M^{-1}. \tag{4.158}$$

Theorem 4.8 is thereby proven.

Theorem 4.8 can be easily applied to various particular functional equations. We illustrate this, first, in the case of the equation

$$x(t) = \int_R k(t-s)x(s)\,ds + (fx)(t), \; t \in R, \tag{4.159}$$

which we dealt with earlier, in connection with the existence of AP_r-solutions.

Condition 2) of Theorem 4.8 is secured by the frequency type assumption

$$det\,[i\omega I - \tilde{k}(i\omega)] \neq 0, \; \omega \in R. \tag{4.160}$$

Assumption 1) of the theorem is assured by $k \in L^1$, a condition which is also needed to construct $\tilde{k}(i\omega)$.

The Lipschitz condition (4.156) remains the same and an estimate like (4.158) must be also imposed.

Therefore, for integro-functional equations of the form (4.159), the application of Theorem 4.8 is warranted.

A second application of Theorem 4.8 is related to the "pure" functional equation

$$x(t) = px(\lambda t) + f(t), \; t \in R, \tag{4.161}$$

to limit our considerations only to the linear case (the extension to nonlinear case, with $f(t)$ replaced by $(fx)(t)$, being obvious).

This case is due to Mr. Peter Jossen, who developed the case in 2006 as he was taking a short course on almost periodic functions at the Central European University in Budapest.

The hypotheses for (4.161) were (i) $f \in AP(R,\mathcal{C})$; (ii) $p \in \mathcal{C}$, $|p| \neq 1$; and (iii) $\lambda \in R$, $\lambda \neq 0$. One should notice that the Banach contraction principle can be applied when $|p| < 1$, in order to obtain unique solution to (4.161).

When $|p| > 1$, one rewrites the equation in the form $x(\lambda t) = p^{-1}x(t) - p^{-1}f(t)$, and denotes $\lambda t = \tau$, keeping also in mind that $x(\lambda^{-1}\tau), f(\lambda^{-1}\tau) \in AP(R,\mathcal{C})$, while $|p|^{-1} < 1$.

We invite the readers to find other applications for Theorem 4.8. For instance, one might consider (linear counterpart) of the form

$$x(t) = Ax(\lambda t) + f(t),$$

in $AP(R,\mathcal{C}^n)$.

4.7 DYNAMICAL SYSTEMS AND ALMOST PERIODICITY

The concept of a dynamical system, as defined by G. D. Birkhoff [66], can be briefly described in the following terms:

One starts with a complete metric space (X,d), and considers a continuous map $f : X \times R \to X$, say $y = f(x,t)$, $x \in X$, $t \in R$, such that the following conditions are satisfied:

$$\begin{aligned} &1) \quad f(x,0) = x,\ x \in X \\ &2) \quad f(x,t+s) = f(f(x,t),s),\ t,s \in R. \end{aligned} \tag{4.162}$$

Property 2) is known as the *group property* of the dynamical system.

For fixed $x \in X$, the set $\{f(x,t); t \in R\}$ is called the *trajectory* of the point x. Based on property 2), we see that it is also the trajectory to each point $f(x,t)$.

The theory of dynamical systems has been developed by several authors, and we encourage readers to consult the books by Nemytskii and Stepanov [416], Sibirskii [499], and Sibirskii and Shuba [498].

In particular, the almost periodicity of dynamical systems is defined, by extending the definition of Bohr, as follows:

Definition 4.1 *The motion/map $t \to f(x,t)$, from R into X is called almost periodic, iff to each $\epsilon > 0$ corresponds $\ell = \ell(\epsilon) > 0$, such that each interval $(a, a+\ell) \subset R$ contains a point τ, with the property*

$$d(f(x,t+\tau), f(x,t)) < \epsilon, t \in R \tag{4.163}$$

where $d : X \times X \to R_+$ is the metric of X.

If this definition holds for an $x \in X$, then x is called an almost periodic point.

It is interesting that most properties of the classical almost periodic functions are valid for the almost periodic motions of a dynamical system.

For instance, for an almost periodic map $t \to f(x,t), t \in R, x \in X$, the uniform continuity on R (for fixed $x \in X$) is valid , as well as its boundedness in the space X. These elementary properties are easy to check.

We will now formulate other basic properties of almost periodicity in dynamical systems. For proof, we encourage readers to consult the references cited in the text.

Theorem 4.9 *The following statements hold for a dynamical system defined by the map* $f : X \times R \to X$, $y = f(x,t)$:

1) $f(x,t)$ *is almost periodic, according to Definition 4.1, for some* $x \in X$.
2) *The trajectory of* $x \in X$, $\{f(x,t); t \in R\}$ *is relatively compact in* X *(in the topology induced by the metric* d*).*
3) *If* $y \in \{f(x,t); t \in R\}$ *(i.e., it is on the trajectory of* x*) then* $\{f(y,t); t \in R\} = \{f(x,t); t \in R\}$.
4) *For every sequence* $\{t_n; n \geq 1\} \subset R$, *there exists a subsequence* $\{t_{n_k}; k \geq 1\}$, *such that* $f(x, t_{n_k})$ *converges with respect to the topology induced by the distance* d *(relative compactness of any trajectory).*

Remark 4.15 *The theory of dynamical systems is a conspicuous division of topological dynamics and deals with other concepts that originate from the classical theory of differential equations. In particular, there are profound results relating almost periodicity and Liapunov stability theory (in the abstract formulation of dynamical systems). In addition to the aforementioned references, see also Gottschalk and Hedlund [227] and Bhatia and Szegö [64].*

Next, we examine how one can treat problems related to spaces of almost periodic functions, like $AP_r(R,R^n)$, in the framework of the theory of almost periodic dynamical systems. These results were personally communicated to us by D. Cheban.

The key result can be stated in the following form:

Theorem 4.10 *Let* $X = AP_r(R,\mathcal{C})$, $r \in [1,2]$, *be the complete metric space in Theorem 4.1, and define the map* $t \to f(x,t)$ *by letting* $f(x,t) = x(\cdot + t)$, $t \in R$, *for each fixed* $x \in AP_r(R,\mathcal{C})$. *Then, the triplet* $[AP_r(R,\mathcal{C}), R, f]$ *is a dynamical system (Birkhoff).*

Proof. First, we notice that $f(x,0) = x(\cdot)$, which is property 1) in the definition of dynamical systems. With regard to property 2), we take into account that each side represents the same trajectory, $\{f(x,t); t \in R\}$. We must prove that the map $(x,t) \to f(x,t)$ is continuous from $AP_r(R,\mathcal{C}) \times R$ into $AP_r(R,\mathcal{C})$. Let us consider an arbitrary sequence $\{t_n; n \geq 1\} \subset R$, and another $\{x_n; n \geq 1\}$ $\subset AP_r(R,\mathcal{C})$, such that $t_n \to t$ and $x_n \to x$, in $AP_r(R,\mathcal{C})$, as $n \to \infty$. Since the distance function on $AP_r(R,\mathcal{C})$ is given by the *norm* of this space (see Section 4.1), while the distance in the product space is $d[(x,t),(y,0)] = |x-y|_r + |t-s|$, one finds the inequalities

$$
\begin{aligned}
d[(x_n,t_n),(x,t)] &\leq d[(x-n,t_n),(x,t_n)] + d[(x,t_n),(x,t)] \\
&= |x-x_n|_r + |t-t_n| < \epsilon,
\end{aligned}
$$

as soon as $|x-x_n|_r < \frac{\epsilon}{2}$, $|t-t_n| < \frac{\epsilon}{2}$. This is accomplished for $n \geq N(\epsilon)$, which means

$$
\lim d[(x_n,t_n),(x,t)] = 0 \text{ as } n \to \infty.
$$

This ends the proof of Theorem 4.10.

Based on the results in Theorems 4.9 and 4.10, one can obtain additional properties of the almost periodic functions in $AP_r(R,\mathcal{C})$, $r \in [1,2]$.

The following is an example in this direction. We know that Bohr's definition of almost periodicity (the classical one) is equivalent to "approximation property" (uniformly on R, by trigonometric polynomials) and to the property of relative compactness (also in uniform convergence on R, of the family $\mathcal{F} = \{f(t+h; h \in R\}$ of translates).

Regarding the functions in $AP_r(R,\mathcal{C})$, we can conclude, on behalf of Theorems 4.9 and 4.10, that the same is true in this case. Of course, if we substitute uniform convergence in R with convergence in the space $AP_r(R,\mathcal{C})$, this means the Minkowski's norm (see Section 4.1).

In this concluding section, we consider an interesting connection between the almost periodicity of a dynamical system (Definition 4.1) and its stability in Liapunov sense. This is the definition of Liapunov stability for dynamical systems.

Definition 4.2 *The motion $f(x,t)$ of the dynamical system on the space (X,d) is called Liapunov stable, if for each $\epsilon > 0$ one can find $\delta = \delta(\epsilon) > 0$, such that*

$$
d(x,y) < \delta \text{ implies } d(f(x,t),f(y,t)) < \epsilon, \tag{4.164}
$$

for each $t \in R$.

Remark 4.16 *Definition 4.2 is not equivalent to the definition of stability we dealt with in Chapter 3. A closer definition similar to that used in Chapter 3* *would require $t \in R_+$, instead of $t \in R$. This kind of stability in dynamical systems has been largely investigated in topological dynamics (see the references mentioned above, in this section). In fact, one distinguishes three types of Liapunov stability: positive ($t \in R_+$), negative ($t \in R$), and the one described in Definition 4.2.*

Definition 4.3 *A dynamical system $f(x,t)$ on the complete metric space (X,d) is called Liapunov stable if each motion ($x \in X$) is Liapunov stable, according to Definition 4.2.*

The result showing the connection between almost periodicity of a dynamical system and its (Liapunov) stability, can be formulated as follows:

Theorem 4.11 *Let $f(x,t)$ denote an almost periodic dynamical system on (X,d), with X a complete metric space. Then, it is also Liapunov stable.*

Proof. Before getting into the details of the proof, let us point out that in any dynamical system $[X,R,f]$, with X a complete metric space, the continuous dependence of $f(x,t)$, with respect to $x = f(x,0)$ =initial point, holds. This means that if we choose a $x \in X$, then for each $\epsilon > 0$, one can find $\delta = \delta(\epsilon) > 0$, such that on a fixed interval $[0,T]$, $T > 0$,

$$d(x,y) < \delta \text{ implies } d(f(x,t),f(y,t)) < \epsilon. \tag{4.165}$$

This property, known since Birkhoff introduced the concept of dynamical system, can be found with its complete proof in the sources quoted earlier in this section.

Let us now proceed to the proof of Theorem 4.11. Consider arbitrary $\epsilon > 0$ and $x \in X$. Due to almost periodicity property, for $\frac{\epsilon}{3}$ one can find $T > 0$, such that any interval $(a, a+T) \subset$ contains an $\frac{\epsilon}{3}$-almost period. On the other hand, for $\frac{\epsilon}{3}$ and $T > 0$, there exists $\delta = \delta(\epsilon, T) > 0$, such that

$$d(f(x,t),f(y,t)) < \frac{\epsilon}{3} \text{ for } d(x,y) < \delta,\ t \in [0,T]. \tag{4.166}$$

If $y \in X$ is any point such that $d(x,y) < \delta$, we will show that

$$d(f(x,t),f(y,t)) < \frac{\epsilon}{3}, \text{ for } t \in R. \tag{4.167}$$

Indeed, from (4.166) one can write

$$d(f(x,t+\tau),f(y,t+\tau)) < \frac{\epsilon}{3}, \tag{4.168}$$

as soon as $0 \leq t+\tau \leq T$, or $\tau \in [-t, -t+T]$. But T is the length, corresponding to $\frac{\epsilon}{3}$, of any interval $(a, a+T) \subset R$, containing an $\frac{\epsilon}{3}$ translation number for f. Hence, we can choose τ as an $\frac{\epsilon}{3}$ almost period for $f(x,t)$, $x \in X$:

$$d(f(x,t+\tau), f(x,t)) < \frac{\epsilon}{3}, \; d(f(y(t+\tau)), y(t)) < \frac{\epsilon}{3}, \tag{4.169}$$

for all $t \in R$. Combining (4.166)–(4.169) and using triangle inequality for d, one obtains

$$d(f(x,t), f(y,t)) < \epsilon, \; t \in R, \tag{4.170}$$

which proves that our dynamical system is Liapunov stable.

This completes the proof of Theorem 4.11.

The theory of dynamical systems, introduced and developed by Birkhoff [66], has been extensively developed throughout the past 50 years. In particular, the concept of semi dynamical system has been created and investigated. Also, instead of maps in the usual sense, the set valued maps have been involved.

Since this book is not focused on the extensions of dynamical systems that have appeared during the last half-century, we send interested readers to sources that discuss such developments: Bhatia and Hájek [63], Sibirskii and Shuba [498], and Szegö and Treccani [514].

4.8 BRIEF COMMENTS ON THE DEFINITION OF $AP_r(R, \mathcal{C})$ SPACES AND RELATED TOPICS

The definition formulated in section 4.1 for $AP_r(R, \mathcal{C})$, $1 \leq r \leq 2$, is as follows: Consider the set of all trigonometric series of the form $\sum_{k=1}^{\infty} A_k e^{i\lambda_k t}$, $t \in R$, $A_k \in \mathcal{C}$, $\lambda_k \in R$, $k \geq 1$. If we limit ourselves to those formal series of this form, with the property $\sum_{k=1}^{\infty} |A_k|^r < \infty$, $1 \leq r \leq 2$, then we can easily prove that this set of formal trigonometric series constitute a Banach space over $\mathcal{C}$. The norm is defined by

$$\left| \sum_{k=1}^{\infty} A_k e^{i\lambda_k t} \right|_r = \left(\sum_{k=1}^{\infty} |A_k|^r \right)^{\frac{1}{r}}, \; 1 \leq r \leq 2. \tag{4.171}$$

Indeed, the Minkowski norm induces a topology of normed space, say $AP_r(R, \mathcal{C})$, $1 \leq r \leq 2$. The completeness required for a normed space, to be a Banach space, is a simple consequence of the following argument: It is known that the completion of a normed linear space is the minimal complete space, hence of Banach type, which contains the normed linear space (generally, not complete). Based on our definition, the normed space contains all the trigonometric series of the form which appear in (4.171), and the coefficients series on the right-hand side is convergent. Consequently, the normed linear space is the *minimal* Banach type space to which it belongs.

On the contrary, if there exists another linear complete space, larger than this set (space we denoted by $AP_r(R,\mathcal{C})$), then it should contain at least one element (series) which converges in the Minkowski norm. This contradicts the definition of the space $AP_r(R,\mathcal{C})$, which contains all the series in (4.171) with the property of convergence.

In Corduneanu [161], other properties of the spaces $AP_r(R,\mathcal{C})$, $1 \leq r \leq 2$ are established, among them the inclusions

$$AP_1(R,\mathcal{C}) \subset AP_r(R,\mathcal{C}) \subset AP_s(R,\mathcal{C}) \subset AP_2(R,\mathcal{C}), \tag{4.172}$$

valid for $1 < r < s < 2$. The inclusion is not only set-theoretic but also topologic (i.e., the topology/convergence in $AP_r(R,\mathcal{C})$ is stronger than that of $AP_s(R,\mathcal{C})$, $r < s$).

Many problems arise in connection with the spaces $AP_r(R,\mathcal{C})$, $r \in [1,2]$. For instance, what is the relationship of such spaces with the classical spaces of almost periodic functions (Bohr, Stepanov, Besicovitch)?

To answer such questions, at least partially, it is appropriate to consider the case of Bohr almost periodic functions, the space denoted by $AP(R,\mathcal{C})$.

The answer is easy, based on the classical theory of almost periodic functions (see our list of references under the names of Bohr, Favard, Fink, Corduneanu, Zhang, Amerio and Prouse, and Levitan and Zhikov).

The only space in the scale $AP_r(R,\mathcal{C})$, $1 \leq r \leq 2$ that is entirely contained in $AP(R,\mathcal{C})$, is $AP_1(R,\mathcal{C})$. One easily sees that the intersection $AP_r(R,\mathcal{C}) \cap AP(R,\mathcal{C}) \neq \emptyset$, $1 < r \leq 2$. These contain at least the trigonometric polynomials, which can be used to construct all spaces in the scale. Moreover, one observes from the definition of Minkowski norms, that the set $\mathcal{T}$ is everywhere dense in each space $AP_r(R,\mathcal{C})$, $1 \leq r \leq 2$ (obvious for $r = 1$).

An important problem, not yet completely solved, is to characterize those series/elements of a space $AP_r(R,\mathcal{C})$ which also belong to $AP(R,\mathcal{C})$, $1 < r \leq 2$. The continuity of the attached function/sum is certainly one of the conditions to be satisfied by the series of $AP_r(R,\mathcal{C})$, $1 < r \leq 2$. Is the continuity also sufficient? Shubin showed in his papers [496] and [497] that elements of $AP_r(R,\mathcal{C})$, or similar spaces, also are in $AP(R,\mathcal{C})$, when they are continuous. The results obtained in these papers are about partial differential operators/equations in higher dimensions.

We encourage the readers to attempt proving similar results for $AP_r(R,\mathcal{C}^n)$ or $AP_r(R,\mathcal{C})$ solutions of equations (4.61) and (4.95).

Hint: One can easily note that an AP_r-solution of such equations has a first order derivative, which also belongs to AP_r-space. It is also worthwhile to notice, based on the fact discussed in Section 4.7, that an AP_r space generates a dynamical system (Birkhoff). The property that serves as the definition for Bohr almost periodic functions is a simple consequence of this fact. Also, one should consider the known result that states: A Stepanov almost periodic function is also Bohr almost periodic, if and only if it is uniformly continuous on R.

Another problem to be clarified is to establish the relationship between AP_r-space and Stepanov's almost periodicity. See Refs. [23] and [24] for helpful material.

We will close this section by considering the concept of an AP_r-space, for Hilbert-space valued series/functions. We stress the fact that, under adequate conditions, for each generalized Fourier series we will consider its "sum," in either the classical sense or in a generalized sense (e.g., a summability procedure).

In Appendix, one can find a more detailed description of the connections between generalized Fourier series and functions. Because of the choice between norm or semi-norm in defining the topology/convergence in the function space associated with a series, we may not have a single-valued series-functions correspondence. Note that most general cases encountered in this book are maps with values in a finite dimensional space. However, any approach for obtaining results for *partial functional differential equations* requires function spaces in which the elements are infinite-dimensional valued maps.

Let us consider a complex Hilbert space H, whose elements are denoted by $h_1, h_2, \ldots$. Each h_j is a sequence in $\ell^2(N, C)$. If the scalar product on $H \times H$ is denoted by $< ., . >$, then $|h|^2 =< h, h >$ is the norm, $h \to |h|$. This makes H a linear complete Banach space of a special form, where the norm is derived from a scalar product!.

To define the space $AP_r(R, H)$, $1 \leq r \leq 2$, one considers all Fourier series with coefficients in H, say

$$\sum_{j=1}^{\infty} h_j e^{i\lambda_j t}, \tag{4.173}$$

with $\lambda_j \in R, j \geq 1$, under the conditions

$$\sum_{j=1}^{\infty} |h_j|^r < \infty, \; 1 \leq r \leq 2. \tag{4.174}$$

Since the map $r \to \left(\sum_{k=1}^{\infty} |A_k|^r\right)^{\frac{1}{r}}$ is decreasing on $(1,2)$, one derives the inclusion $AP_r(R,H) \subset AP_s(R,H)$, $1 \leq r < s \leq 2$. In other words, $AP_2(R,H)$ contains any space $AP_r(R,H)$ for $r < 2$. As noticed in Section 4.2, in the particular case $H = \mathcal{C}$, one can derive the basic properties of $AP_r(R,H)$, known for classical almost periodic functions. On behalf of discussion in Section 4.7, the triplet $[AP_r(R,H), R, f]$, with $f = x(.+t)$, $t \in R$ (Theorem 4.10) is a dynamical system (Birkhoff). Of course, from this last statement, one can derive all other basic properties of $AP_r(R,H)$.

Historically, the space $\ell^2(N,\mathcal{C})$ constitutes the first example of a Hilbert space, which has a countable basis. It is known from elementary Functional Analysis that the function space $L^2([a,b],\mathcal{C})$ is isomorphic to $\ell^2(N,\mathcal{C})$. These spaces are frequently encountered in the literature.

4.9 BIBLIOGRAPHICAL NOTES

In Chapter 4, we have included mostly recent results. The spaces of almost periodic functions, under consideration, constitute a scale of spaces. The first (and smallest) is the Poincaré's space of almost periodic functions, which is represented by absolutely and uniformly convergent (on R) trigonometric series of the form

$$\sum_{k=1}^{\infty} a_k e^{i\lambda_k t},\ t \in R,\ a_k \in \mathcal{C},\ \lambda_k \in R,\ k \geq 1.$$

Poincaré's space, $AP_1(R,\mathcal{C})$, has been used by authors in applications to the oscillatory behavior of solutions to some classes of functional differential equations. It is a convenient feature to obtain the solution as an absolutely and uniformly (on R) convergent Fourier series. This situation can be found in classical books on almost periodic oscillations (of course, the represented solutions can be immediately approximated by trigonometric polynomials in the usual supremum norm).

Such books are the works of many authors, and we shall mention here only some of them: Malkin [355], Levitan [326], Levitan and Zhikov [327], Corduneanu [129, 156], Fink [213], Favard [208], Halanay [237], Persidskii [447], Hino et al. [263], Yoshizawa [540], Zaidman [546, 547], and Zhang [553].

There is extensive literature concerning the work on almost periodicity in classical sense (Bohr, Stepanov, Besicovitch). A great part of it is contained in the list of references to Corduneanu [129] and Fink, who also prepared a long list, which is circulated in Xerox format.

The number of contributions to the development and applications of the theory of almost periodic functions/solutions is also very large. For this reason, we will not even provide a reduced listing. The periodicals Mathematical Reviews, Zentralblatt fur Mathematik, Referativnyi Zhurnal, and other similar publications dedicate sections to almost periodicity.

Our list of references contains names of researchers/contributors to this field from many countries (the United States, France, Italy, Canada, Japan, Peoples Republic China, Taiwan, India, Germany, the United Kingdom, Czeck Republic, Algeria, Russian Federation, Ukraine, Greece, Georgia, Korea, Hungary, Romania, Ireland, Belgium, Moldova Republic, Bulgaria, Morocco, Israel, Poland, Turkey, Mexico, Brazil, Chile, Spain, Austria, Australia, Argentina a.o). Undoubtedly, an interest in Functional Equations is present around the world.

Historically, the Functional Equations, not necessarily the traditional ones (Differential, Integral or Integro-differential) have appeared in the twentieth century. We recognize early contributors like Radon [467] and Riesz [470] who paid attention to this subject and who were among the founders of Modern Analysis. We can say that the theory of Functional Equations has deep roots in the history of mathematics and is spread around the world. We are certain that this field of study will continue to grow.

The topics associated with Functional Equations are, generally speaking, the same as in case of traditional functional equations: existence, uniqueness, estimates, approximations of solutions, stability, or oscillatory properties, their use in Control Theory.

It was Tonelli (1928) who introduced the concept of Volterra operator, also known as *causal* or *non-anticipative*, and proved basic properties of associated functional equations. The following books are totally or partially dedicated to this class of functional equations, which naturally arise in application, when the change of the system described by the equations depends on its past moments: Corduneanu [149], Gripenberg et al. [228], Lakshmikantham V. et al. [311], Azbelev, et al. [33]. Vasundhara Devi, J. [526] generalized the basic concepts to the case of set valued mappings. The book by Vath [527] is also partially dedicated to this topic. The aforementioned books, contain references to journal papers. A brief listing of equations with causal operators is included in Corduneanu [135, 149].

A significant amount of publications deal with functional differential equations in infinite dimensional spaces. In most cases, the investigation is based on semigroups method. Also, the convolution product plays an important role in applications. The solution of abstract differential equations, with the form

$$\dot{x}(t) = Ax(t) + f(t),\ t \in R_+,$$

can be expressed by the formula

$$x(t) = T(t)x^0 + \int_0^t T(t-s)f(s)\,ds,\ t \geq 0,$$

provided A is the infinitetessimal generator of a co-semigroup of linear operators, $T(t)$, $t > 0$, and f is differentiable, with the first-order derivative locally integrable. The statement is true for $x \in E =$ reflexive Banach space and $x^0 \in D(A)$.

This is the key to useful applications to partial differential equations. See, for instance, Barbu [44] and the references therein, Pazy [442].

Many results on almost periodicity in case of the space $AP(R,\mathcal{C})$ should be extended to AP_r-spaces, considered in this chapter. In Sections 4.1–4.4, we limited our presentation to the case of the space $AP(R,\mathcal{C})$. It is interesting to try the extension of these kinds of results to infinite-dimensional spaces $AP_r(R,X)$, with X a Banach or Hilbert Space. This approach will lead to results related to functional partial differential equations or other types of functional equations.

We are proposing here the modified Schroedinger equation $i\,u_t + \Delta\,u + k * u = 0$, where k is representing the potential and the convolution must be performed with respect to the time variable. The variable u, the unknown element, has the usual meaning in quantum mechanics. In one-dimensional case, $u = u(x,t)$ and $k = k(x,t)$.

For existence results in the space $AP_r(R,\mathcal{C})$, the following references contain such results: Corduneanu [159, 161, 162], Corduneanu and Li [171], Corduneanu and Mahdavi [182], and Mahdavi [351]. These references extend and complete some results in the paper [161]. There are many possible extensions when substituting the space $\mathcal{C}$ with other spaces like R^n, $\mathcal{C}^n$ or general Hilbert (or Banach) spaces.

In Appendix A, we briefly present the method for constructing spaces of oscillatory functions, including the classical ones (periodic, almost periodic). Some introductory elements are taken from the papers by Corduneanu [161], [165]. Spaces of oscillatory functions have been constructed and investigated by V. F. Osipov [432] and Zhang [553–556].

The pseudo almost periodic functions have been introduced by Zhang [553], and the theory of these oscillatory functions has been developed by many authors. Also, the Bochner concepts of automorphy and its immediate extension to pseudo-automorphy have generated a vast literature. We will mention here the books by G. M. N'Guerekata [417] and T. Diagana [193]. A significant number of journal papers have been produced that relate to the study of these types of "oscillatory" functions. This problem remains to be clarified by future work. The paper of Bochner, see Academy of Science USA, (1961) gives birth to new generalizations of almost periodicity concept. For recent research in this direction, see the books by N'Guerekata and Diagana, mentioned earlier, and the references therein.

Another generalization of the classical concept of almost periodicity is obtained by considering functions with values in a locally convex space. The paper by Shtern [495] contains basic facts of the theory. It also connects components of this concept with other aspects of the theory (like representations in locally convex spaces). The list of references has more than 200 items; the paper constitutes a main source for the study of extensions of the concept of almost periodicity.

Let us also mention two important contributions, by Besicovitch and Bohr [62] and Bohr and Folner [73], regarding basic concepts related to almost periodicity (connections with trigonometric series and functional spaces). These contributions came before the Functional Analysis became prevalent in research.

Fréchet's concept of asymptotic almost periodicity has been present in many publications. We mention here the recent book by Cheban [99], which is primarily concerned with applications to differential equations.

Books containing results related to almost periodicity (or its generalizations) of solutions of functional equations of varied types are in abundance in the mathematical literature. Of course, we are most concerned with classical types of almost periodicity, and we will mention some of them: Malkin [356], Halanay [237], Hale [240], Hale and Lunel [242], Hino et al. [263], Fink [212], Favard [208], Corduneanu [129, 156], Krasnoselskii et al. [297], Amerio and Prouse [21], Leviton and Zhikov [327], Zaidman [547], and N'Guerekata and Pankov [418].

We will only mention a few of the thousands papers, dedicated to problems in almost periodicity. The paper by Andres et al. [24] caught our attention. They thoroughly investigated the relationships and hierarchy between various spaces/classes of almost periodic functions (classical type).

Zhang [554, 556] built up the space $\mathcal{SLP}$-functions, making new progress in the generalization of almost periodicity and producing a good example of oscillatory function spaces, pertaining to the third stage in Fourier Analysis.

Corduneanu and Mahdavi [182] provided several results on the existence of AP_r-type almost periodic solutions to various classes of functional equations/functional differential equations.

Yi (Yingfei) [537] treated, in some detail, the almost automorphic oscillations, making a valuable contribution to the Almost Automorphic Dynamics. Over 100 references were included.

Kordonis and Dhilosch [296] investigated the oscillating behavior (not almost periodic type) of solutions to classes of advanced–delayed functional equations.

Andres and Bersani [23] treated the problem of almost periodicity as a fixed point problem for evolution inclusions.

Corduneanu [158] surveyed some results obtained by several contributors, related to the almost periodicity of solutions of certain differential or functional differential equations.

Ruess and Summers [476] provided interesting results concerning the connections between weak almost periodicity, ergodicity, and the theory of semigroups of linear contraction operators.

Fink [212] presented the aspect of connections between almost periodicity (Bohr) and various applied problems. He studied how new classes of almost periodic functions appear from specific purposes.

Giesel and Rasmussen [217] provided an example of an almost periodic function (scalar) whose derivative is not almost periodic. They also applied their finding to investigate ordinary differential equations $\dot{x}(t) = F(t,x(t))$, using the variational equations.

Hasil and Vesely [250] treated oscillatory properties of half-linear second order equations including almost periodicity.

Kurihara and Suzuki [302] applied Galarkin's method to quasi-periodic differential–difference equations.

Cheban and Mammana [101] investigated the almost periodicity of solutions of second-order equations $\ddot{x} = f(t,x)$, under monotonicity of f with respect to x. Also, the property of almost automorphy.

Naito and Nguyen [414] investigated the evolution semigroups (Banach space) and provided spectral criteria for almost periodicity.

Corduneanu [154] provided results for semi-linear systems of functional differential equations in the spaces $AP(R,R^n)$ and $AP(R,\mathcal{C})$.

Hamaya [245] found results of existence for almost periodic solutions to a nonlinear Volterra difference equation.

Let us also mention some papers dedicated to applications of the concept of almost periodicity to diverse problems in applied mathematics.

In a lengthy paper, Meyer [392] investigated problems related to almost periodic patterns (other authors use the term *structure* instead of pattern) in connection with the quasi-crystals, mean-periodic functions, and irregular sampling. The mean-periodic functions are continuous solutions of the equation $f * \tau = 0$, where τ denotes a Borel measure with compact support. It is a paper that can inspire further research.

Mickelson and Jaggard [394] studied problems in connection with electromagnetic wave propagation in almost periodic media. The Fourier series were a main tool. It is an engineering-mathematical article.

Mentrup and Luban [391] investigated almost periodic wave packets, including those of invariant shape. This is another interesting engineering-mathematical paper. The wave packets are important in applications.

Mäkilä et al. [353] investigated bounded power signal spaces for robust control and modeling, *SIAM Journal on Control and Optimization.*

In Corduneanu [162] and Corduneanu and Li [169], AP_r-almost periodic solutions were investigated in equations involving convolution.

Henriquez and Vasquez [260] investigated the existence of bounded, almost periodic and asymptotically almost periodic solutions to first- and second-order functional differential equations with unbounded/infinite delay. This is a conspicuous paper containing many results, involving the semigroups theory and applications to biomathematics.

With regard to the application of various types of functional differential equations, requiring advanced functional analysis, the books by Kolmanovskii and Myshkis [291, 292] contain a large variety of results and methods. They contain a list of references with over 500 entries.

5

NEUTRAL FUNCTIONAL DIFFERENTIAL EQUATIONS

The investigation of neutral functional differential equations, both with continuous and discrete argument, has been developed rapidly throughout the past 50–60 years, and is sometimes comparable with the case of ordinary functional equations. The literature abundantly illustrates this situation. Monographs and treatises like Halanay [237], Hale [240], and Agarwal et al. [10] are pertinent examples.

For instance, in the book by Hale and Lunel [242], which deals with functional differential equations of the form,

$$\frac{d}{dt}[D(t)x_t] = L(t)x_t + h(t),\ t \geq 0,\ x_\sigma = \phi,$$

with $(\sigma, \phi) \in C([-\tau, 0], R^n)$, $h \in L^1_{\text{loc}}(R, R^n)$, $D(t) : C \to R^n$, $L(t) : C \to R^n$, and $D(t) = \phi(0) - \int_{-\tau}^{0} d[\mu(t,0)]\,\phi(\theta)$, $L(t)\phi = \int_{-\tau}^{0} d[\eta(t,\theta)]\,\phi(\theta)$, adequate conditions are imposed on η and μ, and one proves existence, uniqueness of the Cauchy problem, while an integral representation formula is obtained. The method for investigating quasilinear systems is thus open.

Functional Differential Equations: Advances and Applications, First Edition.
Constantin Corduneanu, Yizeng Li and Mehran Mahdavi

When dealing with neutral-type functional differential equations, we start from the following point of view. An ordinary differential system is usually written in the form $\dot{x}(t) = f(t, x(t))$: the derivative on the left-hand side measures the speed of the change in time for each coordinate/parameter. In the practical approach, it is actually more convenient to measure the rate of change for a combination of the coordinates. For instance, the energy of the system, described by the equations, could be one of these combinations. Therefore, a more general type of system could be of the form $\frac{d}{dt}f(t, x(t)) = F(t, x(t))$. The word *neutral* is frequently encountered in the literature for equations or systems of the form $\dot{x}(t) = f(t, x(t), \dot{x}(t))$. There are many other types of equations, termed *neutral* in existing literature. These functional equations could also be with delay (finite or infinite), as those mentioned earlier from Hale [240], or even of causal-type (see Corduneanu [149]).

In this chapter, we will consider some types of systems that fall within the categories we have recently investigated.

5.1 SOME GENERALITIES AND EXAMPLES RELATED TO NEUTRAL FUNCTIONAL EQUATIONS

Neutral functional differential equations are of the form

$$\frac{d}{dt}(Vx)(t) = (Wx)(t), \tag{5.1}$$

where V and W are causal operators acting on various function spaces whose elements are continuous maps from R (or an interval $I \in R$), or belong to some measurable function spaces. When we deal with spaces of measurable functions, it is understood that the solutions will be meant in the Carathéodory sense.

By choosing the operator V in different manners, we shall obtain many types of neutral equations, including the "normal" ones

$$\frac{dx}{dt} = (Wx)(t), \tag{5.2}$$

which corresponds to $V = I$. Other encountered choices in the literature are

$$\begin{aligned}
(Vx)(t) &= x(t) + cx(t-h), \\
(Vx)(t) &= x(t) + \sum_{k=1}^{m} c_k x(t-h_k), \\
(Vx)(t) &= x(t) + (\nu_0 x)(t), \\
(Vx)(t) &= x(t) + g(t, x_t),
\end{aligned} \tag{5.3}$$

where x_t stands for the function $x_t(u) = x(t+u)$, $u \in [-h, 0)$. The last case was introduced by Hale [240], and has since diffused into the mathematical literature. In addition, chapters dedicated to these types of equations can be found in books by Corduneanu [149], Kolmanovskii and Myshkis [291, 292], Hale and Lunel [242], Azbelev et al. [33], and Kurbatov [301].

In case we choose V as a Volterra linear operator

$$(Vx)(t) = x(t) + \int_0^t k(t,s)x(s)\,ds, \ t \in [0,T], \tag{5.4}$$

which for differentiable $k(t,s)$, with respect to t, leads to the functional integro-differential equation of the form

$$\frac{dx(t)}{dt} = A(t)x(t) + \int_0^t K(t,s)x(s)\,ds + (Wx)(t), \tag{5.5}$$

the first initial value problem related to such equations is

$$x(0) = x_0, \tag{5.6}$$

while the second type of initial value problem is

$$x(t) = \phi(t), \ 0 \le t < t_0, \ x(t_0) = x_0. \tag{5.7}$$

We generally study such initial value problems either locally or globally.

For ordinary differential equations (ODEs) or delay equations, one can further investigate problems like uniqueness, dependence on data, estimates for solutions, stability, oscillatory solutions, control problems and other general or specific problems (related to the special form of the equation).

When dealing with higher-order neutral equations (we will limit our approach to the second order), we will try reducing these equations to first-order equations.

The existence of the inverse operator V^{-1} leads to results concerning the existence of solutions, assuming that the inverse is also causal, and Equation (5.1) can be rewritten as

$$x(t) = V^{-1}\left(c + \int_0^t (Wx)(s)\,ds\right), \tag{5.8}$$

where c is a constant (possibly a vector in R^n, or even a Banach space).

It is useful to note that the inverse of a causal operator, though existing, may itself not be causal.

A very simple example is provided by

$$(Vx)(t) = x\left(\frac{t}{2}\right), \ x \in C(R_+, R^n). \tag{5.9}$$

The inverse operator is given by

$$(V^{-1}y)(t) = y(2t), \ y \in C(R_+, R^n).$$

Another example is offered by the Volterra integral operator, for which the equation

$$(Vx)(t) = \int_0^t k(t,s)x(s)\,ds = f(t), \tag{5.10}$$

does not possess, in general, solutions. It is sufficient to choose $f(t)$ outside the range of the operator on the right-hand side of (5.10).

On the contrary, the linear Volterra equation of the second kind

$$(Vx)(t) = x(t) + \int_0^t k(t,s)x(s)\,ds = f(t), \ t \in [0,T], \tag{5.11}$$

has always solutions, under fairly general conditions, which implies (because of the uniqueness) the invertibility of the operator V in (5.11). Moreover, the inverse operator is given by the formula

$$y(t) = f(t) + \int_0^t \tilde{k}(t,s)y(s)\,ds, \ t \in [0,T], \tag{5.12}$$

where $\tilde{k}(t,s)$ is the *resolvent* kernel, associated to $k(t,s)$. For details, regarding the validity of formula (5.12) see, for instance, Tricomi [520].

Now we shall rewrite equation (5.1) in the form

$$\frac{d}{dt}[x(t) + (Vx)(t)] = (Wx)(t), \ t \in [0,T], \tag{5.13}$$

noticing the fact that $(Vx)(t)$ and $x(t) + (Vx)(t)$ are simultaneously causal.

The following hypotheses will be made, in order to obtain the existence of solutions of equation (5.13).

$(H1)$ The operators V and W are continuous causal operators on the space $C([0,T], R^n)$.

$(H2)$ V is compact and has the fixed initial value property

$$(Vx)(0) = \theta \in R^n, \ x \in C([0,T], R^n). \tag{5.14}$$

$(H3)$ The operator W takes bounded sets into bounded sets of $C([0,T], R^n)$.

We can state and prove the following existence result, based on Schauder fixed-point theorem, in the underlying space $C([0,T],R^n)$.

Theorem 5.1 *Consider equation (5.13) under hypotheses* $(H1)$, $(H2)$, $(H3)$, *and the initial condition (5.6).*

Then, there exists a solution of (5.13), satisfying the initial condition (5.6), defined on an interval $[0,a]$, *with* $a \leq T$.

Proof. Integrating both sides of (5.13) from 0 to $t \leq T$, we obtain the equation

$$x(t) + (Vx)(t) = x_0 + \int_0^t (Wx)(s)\,ds, \tag{5.15}$$

taking into account (5.14).

Let us now consider, on $C([0,T],R^n)$, the operator

$$(Ux)(t) = -(Vx)(t) + x_0 + \int_0^t (Wx)(s)\,ds. \tag{5.16}$$

Consequently, we can rewrite equation (5.16) in the following form:

$$x(t) = (Ux)(t),\ x \in C([0,T],R^n). \tag{5.17}$$

Hypotheses $(H1)$, $(H2)$, and $(H3)$ assure the compactness of the operator U because V is compact according to $(H2)$, while

$$x(t) \to \int_0^t (Wx)(s)\,ds,$$

is a compact operator on $C([0,T],R^n)$.

The proof will be complete if we prove the local existence of solutions for equation (5.17). This proof can be found in Corduneanu [149, Chapter 3]. Because it is very simple we will provide it here.

If $x^0 \in R^n$ is the fixed initial value of U, $(Ux)(0) = x_0$, for any $x \in C([0,T],R^n)$, equation (5.17) is equivalent to

$$x(t) - x_0 = (Ux)(t) - x_0, \tag{5.18}$$

and due to compactness of U, we can write $\forall \epsilon > 0$, $|(Ux)(t) - x_0|_C < \epsilon$, provided $0 \leq t \leq \delta$, $\delta = \delta(\epsilon)$, based on the equicontinuity in Ascoli–Arzelà criterion of compactness, for all $x(t)$ such that $|x(t) - x_0|_C \leq r$, $t \in [0,T]$. Once r is fixed, δ will depend only on ϵ. Without loss of generality, we can choose $\epsilon < r$. Then, from (5.18) we obtain $|(Ux)(t) - x_0|_C < \epsilon < r$, with $|x(t) - x_0|_C \leq r$, provided $t \in [0,\delta]$.

This ends the proof of Theorem 5.1.

Remark 5.1 *It is possible to obtain a similar global existence result if we add to hypotheses $(H1)$, $(H2)$, and $(H3)$ the following growth conditions on V and W:*

$$|x(t)+(Vx)(t)| \geq \gamma|x(t)|,\ \gamma>0,\ t\in[0,T],$$

and

$$|(Wx)(t)| \leq \alpha|x(t)|+\beta,\ \alpha,\beta>0,\ x\in C([0,T],R^n).$$

From equation (5.13) and the given growth conditions we find the inequality

$$\gamma|x(t)| \leq |x_0|+\beta t+\alpha\int_0^t |x(s)|\,ds, \tag{5.19}$$

$\forall t\in[0,T]$, as long as the solution $x(t)$ is defined. But (5.19) implies that $|x(t)|$ must remain bounded on the whole interval of existence. This implies the fact that $x(t)$ can be extended to the interval $[0,T]$, by using Gronwall's inequality.

Remark 5.2 *The equation (5.17) can be considered in the space $L^p([0,T],R^n)$, $1\leq p<\infty$. Under the same condition of compactness for U, results of existence similar to that found already in the continuous case are valid. Moreover, the singularly perturbed equation $\epsilon\dot{x}_\epsilon(t)=-x_\epsilon(t)+(Ux_\epsilon)(t)$, for small positive ϵ has solutions defined on an interval $[0,\delta]\subset[0,T]$. The solutions $\{x_\epsilon(t);\epsilon>0\}$ can be regarded as regularized (they are absolutely continuous) solutions of equation (5.17), on the interval $[0,\delta]$.*

The statement of the result, as well as its relatively lengthy proof can be found in Corduneanu [[135], Theorem 3.4.1].

Remark 5.3 *Instead of equation (5.1) or (5.3), one can deal with the neutral equation*

$$\frac{d}{dt}[(V_0x)(t)+(Vx)(t)]=(Wx)(t),\ t\in[0,T]. \tag{5.20}$$

Basic assumptions will be the invertibility of the operator V_0, in the underlying space. This space can be $C([0,T],R^n)$ or $L^p([0,T],R^n)$, $1\leq p<\infty$.

We obtain from (5.20) the following functional equation:

$$x(t)=V_0^{-1}\left(-(Vx)(t)+\int_0^t (Wx)(s)\,ds+x_0\right). \tag{5.21}$$

For the operators V and W, we keep the same hypotheses as in Remark 5.2. If we assume the continuity and causality of V_0^{-1}, then the operator on the right-hand side of (5.21) will be compact on $C([0,T],R^n)$.

Since (5.21) can be assimilated with the equation of the form (5.17), with the compact operator on the right-hand side, we can infer the local existence for equation (5.20) with the initial condition (5.6).

The case of measurable solutions can be treated as shown in Remark 5.2.

Let us consider again the functional differential equation of neutral type (5.13) under different assumptions made above. Since we deal with the concept of "large contraction," developed by Burton [82], we provide here the definition: $V : C([0,T],R^n) \to C([0,T],R^n)$ is a *large contraction* if $|Vx - Vy|_C < |x-y|_C$ for $x \neq y$, and if $\forall \epsilon > 0$, there exists $\delta = \delta(\epsilon) < 1$, such that

$$(x,y \in C,\ |x-y| \geq \epsilon) \Longrightarrow |Vx - Vy|_C \leq \delta |x-y|_C.$$

The following hypotheses will be made for the operators V and W:

$(h1)$ $V : C([0,T],R^n) \to C([0,T],R^n)$ is a large contraction.

$(h2)$ For each $f \in C([0,T],R^n)$, there exists $g \in C([0,T],R^n)$, such that $\tilde{V} = V + f$ satisfies $|g - \tilde{V}^n g| \leq K < \infty$ for $n \geq 1$.

$(h3)$ $W : C([0,T],R^n) \to C([0,T],R^n)$ is a continuous causal operator, taking bounded sets into bounded sets of $C([0,T],R^n)$.

The following existence result for solutions of equation (5.13) is given in the

Theorem 5.2 *Consider the neutral functional differential equation (5.13), under initial condition (5.6). Assume that* $(h1)$, $(h2)$, *and* $(h3)$ *are satisfied by the operators* V *and* W.

Then, the problem (5.13), (5.6) has a solution $x = x(t)$, *defined on an interval* $[0,T_1]$, $T_1 \leq T$, *such that* $x(t) + (Vx)(t)$ *is continuously differentiable on* $[0,T_1]$.

Before we proceed to the proof of Theorem 5.2, we notice that in the statement of the theorem we did not mention fixed initial value property for the operator V. This is because without loss of generality, we can assume $(Vx)(0) = \theta$ =the null element of $C([0,T],R^n)$. Indeed, if V does not satisfy this condition, then the operator $(\tilde{V}x)(t) = (Vx)(t) - (Vx)(0)$ will satisfy it.

Proof. We transform equation (5.13) with the initial condition (5.6) into the equivalent problem of solving equation (5.15).

Let us denote

$$(Ux)(t) = x_0 + \int_0^t (Wx)(t)\,ds,\ x \in C([0,T],R^n), \tag{5.22}$$

and

$$(V_1x)(t) = x(t) + (Vx)(t),\ x \in C([0,T],R^n), \tag{5.23}$$

which allows us to write equation (5.17) in the following form:

$$(V_1x)(t) = (Ux)(t),\ x \in C([0,T],R^n). \tag{5.24}$$

Equation (5.24) will be equivalent to the equation

$$x(t) = V_1^{-1}((Ux)(t)),\ x \in C([0,T],R^n), \tag{5.25}$$

if we successfully prove that V_1 is invertible and has a continuous and causal inverse. Actually, we will prove that the map $x(t) \to (V_1x)(t)$ is a homeomorphism of the space $C([0,T],R^n)$. On behalf of assumptions $(h1)$ and $(h2)$, it is obvious that $x(t) \to (V_1x)(t)$ is a continuous map on $C([0,T],R^n)$. We need to show that this map is onto $C([0,T],R^n)$, and is one-to-one. These properties will follow if we examine the auxiliary functional equation

$$x(t) + (Vx)(t) = f(t),\ t \in R, \tag{5.26}$$

in the space $C([0,T],R^n)$. Equation (5.26) is the same as $(V_1x)(t) = f(t)$, and it can also be written as follows:

$$x(t) = -(Vx)(t) + f(t) = (Tx)(t). \tag{5.27}$$

Since V is, by assumption, a large contraction on $C([0,T],R^n)$, we see from (5.27) that T is also a large contraction. Hence, from (5.27), which is $x(t) = (Tx)(t)$, we obtain the existence of a unique solution $x(t) \in C([0,T],R^n)$, $\forall f \in C([0,T],R^n)$, (see Burton [82], Theorem 1). This means that V_1 maps $C([0,T],R^n)$ onto itself. The uniqueness part tells us that the map $x(t) \to (V_1x)(t)$ is one-to-one. Therefore, this map is a homeomorphism of $C([0,T],R^n)$ onto itself.

This ends the proof of Theorem 5.2.

Remark 5.4 *Under assumptions of Theorem 5.2, the uniqueness of the solution may not be true. A simple example is provided if we choose* $(Vx)(t) \equiv \theta$, *and* $(Wx)(t) = f(t,x(t))$, *in (5.13), with* f *continuous, but such that* $\dot{x}(t) = f(t,x(t))$ *is deprived of uniqueness. One could, for instance, impose on* W *a Lipschitz-type condition to assure uniqueness.*

Remark 5.5 *Following the same procedure as in the proof of Theorem 5.2, one can also prove local existence when the underlying space is an* L^p*-space. The solution of equation (5.13) will be such that* $x(t) + (Vx)(t)$ *is almost everywhere differentiable. It will be an absolutely continuous function, while the validity of (5.13) will be assured only almost everywhere on the interval of existence. Of course, any norm used in the proof will be an* L^p*-norm instead of C-norm.*

Remark 5.6 *In the following text we will sketch the procedure to obtain a global existence result for (5.13), and we will choose the space* $L^2([0,T],R^n)$ *as underlying space. Obviously, the solution will be meant in the Carathéodory's sense.*

Preserving the notations introduced in the proof of (5.13), the basic equation to be examined will have the form

$$(V_1x)(t) = c + \int_0^t (Wx)(s)\,ds, \ t \in [0,T], \tag{5.28}$$

with $c \in R^n$ arbitrary. One also has to prove the existence of the inverse operator V_1^{-1} for which one needs to assure continuity and causality.

The readers are invited to carry out the details of the proof under the following assumptions:

1) The operator $V : L^2([0,T],R^n) \to L^2([0,T],R^n)$ is continuous and such that $V_1 = I + V$ has a continuous causal inverse V_1^{-1}.
2) The operator V_1 is a homeomorphism between a bounded subset $B \subset L^2([0,T],R^n)$ and the whole space $L^2([0,T],R^n)$.
3) The operator $W : L^2([0,T],R^n) \to L^2([0,T],R^n)$ is continuous, causal, and takes bounded sets into bounded sets of L^2.

The existence will be obtained by applying Shauder's fixed-point theorem, with the existence being assured on the whole interval $[0,T]$.

5.2 FURTHER EXISTENCE RESULTS CONCERNING NEUTRAL FIRST-ORDER EQUATIONS

In this section, we are primarily concerned with equation (5.13). Our objective is to obtain some applications of the general results, to the special situation when the operators involved are classical Volterra integral operators.

For formulating the results in the general case, we will use the underlying spaces $L^p([0,T],R^n)$ with $p \in [1,\infty)$. The following hypotheses will be considered:

$(H1)$ V and W are causal operators on the space $L^p([0,T],R^n)$, $1 \leq p < \infty$.

$(H2)$ V is a continuous compact operator, such that for each $x \in L^p$ one has

$$\lim_{t\to 0+} (Vx)(t) = \theta, \tag{5.29}$$

where θ stands for the null element of R^n.

$(H3)$ W is a continuous operator which takes bounded sets of L^p into bounded sets.

Before going further, we note that condition (5.29) imposed on V is a slight generalization of the "fixed initial value property," which we have used in the case of $C([0,T],R^n)$. If instead of (5.29), we impose

$$\lim_{t\to 0+} (Vx)(t) = c \in R^n, \tag{5.30}$$

we will not achieve greater generality because one could substitute to V the operator $\tilde{V} = V - c$.

Theorem 5.3 *Consider equation (5.13) under the initial condition (5.6) and assume that hypotheses $(H1)$, $(H2)$, and $(H3)$ hold.*

Then, equation (5.13) with the initial condition (5.6) has a solution $x(t) \in L^p([0,a],R^n)$, $0 < a \leq T$, such that $x(t) + (Vx)(t)$ is absolutely continuous on $[0,a]$ and satisfies a.e. equation (5.13).

Proof. First let us note that the problem of existence is equivalent in our case, with the problem of existence of the functional equation

$$x(t) + (Vx)(t) = x_0 + \int_0^t (Wx)(s)\,ds, \tag{5.31}$$

in the space L^p. Indeed, since $L^p \subset L^1$ on $[0,T]$, we can integrate both sides of equation (5.13) from 0 to $t \leq T$ as long as the solution exists, thus obtaining

(5.31). Since the right-hand side of (5.31) is an absolutely continuous function on the interval of existence of $x(t)$, we can differentiate almost everywhere (5.31) obtaining (5.13). If in (5.31) we let $t \to 0^+$, and rely on $(H2)$, we obtain $\lim x(t) = x_0$, as $t \to 0^+$.

Obviously, (5.31), on the interval of existence of solution (which has to be proven), is equivalent to the functional integral equation

$$x(t) = x_0 + \int_0^t (Wx)(s)\, ds - (Vx)(t), \tag{5.32}$$

which is of the typical form for applying a fixed-point theorem, namely

$$x(t) = (\nu x)(t), \tag{5.33}$$

and when ν is compact on $L^p([0,T],R^n)$, while W is continuous and takes bounded sets into bounded sets of L^p, the application is immediate. We will again rely on Theorem 3.4.1 in Corduneanu [135], because it also provides regularized approximate solutions (absolutely continuous functions). Based on $(H2)$ and $(H3)$, we easily obtain the compactness of ν on $L^p([0,T],R^n)$.

From Theorem 3.4.1, we obtain the existence of a solution for (5.13) on $[0,a] \subset [0,T]$. Moreover, the ϵ-approximate solutions satisfy the equation $\epsilon \dot{x}_\epsilon(t) = -x_\epsilon(t) + (\nu x_\epsilon)(t)$, *a.e.* on $[0,a]$.

This ends the proof of Theorem 5.3.

The validity of the conclusion in Theorem 5.3 can be substantially improved, leading to a global existence theorem, on the interval $[0,T)$, $0 < T \leq \infty$.

The following global existence result can be proved when the right-hand side of (5.13) contains only linear/affine operators. In other words, we will deal with the equations of the form

$$\dot{x}(t) = (Lx)(t) + (fx)(t), \ t \in [0,T), \tag{5.34}$$

where L is a causal, continuous linear operator on $L^p_{\text{loc}}([0,T),R^n)$, and $f : L^p_{\text{loc}}([0,T),R^n) \to L^p_{\text{loc}}([0,T),R^n)$ is an affine operator. The linear counterpart of equation (5.34) is

$$\dot{x}(t) = (Lx)(t) + f(t), \ t \in [0,T). \tag{5.35}$$

It has been investigated by Mahdavi [347], and Mahdavi and Li [352] in the quasilinear case (5.34). The details of the proof are given in the aforementioned works.

In concluding this section, which contains results related to first-order neutral functional differential equations, we consider a special case of operators that appear in equation (5.13). Namely, we consider the functional integro-differential equation

$$\frac{d}{dt}\left[x(t)+\int_0^t K(t,s)x(s)\,ds\right]=(Wx)(t). \tag{5.36}$$

In order to satisfy the requirements in Theorem 5.3, we make the following assumptions on the operator:

$$(\nu x)(t)=\int_0^t K(t,s)x(s)\,ds,\ t\in[0,T), \tag{5.37}$$

considering $p=2$ (p arbitrary has similar treatment), it is known that the condition of continuity of the operator ν is given by

$$\int_0^{\bar{t}} dt\int_0^t |K(t,s)|^2\,ds<\infty, \tag{5.38}$$

for any $t\le\bar{t}<T$, and $|K(t,s)|$ denotes any $n\times n$ matrix norm. Moreover, condition (5.38) assures the compactness of the operator ν, because (5.29) is verified. This leads us to the following.

Proposition 5.1 *Consider equation (5.36) under the following conditions:*

1) *The kernel $K(t,s)$ is measurable for $0\le s\le t<T$, and verifies condition (5.38).*
2) *$W: L^2_{\text{loc}}([0,T),R^n)\to L^2_{\text{loc}}([0,T),R^n)$ is a linear, causal, continuous operator.*

Then, there exists a unique solution $x(t)$ of equation (5.36) satisfying (5.6), such that

$$x(t)+\int_0^t K(t,s)x(s)\,ds,$$

is locally absolutely continuous on $[0,T)$.

Let us provide some examples for possible choices of the operator W occurring in (5.36):

1) $(Wx)(t)=\int_0^t K_1(t-s)x(s)\,ds+f(t)$, where $|K_1|\in L^1_{\text{loc}}([0,T),\mathcal{L}(R^n,R^n))$ and $f\in L^2_{\text{loc}}([0,T),R^n)$.
2) $(Wx)(t)=\sum_{k=1}^m A_k x(t-t_k)+f(t)$, where $0<t_1<t_2<\cdots<t_m<T$ are given numbers, A_k, $k=1,2,\ldots,m$, are $n\times n$ matrices and $f\in L^2_{\text{loc}}([0,T),R^n)$.

In this case, one must assign the initial values not only for $t=0$ but on the whole interval $[-t_m,0)$. In order to assure the continuity of W on $L^2_{\rm loc}([0,T),R^n)$, we will assign $x(t)=x^0(t)\in L^2([-t_m,0),R^n)$, while preserving the pointwise initial condition $x(0)=x_0\in R^n$.

Remark 5.7 *Since equation (5.36) with the initial condition (5.6) is equivalent to the functional integral equation*

$$x(t)+\int_0^t K(t,s)x(s)\,ds = x_0+\int_0^t (Wx)(s)\,ds, \tag{5.39}$$

we can adopt another approach to discuss the solution of the problem. This solution is based on the existence of the associated kernel to $K(t,s)$, say $\tilde{K}(t,s)$, which will also satisfy a condition of the form (5.38). With the help of $\tilde{K}(t,s)$, we can transform equation (5.39) into an equivalent equation

$$x(t)=x_0+\int_0^t (Wx)(s)\,ds+\int_0^t \tilde{K}(t,s)x_0\,ds + \int_0^t\int_0^s \tilde{K}(t,s)(Wx)(u)\,du\,ds.$$

This rather complicated functional integral equation can be investigated as soon as the operator W is assigned. For instance, one can construct a solution by means of the method of successive approximations and by using a Lipschitz-type continuity condition that follows from the continuity of the operator W. See Corduneanu [135, 149] for related results.

We encourage the readers to conduct this procedure, keeping in mind that the existence of solution has already been proven. For the stochastic neutral equations, see Okonkwo et al. [421,422], and the references therein.

5.3 SOME AUXILIARY RESULTS

We begin this section by considering several types of neutral functional differential equations involving abstract Volterra (causal) operators.

A formula that relates abstract operators and classical ones, which we will use later, has the form

$$\int_0^t (Lx)(s)\,ds = \int_0^t k(t,s)x(s)\,ds,\ t\in[0,T), \tag{5.40}$$

where L stands for a linear operator acting on a function space $E([0,T),R^n)$, while $k\in\mathcal{L}(R^n,R^n)$ for $t\in[0,T)$, $s\in[0,t]$. A formula like (5.40) will help us reduce abstract operators to classical/integral ones.

We would like to stress the validity of formula (5.40) and its applications. More information can be found in classical reference sources such as Dunford and Schwartz [204] and Kantorovich and Akilov [274]. Our aim is to provide some illustrations for its validity.

As underlying space, let us use the Lebesgue space $L^2_{\text{loc}}([0,T),R^n)$, $T \leq \infty$, and assume $L : L^2_{\text{loc}} \to L^2_{\text{loc}}$ is a linear, continuous and causal map. We fix $t \in (0,T)$, and consider the linear functional on $L^2[0,t]$,

$$x \to \int_0^t (Lx)(s)\,ds, \tag{5.41}$$

with values in R^n. Due to our assumption of continuity for L, we know from Functional Analysis that there exists an element $k(t,s) \in L^2([0,t],R^n)$, for each $t \in [0,T)$, such that

$$\int_0^t (Lx)(s)\,ds = \int_0^t k(t,s)x(s)\,ds, \tag{5.42}$$

where $k(t,s) \in L^2([0,t],R^n)$, as a function of $s \in [0,t]$. One can prove that $k(t,s)$ is measurable, and if we want, for instance, the operator

$$x \to \int_0^t k(t,s)x(s)\,ds,\ x \in L^2_{\text{loc}}([0,T),R^n),$$

to take values in $L^2([0,T),R^n)$, there results the following condition on $k(t,s)$: For each $\bar{t} \in (0,T)$, one must have

$$\int_0^{\bar{t}} dt \int_0^t |k(t,s)|^2\,ds < +\infty. \tag{5.43}$$

We have met such a condition in the preceding section, in formula (5.38).

Similar conditions to (5.43) can be obtained when the pair $(L^2_{\text{loc}}, L^2_{\text{loc}})$ is substituted by another pair (E_1,E_2) of function spaces on $[0,T)$, whose elements take values in $E_2([0,T),R^n)$.

We will indicate such pairs of spaces when they appear in our considerations.

Another topic, which we would like to discuss, is related to the concept of resolvent kernel, attached to a given kernel appearing in the integral operators of the form

$$(Kx)(t) = \int_0^t k(t,s)x(s)\,ds, \tag{5.44}$$

or even

$$(Kx)(t) = \int_0^T k(t,s)x(s)\,ds, \ T < \infty, \tag{5.45}$$

which correspond to Volterra, respectively, Fredholm types.

Due to the kind of applications we have in mind, we are concerned only with Volterra-type operators (5.44), and constructing the resolvent kernel. The resolvent kernel appears naturally when we use the method of iteration to solve the linear Volterra equation

$$x(t) = f(t) + \int_0^t k(t,s)x(s)\,ds, \ t \in [0,T), \tag{5.46}$$

allowing the case $T = \infty$, which will be encountered when we want to study the asymptotic behavior (as $t \to \infty$) of the solution to our equations.

The iteration procedure, when applied to equation (5.46), leads (formally) to the function sequence

$$x_0(t) = f(t), \ x_m(t) = f(t) + \int_0^t k(t,s)\,x_{m-1}(s)\,ds, \ m \geq 1,$$

which implies

$$x_m(t) = f(t) + \int_0^t \left\{ \sum_{j=1}^{m} k_j(t,s) \right\} f(s)\,ds, \tag{5.47}$$

where we denoted

$$k_1(t,s) = k(t,s), \ k_m(t,s) = \int_s^t k_{m-1}(t,u)\,k(u,s)\,du, \ m \geq 2. \tag{5.48}$$

The functions $k_m(t,s)$, $m \geq 2$ are defined under the conditions stated later There are two cases that interest us: when we can prove that the series $\sum_{m=1}^{\infty} k_m(t,s)$ converges either in $L^{\infty}_{loc}(\Delta, \mathcal{L}(R^n,R^n))$, or, in $L^2(\Delta, \mathcal{L}(R^n,R^n))$, *a.e.*, to the *resolvent kernel*

$$\tilde{k}(t,s) = \sum_{m=1}^{\infty} k_m(t,s), \ (t,s) \in \Delta, \tag{5.49}$$

with

$$\Delta = \{(t,s); 0 \leq s \leq t < T\}. \tag{5.50}$$

From (5.44), taking into account (5.47) and (5.48), we obtain the so-called resolvent formula for equation (5.46):

$$x(t) = f(t) + \int_0^t \tilde{k}(t,s)f(s)\,ds,$$

while for $\tilde{k}(t,s)$, one obtains from (5.48) and (5.49), the equation

$$\tilde{k}(t,s) = k(t,s) + \int_s^t k(t,u)\tilde{k}(u,s)\,du,$$

which is known as the equation of the resolvent kernel, attached to $k(t,s)$.

Other L^p-spaces are acceptable for defining the resolvent kernel associated to the kernel $k(t,s)$ in (5.46), but we will limit our considerations to the cases L^∞ and L^2.

Let us return to the series (5.49) and show its convergence in the case of L^∞-functions. In this case, we can find $M > 0$, such that

$$\operatorname*{ess\text{-}sup}_{\Delta} |k(t,s)| = M < \infty, \tag{5.51}$$

and we consequently have *a.e.*,

$$|k_2(t,s)| \leq \frac{M(t-s)^2}{2!}, \ldots, |k_m(t,s)| \leq \frac{M(t-s)^m}{m!}, \ldots \tag{5.52}$$

The estimates (5.52) prove the convergence of the series (5.49) in $L^\infty_{\text{loc}}(\Delta, \mathcal{L}(R^n, R^n))$, which guarantees the existence of the resolvent kernel $\tilde{k}(t,s) \in L^\infty_{\text{loc}}(\Delta, \mathcal{L}(R^n, R^n))$.

Concerning the case when $k(t,s)$ belongs to $L^2_{\text{loc}}(\Delta, \mathcal{L}(R^n, R^n))$, and thus (5.43) is valid, the convergence of the series (5.49) requires a more difficult argument.

We send the readers to the book Tricomi [520], where a beautiful classical construction is presented. We point out that condition (5.43) is always understood on the set $\{0 \leq s \leq t \leq \bar{t} < T\}$.

Based on Functional Analysis, the existence of the resolvent kernel is treated in a more direct fashion when dealing with the concept of adjoint operator of a linear operator, on a Hilbert space. More information on this topic is presented in Gohberg and Goldberg [224].

The last topic we will discuss in this section concerns an application of the Tychonoff fixed-point theorem in locally convex spaces. This theorem is given in Corduneanu [120]. This book also contains the application, which we provide in the following text. The result is a global existence theorem for the functional equation

$$x(t) = (Vx)(t), \ t \in [0,T). \tag{5.53}$$

Theorem 5.4 *Assume V is an operator from $L^2_{\text{loc}}([0,T),R^n)$ into itself, and satisfies the following conditions:*

1) *V is a compact operator on the space $L^2_{\text{loc}}([0,T),R^n)$.*
2) *There exist two functions $A,B : [0,T) \to R_+$, with A continuous and positive, and B locally integrable, such that $x \in L^2_{\text{loc}}([0,T),R^n)$ and*

$$\int_0^t |x(s)|^2 \, ds \leq A(t), \ t \in [0,T), \tag{5.54}$$

implies

$$|(Vx)(t)|^2 \leq B(t), a.e., \text{ on } [0,T), \tag{5.55}$$

and moreover,

$$\int_0^t B(s) \, ds \leq A(t), \ t \in [0,T). \tag{5.56}$$

Then, there exists a solution of (5.53) for which (5.54) holds true.

Proof. We choose a closed convex set S in $L^2_{\text{loc}}([0,T),R^n)$, defined by

$$S = \left\{ x \in L^2_{\text{loc}}([0,T),R^n); \ \int_0^t |x(s)|^2 \, ds \leq A(t), t \in [0,T) \right\}. \tag{5.57}$$

Taking into account (5.54)–(5.56), we obtain $VS \subset S$. Since V is a compact operator on L^2_{loc} and takes the set S into itself, Tychonoff's theorem applies and yields the existence of a solution.

Finally, we will mention another application of Tychonoff's theorem concerning the functional differential equation

$$\dot{x}(t) = (Vx)(t), \ t \in [0,T), \tag{5.58}$$

under the usual initial condition $x(0) = x_0 \in R^n$.

By integrating both sides of (5.58), one obtains an equation similar to (5.53), namely

$$x(t) = x_0 + \int_0^t (Vx)(s)\, ds, \ t \in [0,T). \tag{5.59}$$

A slight reformulation of the case discussed before leads to the following result.

Assume that the operator V in (5.58) is continuous from $C([0,T),R^n)$ into $L_{\text{loc}}([0,T),R^n)$. Also, assume there exists two functions $A(t)$ and $B(t)$, $t \in [0,T)$, with A continuous and positive and $B(t)$ locally integrable and non-negative, such that $|x(t)| \leq A(t)$, $t \in [0,T)$, implies $|(Vx)(t)| \leq B(t)$, *a.e.*, on $[0,T)$. Moreover, consider the inequality

$$A(t) - A(0) \geq \int_0^t B(s)\, ds, \ t \in [0,T).$$

Then, there exists a solution of our problem $x \in AC_{\text{loc}}([0,T),R^n)$, such that $|x(t)| \leq A(t)$, $t \in [0,T)$, provided $|x_0| < A(0)$.

This ends the proof of the Theorem 5.4.

Readers are invited to apply the result, concerning equation (5.58), to the case of the linear equation with infinite delay

$$\dot{x}(t) = \int_{-\infty}^t k(t,s)x(s)\, ds + f(t), \ t \in R_+,$$

with the functional initial condition $x(t) = \phi(t)$, $t \in R_-$.

It is useful to notice that the equation and the initial condition imply dealing with the equation

$$\dot{x}(t) = \int_0^t k(t,s)x(s)\, ds + \int_{-\infty}^0 k(t,s)\phi(s)\, ds + f(t), t \in R_+.$$

5.4 A CASE STUDY, I

In this section, we will investigate the linear functional differential equation with infinite delay

$$\dot{x}(t) = (Ax)(t) + f(t), \ t \in R_+, \tag{5.60}$$

where the operator A is given formally by

$$(Ax)(t) = \sum_{j=0}^{\infty} A_j x(t - t_j) + \int_0^t B(t-s)x(s)\, ds, \tag{5.61}$$

with $A_j \in \mathcal{L}(R^n, R^n), j \geq 0$, and $B : R_+ \to \mathcal{L}(R^n, R^n), f \in L^1_{\text{loc}}(R_+, R^n)$. We will assume in the sequel that

$$\sum_{j=0}^{\infty} |A_j| < +\infty, \quad |B(t)| \in L^1(R_+, \mathcal{L}(R^n, R^n)). \tag{5.62}$$

The numbers t_j, $j \geq 1$, are positive ($t_j > 0$), while $t_0 = 0$. These assumptions allow us to state that the operator A is a casual operator.

Based on conditions (5.62), one can derive the property of the operator A to leave invariant each space $L^p(R_+, \mathcal{L}(R^n, R^n))$, provided we agree that

$$x(t) = h(t) \in L^p(R_-, \mathcal{L}(R^n, R^n)). \tag{5.63}$$

Since the sequence $\{t_j; j \geq 1\} \subset R_+$ can be unbounded, it becomes possible that equation (5.60) is with infinite delay.

The Cauchy problem attached to equation (5.60) will consist of equation itself, with the initial functional condition (5.63) and the pointwise initial condition

$$x(0) = x_0 \in R^n. \tag{5.64}$$

It is possible, by means of adequate transformations, to rewrite this Cauchy problem in the classical way (equation and pointwise initial condition). If we proceed formally and denote

$$(\tilde{B}h)(t) = \sum_{j=0}^{\infty} A_j h(t - t_j), \ t \in R_+, \tag{5.65}$$

then integrating both sides of (5.60) from 0 to $t > 0$, one obtains

$$x(t) = \int_0^t (Ax)(s)\,ds + \int_0^t [f(s) + (\tilde{B}h)(s)]\,ds, \tag{5.66}$$

and taking into account

$$\int_0^t (Ax)(s)\,ds = \int_0^t G(t-s)x(s)\,ds,$$

where

$$G(t) = \sum_{j=0}^{\infty} A_j H(t - t_j) + \int_0^t B(u)\,du, \tag{5.67}$$

with $H(t)$ denoting the Heaviside function, $H(t) = 1$ for $t \geq 0$ and $H(t) = 0$ for $t < 0$. If one denotes

$$g(t) = \int_0^t [f(s) + (\tilde{B}h)(s)]\, ds, \ t \in R_+, \tag{5.68}$$

then (5.66) can be written as an integral equation

$$x(t) = \int_0^t G(t-s)x(s)\, ds + g(t), \ t \in R_+. \tag{5.69}$$

The motivation of all the steps taken in the earlier procedure can be achieved as shown in Corduneanu [120]. Everything can be based on the following result, provided in the aforementioned reference. Denoting

$$\mathcal{A}(s) = \sum_{j=0}^{\infty} A_j e^{-t_j s} + \int_0^{\infty} B(t) e^{-ts}\, dt, \ Re(s) > 0, \tag{5.70}$$

we can state the result.

Theorem 5.5 *Consider the fundamental matrix $X(t)$, defined by the equations*

$$\dot{X}(t) = (AX)(t), \ t > 0; \ X(0) = I; \ X(t) = O, \ t < 0,$$

under assumption (5.62).

Then, the following conditions are equivalent:

(1) *$det[sI - \mathcal{A}(s)] \neq 0$ for $Re(s) \geq 0$;*
(2) *$|X(t)| \in L^1(R_+, R)$;*
(3) *$|X(t)| \in L^p(R_+, R)$, for any p satisfying $1 \leq p \leq \infty$.*

The most important consequence of Theorem 5.5 is the representation formula for the unique solution of the Cauchy problem (5.60), (5.63), (5.64),

$$x(t) = X(t)x_0 + \int_0^t X(t-s)f(s)\, ds + \int_0^t X(t-s)(\tilde{B}h)(s)\, ds. \tag{5.71}$$

Remark 5.8 *It is interesting to note that for the homogeneous equation $\dot{x}(t) = (Ax)(t)$, where the operator A is defined by (5.61), the solution $x = \theta$ is asymptotically stable. This fact is the consequence of condition (2) of Theorem 5.5, which also implies, according to $\dot{X}(t) = (AX)(t)$, $|\dot{X}(t)| \in L^1(R_+, R^n)$.*

Indeed, the integrability of $\dot{X}(t)$ on R_+ implies the existence of $\lim X(t)$, as $t \to \infty$. But on behalf of condition (2) of Theorem 5.5, this limit cannot be different from $O \in \mathcal{L}(R^n, R^n)$; $\lim |X(t)| = O$, as $t \to +\infty$. We can extend this result to more general equations (neutral, nonlinear).

Let us now consider the *neutral*-type linear auxiliary system, related to (5.60), with the operator A defined by (5.61):

$$\frac{d}{dt}\left[x(t) - \int_0^t C(t-s)x(s)\,ds\right] = (Ax)(t) + f(t), \tag{5.72}$$

on the positive half-axis R_+.

We shall assume that $f \in L^1_{\text{loc}}(R_+, R^n)$, while $C : R_+ \to \mathcal{L}(R^n, R^n)$ is differentiable. These assumptions allow us to rewrite, after differentiating the bracket on the left-hand side of (5.72), the system (5.72) in the equivalent form

$$\dot{x}(t) = (\bar{A}x)(t) + f(t), \tag{5.73}$$

where

$$(\bar{A}x)(t) = C(0)x(t) + (Ax)(t) + \int_0^t [B(t-s) + \dot{C}(t-s)]x(s)\,ds, \tag{5.74}$$

which is similar to the operator A given (5.61).

In order to obtain $\bar{A}$ satisfying condition similar to (5.62), we need to impose the condition $|\dot{C}(t)| \in L^1(R_+, R)$.

In order to formulate another auxiliary result, we shall consider the matrix-valued function (similar to (5.70)),

$$\bar{\mathcal{A}}(s) = [C(0) + A_0] + \sum_{j=1}^{\infty} A_j e^{-t_j s} + \int_0^{\infty} [B(t) + \dot{C}(t)]\, e^{-ts}\,dt, \tag{5.75}$$

defined for $Re(s) \geq 0$ (called the *symbol* attached to the operator $\bar{A}$).

Proposition 5.2 *Consider the system (5.72), under the following assumptions:*

1) *The matrix operator A satisfies (5.62).*
2) *$C(t)$ is locally absolutely continuous on R_+, and*

$$|\dot{C}(t)| \in L^1(R_+, R). \tag{5.76}$$

3)

$$det[sI - \bar{\mathcal{A}}(s)] \neq 0 \text{ for } Re(s) \geq 0. \tag{5.77}$$

4)

$$f \in L^p(R_+, R^n), \ p \geq 1. \tag{5.78}$$

5)

$$h \in L^p(R_-, R^n), \ p \geq 1. \tag{5.79}$$

Then, the unique solution $x(t)$ of (5.72), with initial conditions (5.63) and (5.64), is given by the formula

$$x(t) = \bar{X}(t)x_0 + (Yh)(t) + \int_0^t \bar{X}(t-s)f(s)\,ds, \tag{5.80}$$

with $\bar{X}(t)$ and $(Yh)(t)$ defined by

$$\dot{\bar{X}}(t) = (A\bar{X})(t), \ t > 0, \ \bar{X}(0) = I, \ \bar{X}(t) = O, \ t < 0, \tag{5.81}$$

$$(Yh)(t) = \sum_{j=1}^{\infty} \int_{-t_j}^{0} \bar{X}(t - t_j - s)A_j h(s)\,ds, \ t \in R_+, \tag{5.82}$$

and

$$x(t) \in L^p(R_+, R^n), \ p \geq 1. \tag{5.83}$$

Proposition 5.2 is a variant of the statement of Theorem 5.5, replacing the operator A by the operator $\bar{A}$. Also, instead of $L^1(R_+, R^n)$, it appears the space $L^p(R_+, R^n)$, $p \geq 1$, a result that follows from the inequalities

$$|f * g|_{L^p} \leq |f|_{L^1} \ |g|_{L^p}, \text{when } f \in L^1 \text{ and } g \in L^p, p \geq 1.$$

One has also to take into account condition (5.62) for the operator A when estimating the operator $(Yh)(t)$, on R_-. Let us notice that (5.83) implies

$$\lim |x(t)| = 0, \ \text{as } t \to \infty, \tag{5.84}$$

that is, the asymptotic stability of the system.

We shall investigate in what follows the absolute stability of the system

$$\frac{d}{dt}\left[x(t) - \int_0^t C(t-s)x(s)\,ds\right] = (Ax)(t) + (b\xi)(t), \tag{5.85}$$

$$\xi = \phi(\sigma), \sigma = \langle a, x \rangle,$$

where the operator A is formally given by (5.61), with the conditions in (5.62).

$C(t)$ is like in condition 2) of Proposition 5.2.

The operator b is given by

$$(b\xi)(t) = \sum_{j=0}^{\infty} b_j \xi(t-t_j) + \int_0^{\infty} C(t-s)\xi(s)\,ds, \tag{5.86}$$

and under the conditions

$$t_j \geq 0,\, j \geq 0,\, \sum_{j=0}^{\infty} |b_j| < \infty,\, \int_0^{\infty} |C(t)|\,dt < \infty, \tag{5.87}$$

is acting from $BC(R_+, R)$ into itself, or from $L^p(R_+, R)$ into itself, $p \geq 1$.

The function $\phi : R \to R$ is assumed continuous and bounded, such that $\sigma\phi(\sigma) > 0$ for $\sigma \neq 0$, which implies $\phi(0) = 0$.

The property of absolute stability of the zero solution of the system (5.85) has the following meaning: The solution $x = \theta$ is asymptotically stable regardless of the choice of the continuous function $\phi(\sigma)$, $\sigma \in R$, with $\sigma\,\phi(\sigma) > 0$ for $\sigma \neq 0$, or with σ satisfying another similar condition (for instance, $\sigma\,\phi(\sigma) \leq L\sigma^2$, with $L > 0$ given).

In order to state the result of absolute stability of the system (5.85), we will reduce the problem to the investigation of a nonlinear integral equation and use Theorem 2.2 in the book of Corduneanu [120]. The statement used is as follows:

Theorem 5.6 *Consider the scalar integral equation*

$$\sigma(t) = h_0(t) + \int_0^t k(t-s)\phi(\sigma(s))\,ds,\ t \in R_+, \tag{5.88}$$

under the following conditions:

1) $h_0, h_0' \in L^1(R_+, R)$.
2) $k, k' \in L^1(R_+, R)$.
3) $\phi : R \to R$ *is continuous, bounded and such that* $\sigma\,\phi(\sigma) > 0$ *for* $\sigma \neq 0$.
4) *There exists* $q \geq 0$, *such that*

$$Re[(1+isq)\tilde{k}(s)] \leq 0,\ s \in R, \tag{5.89}$$

where $\tilde{k}(s)$ *denotes the Fourier transform of* $k(t)$,

$$\tilde{k}(s) = \int_0^{\infty} k(t)e^{-its}\,dt. \tag{5.90}$$

Then, there exists a solution $\sigma(t) \in C_0(R_+,R)$, of equation (5.88) and any solution $\sigma(t)$ of (5.88), continuous on R_+, will belong to $C_0(R_+,R)$.

In order to apply Theorem 5.6 to the neutral system (5.85), we shall use the variation of parameters formula (5.80), for the case $f(t) = b(\phi(\sigma(t)))$, with $\sigma(t) = < a, x(t) >$, $a \in R^n$ being fixed, and $x(t)$ the solution of (5.85). One obtains an integral equation, preliminary to (5.88), with the following data:

$$\begin{aligned}\sigma(t) = < a, x(t) > = < a, \bar{X}(t)x_0 + (Yh)(t) > \\ + < a, \int_0^t \bar{X}(t-s)(b\xi)(s)\,ds >, \ \xi = \phi(\sigma).\end{aligned} \tag{5.91}$$

In order to bring (5.91) to the form (5.88), we need to perform some transformations involving the convolution products that appeared in (5.91), such that ξ will be a factor, outside brackets.

One obtains, first,

$$(Yh)(t) = \sum_{j=1}^{\infty} \int_{-t_j}^{0} \bar{X}(t-t_j-s)A_j h(s)\,ds, \ t \in R_+. \tag{5.92}$$

Then, from (5.85), one obtains

$$x(t) = \bar{X}(t)x_0 + (Yh)(t) + \int_0^t \bar{X}(t-s)\,(b\xi)(s)\,ds, \tag{5.93}$$

and, taking into account

$$\begin{aligned}\int_0^t \bar{X}(t-s)b_j\xi(s-t_j)\,ds &= \int_{-t_j}^{t-t_j} \bar{X}(t-t_j-s)\,b_j(s)\,ds \\ &= \int_0^t \bar{X}(t-t_j-s)\,b_j\xi(s)\,ds + \int_{-t_j}^{0} \bar{X}(t-t_j-s)\,b_j\xi(s)\,ds,\end{aligned}$$

and

$$\begin{aligned}\int_0^t \bar{X}(t-u)\int_0^u C(u-s)\xi(s)dsdu &= \int_0^t \left\{\int_s^t \bar{X}(t-u)C(u-s)du\right\}\xi(s)\,ds \\ &= \int_0^t \left\{\int_0^{t-s} \bar{X}(t-s-u)C(u)\,du\right\}\xi(s)\,ds.\end{aligned}$$

Let us denote now

$$(y\xi)(x) = \sum_{j=1}^{\infty} \int_{-t_j}^{0} \bar{X}(t-t_j-s)\,b_j\xi(s)\,ds, \ t \in R_+, \tag{5.94}$$

and

$$\tilde{X}(t) = \sum_{j=0}^{\infty} \bar{X}(t - t_j)\, b_j + \int_0^t \bar{X}(t-u)\, C(u)\, du, \ t \in R_+. \tag{5.95}$$

Now, formula (5.93) can be rewritten as follows:

$$x(t) = \bar{X}(t)x_0 + (Yh)(t) + (y\xi)(t) + \int_0^t \tilde{X}(t-s)\xi(s)\, ds, \ t \in R_+. \tag{5.96}$$

We notice that $\bar{X} \in L^1$ implies $\tilde{X} \in L^1$. This statement is a direct consequence of the properties of the convolution product.

From (5.96), taking into account $\xi(t) = \phi(\sigma(t))$ and $\sigma = < a, x >$, we obtain

$$\begin{aligned} \sigma(t) &= < a, \bar{X}(t)x_0 + (Yh)(t) + (y\xi)(t) > \\ &+ \int_0^t < a, \tilde{X}(t-s) > \phi(\sigma(s))\, ds, \ t \in R_+, \end{aligned} \tag{5.97}$$

which is an integral equation of the form (5.88). In order to reach the conclusion of Theorem 5.6, $\sigma(t) \in C_0(R_+, R)$, we need to check the validity of hypotheses 1) and 2) because 3) and 4) represent conditions imposed on the data.

Concerning $h_0(t)$, which according to (5.97) is given by

$$h_0(t) = \bar{X}(t)x_0 + (Yh)(t) + (y\xi)(t), \ t \in R_+,$$

we notice that, the conditions already proven, $|\bar{X}(t)| \in L^1(R_+, R)$ implying $|\dot{\bar{X}}(t)| \in L^1(R_+, R)$, lead to $h_0(t) \in C_0(R_+, R)$. We have clarified this fact in Remark 5.8. Similar considerations lead to the conclusion that both $(Yh)(t)$ and $(y\xi)(t)$ belong to $C_0(R_+, R^n)$.

Let us now consider the kernel of (5.97)

$$k(t) = < a, \tilde{X}(t) >, \tag{5.98}$$

where $\tilde{X}(t)$ is given by (5.95), and prove condition 2) of Theorem 5.6, which means that both k and k' must be in $L^1(R_+, R)$. The condition $k \in L^1(R_+, R)$ is easily obtained from (5.95), taking into account $\{b_j; j \geq 1\} \in \ell^1(N, R^n)$, $C \in L^1(R_+, R)$ and the integrability of $\bar{X}(t)$ on R_+. The same conclusion is reached for $k'(t)$ if we differentiate $k(t)$ and take into account that $\dot{\bar{X}}(t) \in L^1(R_+, \mathcal{L}(R^n, R^n))$, while $\dot{C}(t) \in L^1(R_+, R)$.

With conditions 1) and 2) of theorem 5.6 verified, we write the frequency stability condition (5.89) in the form

$$Re[(1+isq) < a, \tilde{\tilde{X}}(is) >] \leq 0, \ s \in R, \tag{5.99}$$

where $\tilde{\tilde{X}}(is)$ is the Fourier transform of $\tilde{X}(t)$. Condition 3) in Theorem 5.6 is imposed on the function $\phi(\sigma)$ and is preserved for equation (5.97), which allows us to formulate the conclusion; $\sigma(t) \in C_0(R_+, R)$. This implies (5.84), $|x(t)| \to 0 \ as \ t \to \infty$.

To summarize the earlier discussion regarding the *absolute stability* of the neutral system (5.85), we emphasize the following conditions:

a) The operator A is given by (5.61), with $A_j, j \geq 0$, and $B : R_+ \to \mathcal{L}(R^n, R^n)$ satisfying (5.62).
b) $C(t), \dot{C}(t) \in L^1(R_+, \mathcal{L}(R^n, R^n)$.
c) The operator $b : R_+ \to R^n$ is defined by (5.86), under conditions (5.87).
d) $\phi : R \to R$ is continuous and bounded, such that $\phi(\sigma)\,\sigma > 0$ for $\sigma \neq 0$.
e) There exists $q \geq 0$, such that (5.99) is satisfied, where $\tilde{\tilde{X}}$ is the Fourier transform of $\tilde{X}$, as defined by (5.95).

The reduction of the absolute stability problem to the investigation of the scalar integral equation (5.88) constituted the key in the proof.

5.5 ANOTHER CASE STUDY, II

In this section, we are concerned with the neutral functional equation of the form

$$\dot{x}(t) = (Lx)(t) + (Nx)(t), \ t \in [0, T), \tag{5.100}$$

where L denotes a linear, causal, and continuous operator on $L^2_{loc}([0, T), R^n)$, while N is, in general, a nonlinear operator acting on $L^2_{loc}([0, T), R^n)$. We will obtain the existence of a solution, in $L^2_{loc}([0, T), R^n)$, by means of special case of Theorem 5.4.

First, let us discuss some preliminaries, which are necessary to further consider equation (5.100).

As seen in Chapter 2, the solution of the linear equation

$$\dot{x}(t) = (Lx)(t) + f(t), \ t \in [0, T), \tag{5.101}$$

for $L : L^2_{loc}([0,T),R^n) \to L^2_{loc}([0,T),R^n)$ a linear, causal, and continuous operator, with $f \in L^2_{loc}([0,T),R^n)$, can be represented by the formula

$$x(t) = X(t,0)x_0 + \int_0^t X(t,s)f(s)\,ds, \ t \in [0,T), \tag{5.102}$$

where $x_0 \in R^n$ is the initial data. The Cauchy matrix $X(t,s)$, $0 \le s \le t < T$, is given by the formula

$$X(t,s) = I + \int_s^t \tilde{k}(t,u)\,du, \tag{5.103}$$

where $\tilde{k}(t,s)$ is the resolvent kernel associated to the kernel $k(t,s)$, from the relationship

$$\int_0^t (Lx)(s)\,ds = \int_0^t k(t,s)x(s)\,ds, \tag{5.104}$$

valid for $x \in L^2_{loc}([0,T),R^n)$. These considerations are detailed in Section 5.3.

Cauchy's matrix $X(t,s)$ has several properties derived from the formula (5.103), such as

$$\int_0^{\bar{t}} dt \int_0^t |\tilde{k}(t,s)|^2\,ds < \infty, \tag{5.105}$$

for any $\bar{t} \in (0,T)$.

On behalf of formula (5.102), we can transform the nonlinear neutral equation (5.100), with the usual initial condition $x(0) = x_0 \in R^n$, into the functional integral equation

$$x(t) = X(t,0)x_0 + \int_0^t X(t,s)(Nx)(s)\,ds, \ t \in [0,T). \tag{5.106}$$

Equation (5.106) can be interpreted as a result of perturbing the linear equation (5.101). Its neutral character results from the fact that the nonlinear operator N may involve the derivative $\dot{x}(t)$. The case of Niemytzki's type operator $(Nx)(t) = f(t,x(t))$, with f a function defined on $[0,T) \times R^n$, leads to an integral equation of Hammerstein type. This topic has been widely discussed in the mathematical literature.

In order to obtain an existence theorem for equation (5.106), which implicitly solves the Cauchy problem for equation (5.100), we shall use a slightly varied form of, Theorem 5.4, namely:

Consider the equation $x(t) = (Vx)(t), t \in [0,T)$, under the following assumptions:

1) V is causal, continuous and compact on $L^2_{\text{loc}}([0,T),R^n)$.
2) There exists a continuous positive function $A : [0,T) \to (0,\infty)$ with the property

$$\int_0^t |x(s)|^2 ds \leq A(t),\ t \in [0,T), \tag{5.107}$$

implying

$$\int_0^t |(Vx)(s)|^2 ds \leq A(t),\ t \in [0,T). \tag{5.108}$$

Then, there exists a solution $x(t) \in L^2_{\text{loc}}([0,T),R^n)$, of the equation $x(t) = (Vx)(t)$.

Returning to equation (5.106), we must choose in the previous theorem

$$(Vx)(t) = X(t,0)x_0 + \int_0^t X(t,s)(Nx)(s)\,ds,\ t \in [0,T), \tag{5.109}$$

and assume the following growth condition for N:

$$\int_0^t |(Nx)(s)|^2 ds \leq \lambda(t) \int_0^t |x(s)|^2 ds + \mu(t), \tag{5.110}$$

where $\lambda(t)$ and $\mu(t)$ are positive and nondecreasing on $[0,T)$. This is a growth condition on the operator N, and it tells us that N has a sort of sublinear growth in $L^2_{\text{loc}}([0,T),R^n)$.

We rely on conditions (5.105) and (5.110) to conclude the continuity of the operator V. We notice that (5.105) implies a similar condition on $X(t,s)$.

From (5.109), we derive the inequalities

$$\begin{aligned}
|(Vx)(t)|^2 &\leq 2|X(t,0)|^2|x_0|^2 + 2\int_0^t |X(t,s)|^2 ds \int_0^t |(Nx)(s)|^2 ds \\
&\leq 2|X(t,0)|^2|x_0|^2 + 2\gamma(t)\left[\lambda(t)\int_0^t |X(t,s)|^2 ds + \mu(t)\right],
\end{aligned}$$

which, by integration with respect to t, leads to

$$\int_0^t |(Vx)(s)|^2\,ds \leq 2|x_0|^2 \int_0^t |X(s,0)|^2\,ds + 2\int_0^t [\gamma(s)\lambda(s)A(s)+\gamma(s)\mu(s)]\,ds, \qquad (5.111)$$

where

$$\gamma(t) = \int_0^t |X(t,s)|^2\,ds,\ \ t\in[0,T),$$

and x satisfies (5.107). In order to satisfy condition 2) in the previous theorem, we are lead to the following inequality

$$2|x_0|^2 \int_0^t |X(s,0)|^2\,ds + 2\int_0^t \gamma(s)\mu(s)\,ds + 2\int_0^t \gamma(s)\lambda(s)A(s)\,ds \leq A(t),$$

which has the form

$$\alpha(t) + \int_0^t \beta(s)A(s)\,ds \leq A(t),\ \ t\in[0,T), \qquad (5.112)$$

where $\alpha(t)$ and $\beta(t)$ are nonnegative increasing functions on $[0,T)$.

In order to conclude the proof of the existence of a solution to equation (5.109), we must show that inequality (5.112) has a positive solution $A(t)$ on the interval $[0,T)$.

Let us notice that, for $\epsilon > 0$, we strengthen inequality (5.112) if we consider the inequality

$$A(t) \geq \int_0^t [\beta(s)+\epsilon]A(s)\,ds + \alpha(t) + \epsilon,\ \ t\in[0,T). \qquad (5.113)$$

Therefore, a positive solution of (5.113) will be also a solution of (5.112). Let us construct a positive solution for the inequality (5.113). Denote $y(t) = \int_0^t A(s)[\beta(s)+\epsilon]\,ds$, and notice that $\dot{y}(t) = A(t)[\beta(t)+\epsilon]$, $y(0)=0$.

From (5.112) and the given notation, we derive the differential inequality

$$\dot{y}(t) \geq [\beta(t)+\epsilon]y(t) + [\alpha(t)+\epsilon][\beta(t)+\epsilon],\ \ t\in[0,T), \qquad (5.114)$$

for which a solution is obtained when we use only the $=$ sign. This solution is given by the formula [with $y(0)=0$!] known for linear first-order ODEs. Since we obtain $\dot{y}(t) \geq \epsilon^2$ for $t\in[0,T)$ by using (5.114), and also $A(t) = \dot{y}(t)[\beta(t)+\epsilon]^{-1}$, we have $A(t) > 0$ for $t\in[0,T)$. In other words, we have constructed the function $A(t)$ whose existence is postulated in the fixed-point theorem stated before.

To summarize the above discussion, we will state the following theorem.

Theorem 5.7 *Consider the functional equation (5.100) under the following assumptions:*

1. $L : L^2_{\text{loc}}([0,T),R^n) \to L^2_{\text{loc}}([0,T),R^n)$ *is a linear causal operator, continuous on the whole space.*
2. $N : L^2_{\text{loc}}([0,T),R^n) \to L^2_{\text{loc}}([0,T),R^n)$ *is a continuous causal operator, satisfying the growth condition (5.110).*

Then, there exists a solution to equation (5.100), satisfying the condition $x(0) = x_0 \in R^n$, *defined on the interval* $[0,T)$. *This solution is absolutely continuous on each interval* $[0,\bar{t}] \subset [0,T)$.

Remark 5.9 *Since one can use* $(Nx)(t) = f(\dot{x})$ *in (5.100), the equation becomes*

$$\dot{x}(t) = (Lx)(t) + f(\dot{x}), \quad t \in [0,T),$$

where $f : R^n \to R^n$. *This equation is clearly a neutral equation in x. As seen earlier in this section, together with the initial condition, this equation is equivalent to the integro-differential equation*

$$x(t) = X(t,0)x_0 + \int_0^t X(t,s)f(\dot{x}(s))\,ds.$$

An investigation of equations of this type may be of interest to those in the field of mechanics. If one regards the system $\dot{x}(t) = (Lx)(t)$ *as describing a motion, then the term* $f(\dot{x})$ *could be interpreted as a perturbation caused by velocity.*

Remark 5.10 *If we want to investigate an equation of the form*

$$x(t) = (Lx)(t) + (Nx)(t), \quad t \in [0,T), \tag{5.115}$$

then the invertibility of the operator $x(t) - (Lx)(t)$, *which can be obtained under fairly mild assumptions, can help reduce equation (5.115) to the form*

$$x(t) = (I - L)^{-1}(Nx)(t).$$

This form is the preferred one for applying a fixed-point theorem to obtain an existence result.

Equation (5.115) is regarded as a perturbed equation of the linear one ($x(t) = (Lx)(t) + f(t)$). For some results in this category, see the papers of Corduneanu [131] and Mahdavi and Li [352].

5.6 SECOND-ORDER CAUSAL NEUTRAL FUNCTIONAL DIFFERENTIAL EQUATIONS, I

This section is dedicated to the local or global study of neutral second-order functional differential equations of the form

$$\frac{d}{dt}\left[\frac{dx(t)}{dt} - (Lx)(t)\right] = (Vx)(t),\ t \in R_+, \tag{5.116}$$

with L and V causal operators acting on $L^2_{\text{loc}}(R_+, R^n)$, L being assumed linear. The operator V could be also linear, in which case we obtain global existence results, but the nonlinear V is also in our objective.

We attach, as usual, the Cauchy-type initial conditions

$$x(0) = x_0 \in R^n,\ \dot{x}(0) = v_0 \in R^n, \tag{5.117}$$

and search for solutions to the problem (5.116), (5.117) on the real semi-axis R_+.

Since we have chosen the space $L^2_{\text{loc}}(R_+, R^n)$ as underlying space, the expected solutions will, in general, satisfy the equation (5.116) only *a.e.* on R_+. Generally speaking, the solutions will be such that $\dot{x}(t) - (Lx)(t)$ is absolutely continuous, which does not imply the absolute continuity of $\dot{x}(t)$ itself.

We will operate some transformations of the given data in order to apply known/classical methods for proving the existence of solutions.

Let us integrate both sides of (5.116), from 0 to $t > 0$, which is possible for each $t > 0$ due to the fact that the elements of $L^2_{\text{loc}}(R_+, R^n)$ are locally integrable on R_+. One obtains, after taking the second equation in (5.117) into account,

$$\dot{x}(t) - (Lx)(t) = v_0 + \int_0^t (Vx)(s)\,ds, \tag{5.118}$$

and the fixed initial property for the operator L,

$$(Lx)(0) = \theta,\ x \in L^2_{\text{loc}}. \tag{5.119}$$

We cannot differentiate both sides in (5.118) to introduce $\ddot{x}(t)$, because we know from (5.116) that only $\dot{x}(t) - (Lx)(t)$ is *a.e.* differentiable.

Therefore, we need to rely on the results from section 2.2, to apply the representation of the solution of the first-order functional differential equation (5.118) (i.e., the formula of variation of parameters). We obtain

$$x(t) = X(t,0)x_0 + \int_0^t X(t,s)v_0\,ds + \int_0^t X(t,s)\int_0^s (Vx)(u)\,du\,ds, \qquad (5.120)$$

with $X(t,s)$, $0 \leq s \leq t$, the Cauchy matrix corresponding to the operator L, in the manner shown in Section 2.2.

We distinguish two different situations, in accordance with the linearity or nonlinearity of the operator V.

When V is nonlinear, we deal with equation (5.120) as is, using for V a Lipschitz-type condition.

When V *is linear*, we rely again on the formula connecting abstract causal operators with classical Volterra operators, namely (5.40):

$$\int_0^t (Vx)(s)\,ds = \int_0^t k_0(t,s)x(s)\,ds, \; t \in R_+. \qquad (5.121)$$

Substituting the first term of (5.121) on the right-hand side of (5.120), we obtain a classical Volterra linear integral equation

$$x(t) = f(t) + \int_0^t k_1(t,s)x(s)\,ds, \; t \in R_+, \qquad (5.122)$$

where

$$f(t) = X(t,0)x_0 + \int_0^t X(t,s)v_0\,ds, \qquad (5.123)$$

and

$$k_1(t,s) = \int_s^t X(t,u)k_0(u,s)\,du, \; 0 \leq s \leq t. \qquad (5.124)$$

In the sequel, we shall use the notation

$$\Delta = \{(s,t);\; 0 \leq s \leq t\}.$$

With regard to the integral equation (5.122), we know that the unique solution is given by the resolvent formula

$$x(t) = f(t) + \int_0^t \tilde{k}_1(t,s)f(s)\,ds, \; t \in R_+, \qquad (5.125)$$

where $\tilde{k}_1(t,s)$ is the resolvent kernel associated to $k_1(t,s)$ from (5.124). As we have seen in Section 5.3, the existence of the resolvent kernel is assured in the following two situations:

1) $k_1 \in L^\infty_{\text{loc}}(R_+, \mathcal{L}(R^n, R^n))$;
2) $k_1 \in L^2_{\text{loc}}(R_+, \mathcal{L}(R^n, R^n))$.

We notice that the first situation implies the second, but we have $\tilde{k}_1 \in L^\infty_{\text{loc}}$ in the first case, while in the second case $\tilde{k}_1 \in L^2_{\text{loc}}$.

Before we draw conclusions about the existence of solutions to (5.122) or the underlying space, we need to clarify which one of the situations 1) or 2) occurs.

First, taking into account (5.124) and (5.103), we obtain

$$k_1(t,s) = \int_s^t \left[I + \int_u^t \tilde{k}(t,v)\,dv \right] k_0(u,s)\,du. \tag{5.126}$$

If we take into account that both $\tilde{k}(t,s)$ and $k_0(u,s)$ belong to L^2_{loc}, according to our assumption, we must still show that for each $T > 0$

$$\int_0^T \int_0^T |k_1(t,s)|^2\,dt\,ds < \infty. \tag{5.127}$$

Condition (5.127) will be secured if we prove

$$\int_s^t \int_u^t |\tilde{k}(t,v)|\,|k_0(u,s)|\,dv\,du \in L^2_{\text{loc}}(\Delta, \mathcal{L}(R^n, R^n)), \tag{5.128}$$

because it can be easily seen that

$$\int_s^t k_0(u,s)\,du \in L^2_{\text{loc}}(\Delta, \mathcal{L}(R^n, R^n)).$$

In order to assure the validity of (5.128), we shall rely on the elementary integral inequality,

$$\left(\int_s^t \int_u^t |\tilde{k}(t,v)|\,|k_0(u,s)|\,dv\,du \right)^2 \leq$$
$$\left(\int_s^t \int_u^t |\tilde{k}(t,v)|^2\,dv\,du \right) \left(\int_s^t \int_u^t |k_0(u,s)|^2\,dv\,du \right),$$

and since,

$$\left(\int_s^t \int_u^t |\tilde{k}(t,v)|^2 \, dv \, du\right) \left(\int_s^t \int_u^t |k_0(u,s)|^2 \, dv \, du\right) \leq$$
$$\left(\int_0^T \int_0^T |\tilde{k}(t,v)|^2 \, dv \, du\right) \left(\int_0^T \int_0^T |k_0(u,s)|^2 \, dv \, du\right),$$

one obtains

$$\left(\int_s^t \int_u^t |\tilde{k}(t,v)|\,|k_0(u,s)| \, dv \, du\right)^2 \leq$$
$$\left(\int_0^T \int_0^T |\tilde{k}(t,v)|^2 \, dv \, du\right) \left(\int_0^T \int_0^T |k_0(u,s)|^2 \, dv \, du\right). \tag{5.129}$$

It is obvious that the last term in (5.129) is dominated by

$$T^2 \left(\int_0^T |\tilde{k}(t,v)|^2 \, dv\right) \left(\int_0^T |k_0(u,s)|^2 \, du\right). \tag{5.130}$$

When we integrate on Δ the function (of t and s) appearing in (5.130), we certainly obtain a finite number, because both $\tilde{k}$ and k_0 are locally square integrable on Δ.

Therefore, the earlier discussion leads to the conclusion that the L^2-theory can be applied to equation (5.122) in Δ, due to the fact that $T > 0$ is arbitrary in our earlier considerations, which means that our problem is uniquely solvable in $L^2_{\text{loc}}(R_+, R^n)$.

Second, in case we change the space $C(R, R^n)$ with the space $L^2_{\text{loc}}(R_+, R^n)$, the classical theory or even the L^∞-theory (kernel in L^∞_{loc}) can be applied to obtain the existence and uniqueness of solution to (5.122), on R_+. These cases are frequently discussed in the literature (see, Tricomi [520], Corduneanu [135], and Hochstadt [264]).

We shall formulate the basic existence and uniqueness theorem for problem (5.116) and (5.117), specifying various hypotheses under which the conclusion is valid:

Theorem 5.8 *Consider the neutral functional differential equation (5.116) under the following conditions:*

(*a*) *Operators L and V are causal, continuous, and bounded on the space* $L^2_{\text{loc}}(R_+, R^n)$*; then, there exists a unique solution, in* $L^2_{\text{loc}}(R_+, R^n)$*, to the problem (5.116), (5.117), such that* $\dot{x}(t) - (Lx)(t)$ *is locally absolutely continuous on* R_+.

(*b*) *Operators L and V are causal, continuous, and bounded on the space $C(R_+, R^n)$, in which the convergence is the uniform convergence on each finite interval of R_+; then, there exists a unique solution of our problem, such that $\dot{x}(t) - (Lx)(t)$ is locally absolutely continuous on R_+.*

(*c*) *The same statement as in* (*b*) *when $C(R_+, R^n)$ is replaced by $L^\infty_{\text{loc}}(R_+, R^n)$.*

As mentioned at the beginning of this section, we have to treat the nonlinear case for the operator V, which requires us to return to equation (5.120). We will rewrite this equation in the form

$$x(t) = f(t) + \int_0^t X(t,s) \int_0^s (Vx)(u)\,du\,ds, \ t \in R_+, \tag{5.131}$$

with

$$f(t) = X(t,0)x_0 + \int_0^t X(t,s)v_0\,ds, \ t \in R_+. \tag{5.132}$$

Equation (5.132) is nonlinear only in case the operator V is nonlinear on $L^2_{\text{loc}}(R_+, R^n)$. On the operator $V : L^2_{\text{loc}}(R_+, R^n) \to L^2_{\text{loc}}(R_+, R^n)$, we shall impose the Lipschitz-type condition

$$|(Vx)(t) - (Vy)(t)| \leq \lambda(t)|x(t) - y(t)|, \ t \in R_+, \tag{5.133}$$

for $x, y \in L^2_{\text{loc}}$. Because we used the norm of R^n in (5.133), the inequality must be understood to hold *a.e.* on R_+. The function $\lambda(t)$ is assumed non-negative and nondecreasing on R_+.

The method of proof will be the iteration process, applied to equation (5.131), according to the usual scheme:

$$x_{k+1}(t) = f(t) + \int_0^t X(t,s) \int_0^s (Vx_k)(u)\,du\,ds, \ t \in R_+, \tag{5.134}$$

for $k \geq 0$, letting $x_0(t) = f(t)$, $t \in R_+$.

It is appropriate to note that we have already encountered the sequence $\{x_k(t); k \geq 0\} \subset C(R_+, R^n)$ in this section. It is obvious from (5.132) that $x_0(t) = f(t)$ is a locally absolutely continuous function on R_+, which represents the solution of the linear equation $\dot{x}(t) = (Lx)(t) + v_0$, with the initial condition $x(0) = x_0$. By induction on k we obtain the same property of local absolute continuity, for every $x_k(t)$, $k \geq 1$.

Let us consider the differences

$$x_{k+1}(t) - x_k(t) = \int_0^t X(t,s) \int_0^s [(Vx_k)(u) - (Vx_{k-1})(u)]\, du\, ds, \tag{5.135}$$

and estimate the absolute value on the left-hand side.

The following hypothesis will be imposed: For any interval $[0,T]$, $T > 0$, there exists $M = M(T) > 0$, such that

$$\int_0^t |X(t,s)|\, ds \leq M < \infty, \text{ for } t \in [0,T]. \tag{5.136}$$

We notice that condition (5.136) represents the condition for continuity or boundedness of the integral operator

$$x(t) \to \int_0^t X(t,s)x(s)\, ds,$$

on the space $C(R_+, R^n)$.

Applying (5.136) to (5.135), one obtains

$$|x_{k+1}(t) - x_k(t)| \leq M \int_0^t |(Vx_k)(u) - (Vx_{k-1})(u)|\, du,$$

and keeping in mind (5.133), one gets, for $k \geq 1$,

$$|x_{k+1}(t) - x_k(t)| \leq M \int_0^t \lambda(u)|x_k(u) - x_{k-1}(u)|\, du, \ t \in [0,T]. \tag{5.137}$$

If one denotes $A = \sup\{|x_1(t) - x_0(t)|; t \in [0,T]\}$, then (5.137) implies

$$|x_2(t) - x_1(t)| \leq AM \int_0^t \lambda(u)\, du, \ t \in [0,T].$$

We can continue this process, and obtain

$$|x_{k+1}(t) - x_k(t)| \leq \frac{AM^k}{k!} \left(\int_0^t \lambda(u)\, du \right)^k, \ t \in [0,T]. \tag{5.138}$$

Inequality (5.138) shows the uniform convergence of the sequence $\{x_k; k \geq 0\}$ on the interval $[0,T]$. Since $T > 0$ is arbitrarily large, we conclude the uniform convergence of $\{x_k; k \geq 0\}$ on each finite interval of R_+, which shows the convergence in $C(R_+, R^n)$. Taking the limit as $k \to \infty$, in (5.134), we obtain equation (5.131).

Therefore,

$$x(t) = \lim_{k \to \infty} x_k(t) \in C(R_+, R^n) \tag{5.139}$$

represents the solution of equation (5.131), which is equivalent to problem (5.116) and (5.117).

The uniqueness of this solution in $C(R_+, R^n)$ can be proven by a similar iteration procedure as used before. Indeed, if we assume the existence of two solutions $x(t)$ and $y(t)$, we find from (5.120),

$$x(t) - y(t) = \int_0^t X(t,s) \int_0^s [(Vx)(u) - (Vy)(u)]\, du\, ds.$$

This can be processed in the same way as (5.135). After k steps in the process, one obtains on each fixed interval $[0,T]$, the inequality

$$|x(t) - y(t)| \leq \frac{AM^k}{k!} \left(\int_0^t \lambda(u)\, du \right)^k, \ k \geq 1,$$

where $A = \sup|x(t) - y(t)|$. In this inequality, the right-hand side tends to zero as k tends to infinity on any finite interval $[0,T]$. Hence, $x(t) - y(t) = 0$, $\forall t \in R_+$.

These considerations allow us to state the following result.

Theorem 5.9 *Problem (5.116) and (5.117) has a unique solution $x(t) \in C(R_+, R^n)$ if the conditions stated in the following are satisfied:*

- (a) *The operator L on $L^2_{\text{loc}}(R_+, R^n)$ is a linear, causal and continuous operator;*
- (b) *The operator V is acting on $L^2_{\text{loc}}(R_+, R^n)$ continuously and satisfies a.e. the Lipschitz-type condition (5.133) (this condition actually implies the continuity of V).*

Then, there exists a unique solution of the Cauchy problem, $x(t) \in C(R_+, R^n)$, such that $\dot{x}(t) - (Lx)(t)$ is locally absolutely continuous on R_+.

This theorem concludes the investigation of equation (5.116) in this section.

In concluding this section, we refer to an equation related to (5.116), namely

$$\ddot{x}(t) + (L\dot{x})(t) = (Vx)(t), \ t \in [0,T). \tag{5.140}$$

The following approach can be adopted for this equation. Denoting $\dot{x} = y$, the equation can be rewritten as follows:

$$\dot{y}(t) + (Ly)(t) = (Vx)(t), \ t \in [0, T). \tag{5.141}$$

Since

$$x(t) = x_0 + \int_0^t y(s)\, ds, \tag{5.142}$$

we get y from (5.141), using the formula of variation of parameters for (5.141), and substituting in (5.142), thus obtaining an equation in x alone.

This formula for y is

$$y(t) = X(t,0)x_0 + \int_0^t X(t,s)(Vx)(s)\, ds,$$

and after taking (5.142) into account, one obtains

$$x(t) = x_0 + \int_0^t X(s,0)x_0\, ds + \int_0^t \int_0^s X(s,u)(Vx)(u)\, du\, ds.$$

This equation for $x(t)$ has the same form as equation (5.131) and can be treated by the same method of iteration. We leave the readers the task of formulating the conditions which assure the convergence of the procedure.

5.7 SECOND-ORDER CAUSAL NEUTRAL FUNCTIONAL DIFFERENTIAL EQUATIONS, II

We begin this section with the study of the neutral equation

$$\frac{d}{dt}[\dot{x}(t) + (L\dot{x})(t)] = (Vx)(t), \ t \in [0, T), \tag{5.143}$$

under various kinds of hypotheses. As usual, we will impose the initial conditions (5.117). The linear operator $V : L^2_{\text{loc}}([0,T), R^n) \to L^2_{\text{loc}}([0,T), R^n)$ will be assumed to possess the fixed initial value property (5.119), that is, $(Vx)(0) = \theta$ for any $x \in L^2_{\text{loc}}([0,T), R^n)$. Sometimes, we may substitute another space for $L^2_{\text{loc}}([0,T), R^n)$.

If we make the substitution $\dot{x} = y$, then $x(t) = x_0 + \int_0^t y(s)\, ds$, $t \in [0, T)$, and equation (5.143) can be rewritten in the form

$$\frac{d}{dt}[y(t) + (Ly)(t)] = \left(V\left(x_0 + \int_0^t y(s)\, ds\right)\right)(t), \tag{5.144}$$

which involves only the function $y(t)$, $t \in [0,T)$. If one denotes

$$\left(V\left(x_0 + \int_0^t y(s)\,ds\right)\right)(t) = (Wy)(t),\ t \in [0,T), \tag{5.145}$$

then (5.144) takes the usual form

$$\frac{d}{dt}[y(t) + (Ly)(t)] = (Wy)(t),\ t \in [0,T), \tag{5.146}$$

which represents a first order neutral functional differential equation. In other words, by means of the substitution $x(t) = x_0 + \int_0^t y(s)\,ds$, $t \in [0,T)$, we reduced the order of equation (5.143) from 2 to 1.

Moreover, we emphasize the fact that (5.146) is a causal equation in any instance where (5.143) or the operator V is causal.

Returning to equation (5.146), we note that these results were proven in various papers. Readers may refer to Corduneanu [145, 148], Mahdavi [348], and Mahdavi and Li [352], for functional equations of the form

$$x(t) + (Lx)(t) = (Nx)(t),\ t \in [0,T). \tag{5.147}$$

Indeed, after integrating both sides of (5.146) from 0 to $t \in (0,T)$, and taking into account $y(0) = \dot{x}_0 = v_0 \in R^n$, one obtains the

$$y(t) + (Ly)(t) = \int_0^t (Wy)(s)\,ds + y_0,\ t \in [0,T), \tag{5.148}$$

which obviously has the form (5.147). Since W may be a causal operator, the same property will be true for the "integrated" operator on the right-hand side of (5.148).

To conclude the discussion on intervening equations, we can state that any result that is valid for equation (5.147) generates a result for equation (5.143). In particular, it generates existence or behavior-type results, considering the connection given by $y(t) = \dot{x}(t)$.

Therefore, it will be convenient for us to use/establish results valid for equation (5.148).

For instance, let us consider equation (5.148) under the following assumptions, which were formulated in Corduneanu [[135], p. 97]:

1. The operator $L : C([0,T],R^n) \to C([0,T],R^n)$ is linear, causal, continuous, and compact, and such that $(L\theta)(t) = \theta$ on $[0,T]$.
2. The operator $N : C([0,T],R^n) \to C([0,T],R^n)$ is continuous and compact.

3. If we denote $\phi(r) = \sup\{|Nx|_C;\ |x|_C \le r\}$, for $r > 0$, then $\phi(r)$ satisfies

$$\limsup_{r\to\infty} \frac{\phi(r)}{r} = \lambda, \tag{5.149}$$

with λ sufficiently small.

Then, there exists a solution, $x(t) \in C([0,T],R^n)$, to equation (5.147).

The proof is provided in full, in the book referenced earlier.

Let us now apply the above-stated existence result to the equation (5.146), which will be carried on to equation (5.143).

With regards to equation (5.146), we will make the following assumptions:

1. The same, as formulated before when referring to the equation (5.148);
2. The operator $W : C([0,T],R^n) \to C([0,T],R^n)$ is continuous and takes bounded sets into bounded sets;
3. If we denote $\phi(r) = \sup\left\{\left|\int_0^t (Wy)(s)\,ds\right|_C;\ |y|_C \le r\right\}$, $r > 0$, then condition (5.149) is satisfied.

We can now state the following existence result for (5.148).

Theorem 5.10 *Under assumptions 1, 2, and 3, formulated in the text, there exists a solution $y(t) \in C([0,T],R^n)$, to equation (5.148).*

Corollary 5.1 *Consider equation (5.143) under assumption 1, for the operator L, as stated before. The operator $V : C([0,T],R^n) \to C([0,T],R^n)$ is continuous and causal, taking bounded sets into bounded sets.*

Then, equation (5.143) has a solution $x(t) \in C^{(1)}([0,T],R^n)$.

The only remark to be made is obtained from the relationship $x(t) = x_0 + \int_0^t y(s)\,ds$.

This ends the proof of Corollary 5.1, thus providing the existence of a solution for equation (5.143).

We will now consider an application of Theorem 5.10 to a case that we discussed in Section 5.4.

Namely, we will deal with the neutral equation, on R_+,

$$\frac{d}{dt}\left[\dot{x}(t) + \int_0^t k(t-s)\dot{x}(s)\,ds\right] = (Ax)(t) + f(t), \tag{5.150}$$

where the operator A is given by formula (5.61), under the assumptions (5.62), and $k \in L^1(R_+, \mathcal{L}(R^n, R^n))$. Due to the nature of operator A, which implies delays, one has to associate initial conditions with (5.150). These are

$$x(t) = h(t),\ \dot{x}(t) = h_1(t),\ t \in (-\infty, 0),\qquad x(0) = x_0 \in R^n,\ \dot{x}(0) = v_0 \in R^n, \tag{5.151}$$

with h, h_1 belonging to some $L^p((-\infty, 0], R^n)$, $p \geq 1$, the case $p = 1$ received special attention in Section 5.4.

If one denotes

$$\dot{x} + \int_0^t k(t-s)\dot{x}(s)\,ds = y(t), \tag{5.152}$$

one obtains for $y(t)$ the initial condition

$$y(0) = v_0 \in R^n. \tag{5.153}$$

But under frequency-type condition

$$det[I + \tilde{k}(is)] \neq 0,\ \ s \in R, \tag{5.154}$$

where $\tilde{k}$ denotes the Fourier transform of k, one obtains from (5.152)

$$\dot{x}(t) = y(t) + \int_0^t \tilde{k}(t-s)y(s)\,ds,\ \ t \in R_+. \tag{5.155}$$

See, for instance, Corduneanu [120], where the existence of the kernel $\tilde{k} \in L^1(R_+, \mathcal{L}(R^n, R^n))$ is also proven.

Equations (5.150) and (5.152) lead to the following system for x, y:

$$\dot{y}(t) = (Ax)(t) + f(t),\ \ t \in R_+. \tag{5.156}$$

Together with the (vector) equation (5.155), the system represents a system of the form (5.150), for the unknown vector valued $z = \mathrm{col}(x, y)$, with all conditions, used in proving the existence and uniqueness of solution are satisfied.

We leave the readers with the task of writing and investigating the system for $z = \mathrm{col}(x, y)$. Let us note that the delay (infinite!) system thus obtained, is equivalent to the neutral system (5.150) considered above. Related problems, such as those encountered in Section 5.4 regarding the absolute stability, can be also treated in the framework of the systems described by means of operator of type A, given by (5.61).

Let us return now to equation (5.143) to investigate the existence of solutions in L^p-spaces.

Theorem 5.11 *Consider equation (5.143) under the following assumptions:*

1. *L is a linear, causal, continuous and compact operator on the space $L^p([0,T],R^n)$, $1 < p < \infty$.*
2. *V is acting on $L^p([0,T],R^n)$ and is causal, continuous, taking bounded sets into bounded sets.*

Then, there exists $\delta > 0$, $\delta \leq T$, such that the equation (5.143), with initial conditions $x(0) = x_0 \in R^n$, $\dot{x}(0) = v_0 \in R^n$, has a solution $x(t) \in AC([0,\delta],R^n)$, satisfying a.e. on $[0,\delta]$ the equation, such that $\dot{x}(t) \in L^p([0,T],R^n)$, while $\dot{x}(t) + (L\dot{x})(t) \in AC([0,\delta],R^n)$.

Proof. The proof will be conducted based on Theorem 3.3 in Corduneanu [120], where the Schauder fixed-point theorem is used. One must notice that (5.143) can be written as follows:

$$y(t) = -(Ly)(t) + \int_0^t \left(V\left(x_0 + \int_0^t y(s)\, ds \right) \right)(s)\, ds,\ t \in [0,T]. \tag{5.157}$$

The operator on the right-hand side of (5.157) is causal, continuous, and compact on $L^p([0,T],R^n)$, according to our assumption. Consequently, one can directly proceed to the proof of existence for (5.157) by applying the fixed point theorem of Schauder, as indicated in the earlier reference.

The number δ, $0 < \delta \leq T$ is determined from the compactness property of the operator on the right-hand side of (5.157).

From $\dot{x}(t) = x_0 + \int_0^t y(s)\, ds$, $t \in [0,\delta]$, one derives properties for $x(t)$, that is, the solution of (5.143). Namely, denoting

$$(Uy)(t) = x_0 + \int_0^t y(s)\, ds,\ t \in [0,\delta], \tag{5.158}$$

and finding $q > 1$ from $p^{-1} + q^{-1} = 1$, one can write for the operator $U : L^p \to C$, the following inequality:

$$|(Uy)(t) - (Uy)(s)| \leq \left| \int_s^t |y(u)|\, du \right| \leq |t-s|^{\frac{1}{q}} \left| \int_s^t |y(u)|^p\, du \right|^{\frac{1}{p}},$$

and for $y \in B =$ bounded set in $L^p([0,\delta],R^n)$, which means that there exists $A > 0$, such that

$$|(Uy)(t) - (Uy)(s)| \leq A\,|t-s|^{\frac{1}{q}}. \tag{5.159}$$

Inequality (5.159) shows that the set $\{Uy; y \in B \subset L^p\} \subset C([0,\delta],R^n)$ satisfies the compactness condition for that space; and therefore, U is a compact operator from L^p into C. A fortiori, U is compact from L^p into itself, which implies the compactness of the product operator VU, necessary in the fixed point theorem.

We observe that the result will be of global nature, that is, $x(t)$ will be defined on $[0,T]$, where the operator V is also linear on L^p.

This ends the proof of the Theorem 5.4.

Let us now consider the (non-differential) functional equation similar to (5.143), namely

$$x(t) + (Lx)(t) = (Vx)(t), \ t \in [0,T]. \tag{5.160}$$

In fact, the auxiliary equation (5.157) is of the form (5.160), and we dealt with it in the space $L^p([0,T],R^n)$, $p > 1$. Then, we applied the result to equation (5.143).

With regards to equation (5.160), we will adopt an alternate approach that is also based on fixed point theorem. Namely, we notice that the linear equation associated to (5.160),

$$x(t) + (Lx)(t) = f(t), \ t \in [0,T], \tag{5.161}$$

with L causal, continuous and compact on $C([0,T],R^n)$, is uniquely solvable in $C([0,T],R^n)$. Using operator notation, from (5.161) one finds $(I+L)x = f$, which means $x = (I+L)^{-1}f$. If we define the resolvent operator associated to L by

$$R = (I+L)^{-1} - I, \tag{5.162}$$

then the unique solution x of (5.161) can be represented by the formula

$$x = f + Rf, \ f \in C([0,T],R^n). \tag{5.163}$$

Since $x + Lx$ is onto C, there results the continuity of $(I+L)^{-1}$, hence the continuity of the resolvent R.

Returning to equation (5.160), we can rewrite it (in an equivalent form) as

$$x = Vx + RVx. \tag{5.164}$$

Obviously, these considerations are valid for equations like (5.160) in other function spaces on $[0,T]$.

For equation (5.164), one can formulate and prove (rather routinely) two existence theorems, usually encountered with quasilinear equations of the form (5.160). We will limit our statements to the case of the space $C([0,T],R^n)$, this case is more frequently encountered in applications.

Proposition 5.3 *Consider equation (5.160), where L and V are operators acting on the space $C([0,T],R^n)$, with L linear, continuous, and compact, and with V satisfying a Lipschitz-type condition*

$$|(Vx)(t)-(Vy)(t)| \le \lambda |x(t)-y(t)|, \tag{5.165}$$

for $t\in[0,T]$, and any $x,y\in C([0,T],R^n)$, with sufficiently small λ.

Then, there exists a unique solution $x\in C([0,T],R^n)$ to (5.164).

Proof. We notice first that (5.165) implies both continuity and causality for the operator V. Also, $|Vx-Vy|_C \le \lambda |x-y|_C$.

From equation (5.164), one obtains

$$|Vx+RVx-Vy-RVy|_C \le \lambda (1+|R|)\, |x-y|_C, \tag{5.166}$$

which implies the contraction of the operator $V+RV$, on $C([0,T],R^n)$, when $\lambda<(1+|R|)^{-1}$.

Proposition 5.3 is thus proven.

Remark 5.11 *The norm $|R|$ can be estimated, depending on the linear operator L. For instance, for*

$$(Lx)(t) = \int_0^t k(t,s)x(s)\,ds,\ t\in[0,T], \tag{5.167}$$

with k continuous on $0\le s\le t\le T$, the norm will be given by

$$|R| = \sup\left\{\int_0^t |\tilde{k}(t,s)|\,ds;\ t\in[0,T]\right\}.$$

See, for instance, Corduneanu [149], for other cases.

The next statement, regarding equation (5.160), is based on the Schauder fixed-point theorem.

Proposition 5.4 *Consider equation (5.160) under the following assumptions on the operators L and V, acting on* $C([0,T],R^n)$:

1. *L is linear, continuous, and compact.*
2. *V is continuous and compact, and denoting*

$$\phi(r) = \sup\{|Vx|_C; x \in C, \ |x|_C \le r\}, \tag{5.168}$$

one has

$$\lim_{r\to\infty} \frac{\phi(r)}{r} = \lambda, \tag{5.169}$$

where λ *is sufficiently small.*

Then, there exists a solution $x \in C([0,T],R^n)$ *of equation (5.160).*

Proof. Indeed, from equation (5.164), one derives

$$|(I+R)V|_C \le (1+|R|_C)\,\phi(r), \tag{5.170}$$

for all $x \in B_r$ =the ball of radius r, centered at θ. In order to assure the inclusion

$$(I+R)\,V B_r \subset B_r, \tag{5.171}$$

for some $r > 0$, one has to impose the condition

$$(1+|R|)\,\phi(r) \le r, \ \ r > 0,$$

which is always possible on behalf of assumption (5.169), for $\lambda < (1+|R|)^{-1}$.
This ends the proof of Proposition 5.4.

Remark 5.12 *The results of Propositions 5.3 and 5.4 can be used to find existence for the second-order neutral functional differential equation (5.143), in the same manner as with equation (5.157), but under different assumptions, which lead to local existence results.*

Remark 5.13 *Results similar to those given in Propositions 5.3 and 5.4 are available in the literature. We would like to mention O'Regan and Precup [430], based on another fixed-point theorem (Leray–Schauder). The result is also included in Corduneanu [149], and the underlying space is* $L^p([0,T],R^n)$. *Other sources for this type of results are V. Lakshmikantham and Leela [309], V. Kolmanovskii and A. Myshkis [292], N. V. Azbelev et al. [33], and J. K. Hale [240].*

5.8 A NEUTRAL FUNCTIONAL EQUATION WITH CONVOLUTION

In concluding this chapter, we will consider a case that involves $AP_r(R,\mathcal{C})$ type of almost periodicity and convolution product.

The equation we will investigate has the form

$$\frac{d}{dt}[\dot{x}(t)+(k*x)(t)]=f(t),\ t\in R, \tag{5.172}$$

under conditions similar to those used in Chapter 4. Namely, $x,f: R\to\mathcal{C}$, $|k|\in L^1(R,\mathcal{C})$ and a frequency domain type, condition, to be specified later.

The function space $AP_r(R,\mathcal{C})$ is taken as underlying space (see definition in Chapter 4).

Obviously, (5.172) is a neutral functional differential equation of second order, while the convolution operator $x\to k*x$ is also defined in Section 4.4.

The first step in investigating the existence of $AP_r(R,\mathcal{C})$ solutions to (5.172) consists of replacing this equation with an equivalent one, which is obtained by integrating both terms. In this way, one obtains the equation (of first order)

$$\dot{x}(t)+(k*x)(t)=\int^t f(s)\,ds. \tag{5.173}$$

The right-hand side in (5.173) stands for a primitive of the function $f(s)$. We notice that the primitive exists (Riemann or Lebesgue) for functions in $AP_r(R,\mathcal{C})$. Below, we consider a hypothesis that guarantees it also belongs to $AP_r(R,\mathcal{C})$.

To find conditions for the existence of an $AP_r(R,\mathcal{C})$ solution to (5.173), we apply the generalized Fourier transform to both sides of this equation and acquire the result after the transform. Of course, this procedure would be valid only if both sides of (5.173) belong to $AP_r(R,\mathcal{C})$, for a given r, $1\le r\le 2$. We will see that this situation occurs in case $k\in L^1(R,\mathcal{C})$ and $\int^t f(s)\,ds\in AP_r(R,\mathcal{C})$. Indeed, with $x\in AP_r(R,\mathcal{C})$, $k*x\in AP_r(R,\mathcal{C})$. If $\int^t f(s)\,ds\in AP_r(R,\mathcal{C})$, then (5.173) implies $\dot{x}(t)\in AP_r(R,\mathcal{C})$. Therefore, from (5.173), we obtain by using the transform

$$\sum_{j=1}^{\infty} i\lambda_j x_j e^{i\lambda_j t}+\sum_{j=1}^{\infty} x_j\tilde{k}(-\lambda_j)\,e^{i\lambda_j t}\equiv\sum_{j=1}^{\infty} f_j\lambda_j^{-1}\,e^{i\lambda_j t}, \tag{5.174}$$

which implies, under assumption $\lambda_j\neq 0$,

$$\sum_{j=1}^{\infty}[i\lambda_j+\tilde{k}(-\lambda_j)]\,x_j=f_j\lambda_j^{-1},\ j\ge 1. \tag{5.175}$$

Therefore, the coefficients x_j of $x(t)$ must be determined from equation (5.175), provided

$$|i\lambda_j+\tilde{k}(-\lambda_j)|>0,\ j\geq 1. \tag{5.176}$$

Since λ_j cannot be easily determined in applications, one substitutes (5.176) with the apparently stronger condition

$$|is+\tilde{k}(-s)|>0,\ s\in R. \tag{5.177}$$

Under assumption (5.177), one can determine the coefficients x_j, $j\geq 1$, uniquely, which proves the existence and uniqueness of the solution $x\in AP_r(R,\mathcal{C})$.

Let us summarize this discussion with the following statement:

Theorem 5.12 *Assume the following conditions hold for equation (5.173):*

a) $|\lambda_j|\geq m>0,\ j\geq 1;$
b) $|k|\in L^1(R,\mathcal{C})$ *and* $\tilde{k}$=*the Fourier transform of* k;
c) Condition (5.177) is satisfied.

Then, for each $f\in AP_r(R,\mathcal{C})$, *there exists a unique solution* $x=x(t)\in AP_r(R,\mathcal{C})$, $1\leq r\leq 2$.

Proof. First, let us notice that condition $a)$ implies the fact that the primitive of $f=\sum_{j=1}^{\infty}f_je^{i\lambda_jt}$ belongs to $AP_r(R,\mathcal{C})$.

Second, by accepting (5.177), one obtains from (5.175)

$$x_j=\left[i\lambda_j+\tilde{k}(-\lambda_j)\right]^{-1}f_j\lambda_j^{-1},\ j\geq 1, \tag{5.178}$$

for any $f\in AP_r(R,\mathcal{C})$. We only need to show that (5.176), which is equivalent to (5.177) because $\{\lambda_j;j\geq 1\}\subset R$ is an arbitrary sequence, is also equivalent to the stronger inequality

$$|is+\tilde{k}(-s)|\geq\alpha>0,\ s\in R. \tag{5.179}$$

Since (5.179) implies (5.177), we need only to prove that the contrary statement is true. However, if we take into account the property $|\tilde{k}(s)|\to 0$ as $|s|\to\infty$, then (5.177) leads to $|is+\tilde{k}(-s)|>1$, for $|s|>A$, $A>0$. On the compact interval $[-A,A]$, $|is+\tilde{k}(-s)|$ is continuous and bounded. Moreover, it is taking a maximum value at some point in this interval: $|is_0+\tilde{k}(-s_0)|=M>0$. Therefore, one has $|is+\tilde{k}(-s)|^{-1}<1$ for $s\in(-\infty,-A)\cup(A\cup\infty)$, and $|is+\tilde{k}(-s)|^{-1} < M^{-1}$ on $[-A,A]$, which means $|is+\tilde{k}(-s)|^{-1} < M_0 = \min\{1,M^{-1}\}$ on R.

With this information at hand, we return to (5.176), which leads to

$$|x_j| \leq M_0 |f_j| |\lambda_j^{-1}| \leq \frac{M_0}{m} |f_j|, \, j \geq 1. \tag{5.180}$$

These inequalities show that $x = x(t) \in AP_r(R,\mathcal{C})$, because f_j are the Fourier coefficients of $f \in AP_r(R,\mathcal{C})$, $1 \leq r \leq 2$.

In conclusion, the function $x = x(t)$, whose Fourier series is $\sum_{j=1}^{\infty} x_j e^{i\lambda_j t}$, with x_j the Fourier coefficients of $f(t)$ in (5.172), represents the solution of that equation, determined up to an additive arbitrary constant. For equation (5.173), the solution is unique in the space $AP_r(R,\mathcal{C})$ of almost periodic functions.

This ends the proof of Theorem 5.12.

The approach used in dealing with equation (5.172) can also be useful in investigating existence and uniqueness for larger classes of neutral equations. We invite the readers to apply the same method to such equations as

$$\frac{d}{dt}\left[x^{(i)}(t) + (k*x)(t)\right] = f(t), \, t \in R, \, i > 1,$$

or even more general, nonlinear, equations of the form

$$\frac{d}{dt}[\dot{x}(t) + (k*x)(t)] = (fx)(t), \, t \in R,$$

following the same pattern.

5.9 BIBLIOGRAPHICAL NOTES

Neutral functional equations display a large variety of types and have been encountered in mathematical research long time ago. For instance, the very simple system $\dot{x} = y - z$, $\dot{y} = z - x$, $\dot{z} = x - y$, which is in normal form (solved with respect to the derivatives), leads to the integral combination $x\dot{x} + y\dot{y} + z\dot{z} = 0$, or $\frac{d}{dt}(x^2 + y^2 + z^2) = 0$, from which we get the first integral $x^2 + y^2 + z^2 = C$, $C \geq 0$; but the system that resulted from above, namely $\dot{x} = y - z$, $\dot{y} = z - x$, $(x^2 + y^2 + z^2)^{\cdot} = 0$ will be termed, necessarily, as a *neutral* system of ODEs.

When investigating problems by means of functional differential equations in applied areas, it is possible that a combination of unknown functions could be easier observed and its change investigated. This would certainly lead to a neutral equation/system. A common example is from mechanics, when the energy of a system is investigated.

The material in Chapter 5 covers only a *limited* number of neutral systems of functional equations. It is mostly extracted from authors' papers: Corduneanu [142, 144–146, 149, 151]; Corduneanu and Li [171]; Corduneanu and Mahdavi [175–178, 180, 181]; and Mahdavi [348].

Existence results have been obtained by Benchohra and Ntouyas [55], Okonkwo and Turner [422] in case of stochastic operator equations, and Okonkwo et al. [421] with particular regard to stability. Additional results concerning neutral equations can be found in Corduneanu [157], Hale and Lunel [242], Hale and Meyer [243], and Gripenberg et al. [228]. For neutral equations of the form $(Vx)(t) = (Wx)(t)$, or the differential variant $\frac{d}{dt}(Vx)(t) = (Wx)(t)$, see Corduneanu [142] and the references therein. Also, see the survey papers Corduneanu [144] and Corduneanu and Lakshmikantham [167].

Let us point out that in the general form mentioned above, $(Vx)(t) = (Wx)(t)$, $t \in J \subset R$, $J =$interval, this problem is equivalent to the existence of coincidence points for the operators V and W.

We do not possess any results concerning the geometry of solutions of these equations, similar to those obtained by many authors. The book by Górniewicz [226] seems to be a good source of results and methods for advancing this field.

Global existence results for neutral functional equations/inclusions can be found in Benchohra and Ntouyas [55]. Their basic problems can be formulated for inclusions of the form $\frac{d}{dt}[y(t) - f(t,y_t)] \in Ay(t) + F(t,y_t), t \in [-h,\infty)$, $h > 0$, $y_0 = \phi$, with A an infinitesimal generator of a strongly continuous semigroup of bounded linear operators, and $T(t)$ being multivalued. The paper has an adequate list of references.

Also, in regard to global existence results on a semiaxis, see the papers by Corduneanu and Mahdavi [175–178], [181], where the second-order neutral equations are investigated. One form of second-order neutral equations looks like

$$\frac{d}{dt}\left[\frac{dx(t)}{dt} - (Lx)(t)\right] = (Vx)(t),\ t \in R_+,$$

where L is a linear operator on convenient function spaces, for example, $C(R_+,R^n)$, $L^2_{\text{loc}}(R_+,R^n)$, and while V is generally nonlinear. A reduction method to first-order functional differential equations is applied, and the use of variation of parameters formula allows the proof of existence.

The case of stochastic neutral equations, of the form

$$d[x(t,\omega) + V(x(t,\omega))] = (Fx)(t,\omega)\,dt + G(t,x(t,\omega))\,dx(t,\omega)$$

has been considered in several papers by Okonkwo and Okonkwo et al. See Refs. [420, 422], where the results of admissibility and stability are obtained.

Baker et al. [36] analyzes the modeling of natural phenomena and proposes a hierarchy for the use and development of existing models.

In Corduneanu and Li [171], a convolution-type neutral equation is investigated, in relationship with almost periodicity in $AP_r(R,\mathcal{C})$-spaces, $1 \leq r \leq 2$.

A type of neutral equation, throughly investigated by many authors, has been introduced by Hale. The basic facts are given in the book [242] by Hale and Lunel. They are of the form $\frac{d}{dt}[D(t)x_t] = L(t)x_t + h(t)$, $t \geq t_0 \geq 0$, under conditions $x_{t_0} = \phi$, with $D(t)\phi = \phi(0) - \int_{-r}^{0} d[\mu(t,\theta)]\phi(\theta)$, $L(t)\phi = \int_{-r}^{0} d[\eta(t,\theta)]\phi(\theta)$ and several assumptions on data. Existence (both local and global) is established and in the case $x \in B$ =Banach space, the semigroups theory is involved (linear case). Stability problems are also investigated, and results comparable to the classical case (ODE in R^n) are obtained. In particular, Liapunov's method is used. A rich literature is referenced in this book.

Among the multiple applications of this theory, we mention here the papers by R. Yuan [544, 545], who obtained interesting results regarding existence of almost periodic solutions.

More references for neutral functional differential equations are available in *Mathematical Reviews*, *Zentralblatt fur Mathematik*, and *Referativnyi Zhurnal*.

APPENDIX A

ON THE THIRD STAGE OF FOURIER ANALYSIS

Constantin Corduneanu

A.1 INTRODUCTION

The Fourier analysis is a vast field of knowledge, with many connections in recent development of the mathematical theory of vibrations/oscillations and waves.

The *first stage*, amply illustrated by the work of Euler, Fourier, Riemann, Dini, Drichlet, Fejér and many other distinguished mathematicians of the past (see, e.g., Bary [47] or Zygmund [562]), is still in development and is related, mainly, to the investigation of cases involving *periodicity* of phenomena.

The *second stage* in this development of vibratory/oscillating processes is related to cases of *almost periodicity*, a concept due to H. Bohr and developed by many followers. See the Bibliographic entries, under the names of Bohr [72], Besicovitch [61], Favard [208], Levitan [326], Fink [213], Levitan and Zhikov [327], Corduneanu [129, 156], and Ch. Zhang [553]. The almost periodicity concept, obviously more comprehensive than that of periodicity, does not suffice in describing the vibratory or wave phenomena, which are neither periodic nor almost periodic.

Functional Differential Equations: Advances and Applications, First Edition.
Constantin Corduneanu, Yizeng Li and Mehran Mahdavi

During the past few decades, the engineering literature is displaying examples of oscillations, which do not pertain to the classical cases of periodicity or almost periodicity. Such type of oscillations may appear even in man-made machines. Such examples can be found in various IEEE Transactions publications. In Zhang's papers [557, 558], one can find several references to examples of this type of oscillations, as well as an interesting construction of a Banach space containing functions describing such kind of oscillations. The corresponding generalized Fourier series are of the form

$$\sum_{k=1}^{\infty} a_k e^{i\lambda_k(t)}, \tag{A1}$$

with $a_k \in \mathcal{C}$, $\lambda_k : R \to R$, these functions being at least locally integrable. The case when $\lambda_k(t)$ are linear functions in t leads to the series related to the almost periodic functions. The periodic case occurs only when $\lambda_k(t) = \lambda_k t$, $k \geq 1$, $\lambda_k \in R$, $\lambda_k = k\omega$, $k \in Z$, $\omega \neq 0$.

The *third stage* of Fourier analysis has in view, primarily, those phenomena of oscillation/vibration whose description involves series of the form (A1) and the functions (possibly generalized) which are characterized by those type of series. So far, we know only the results due to Osipov [432], who published a book at the University Press of Sankt Petersburg (1992), whose title *Bohr-Fresnel Almost Periodic Functions* suggests inspiration from Optics and several papers (see Bibliography) due to Zhang and some of his students. These contributions display various series of the form (A1) and their properties. Also in 2003, Zhang [553] published his book dedicated to almost periodic functions, with results concerning a new class of oscillatory functions he called *pseudo-almost periodic*. Each function in this space has an almost periodic component, which allows to attach to each pseudo-almost periodic function a Fourier series, while they appear as perturbation of the almost periodic functions (Bohr). These functions have been investigated and generalized by many authors see, for instance, Diagana [193]. The quoted books by Osipov and Zhang are, likely, the first announcing the entrance in the third stage of Fourier analysis. Indeed, relying on the classic theory of almost periodic functions, they contain results related to oscillatory functions (solutions of some equations), which do not belong to the space of Bohr.

A.2 RECONSTRUCTION OF SOME CLASSICAL SPACES

We will now consider an approach to reconstruct (provide new definitions) of the classical spaces of almost periodic functions, starting with formal trigonometric series of the form

$$\sum_{k=1}^{\infty} a_k e^{i\lambda_k t}, \tag{A2}$$

where $a_k \in \mathcal{C}$, $\lambda_k \in R$, $k \geq 1$. It is assumed that $\lambda_k{}'s$ are distinct.

Usually, the function attached to series (A2), in case of convergence, is its sum

$$f(t) = \sum_{k=1}^{\infty} a_k e^{i\lambda_k t}, \quad t \in R. \tag{A3}$$

The convergence is a rare occurrence, but when it takes place uniformly (on R) with respect to t, this is the most natural way of establishing the connection between series of the form (A2), and functions with oscillatory properties. Even in more general situations, when the series (A1) is summable, with respect to the uniform convergence on R, the series associated to a given function f, according to the following formula

$$a_k = \lim_{\ell\to\infty} (2\ell)^{-1} \int_{-\ell}^{\ell} f(t) e^{-i\lambda_k t} dt, \quad k \geq 1, \tag{A4}$$

allows the determination of the function f (see, e.g., the references, Corduneanu [129, 156, 163]).

There is a delicate aspect in this regard, namely, the correspondence function $\to$ series is single-valued, but the inverse is a multi-valued map (see the aforementioned references).

In order to provide some unity in the construction of classical spaces of almost periodic functions, it is convenient to consider trigonometric series of the form

$$\sum_{k=1}^{\infty} \phi(|a_k|) e^{i\lambda_k t}, \quad t \in R, \tag{A5}$$

where, as usual, $a_k \in \mathcal{C}$, $\lambda_k \in R$, $k \geq 1$. Of course, we attach such kind of series, to each series of the type (A2), as a tool which will allow us to obtain various classes/spaces of almost periodic functions.

The function $\phi : R_+ \to R_+$, $R_+ = [0, \infty)$, is assumed vanishing at the origin, $\phi(0) = 0$, increasing at the right of the origin, say on $[0, A)$, $A \leq \infty$, possibly continuous. Such kind of functions are often used in defining various types of stability (see Chapter 3).

It is obvious that we subject the coefficients in (A2) to a transformation, in order to distinguish between various types of almost periodicity.

A typical assumption will be the convergence of the following numerical series:

$$\sum_{k=1}^{\infty} \phi(|a_k|) < \infty. \tag{A6}$$

In particular, for $\phi(u) = u$, $u \in [0,A)$, one obtains the condition that leads to the absolute and uniform convergence (on R) of the series (A2). It is well known that the set of all these series can be organized as a Banach algebra. See, for instance, the author's book [156]. The multiplication is defined pointwise and the rule of Cauchy is considered. We have denoted by $AP_r(R,\mathcal{C})$ the space of all almost periodic functions, which can be represented by series of the form (A2), with coefficients calculated by (A4). The space $AP_1(R,\mathcal{C})$ is the first in the scale of spaces $AP_r(R,\mathcal{C})$, $1 \leq r \leq 2$, encountered in Chapter 4. The functions represented by the series (A2), under condition (A6), for $\phi(u) = u^r$, $1 < r < 2$, are also almost periodic in Besicovitch space B^2, denoted by $AP_2(R,\mathcal{C})$. We shall deal below with this case, which needs special considerations.

Let us consider now the last space in the scale mentioned in Corduneanu [161, 165], corresponding to $r = 2$. The condition (A6) becomes

$$\sum_{k=1}^{\infty} |a_k|^2 < \infty, \tag{A7}$$

which means $\{a_k; k \geq 1\} \subset \ell^2(N,\mathcal{C})$.

This space, $AP_2(R,\mathcal{C})$, also denoted by $B^2(R,\mathcal{C})$, is exactly the Besicovitch space of almost periodic functions. But the condition (A7) does not imply the usual convergence of the series (A2).

A complete examination of this case, including the reconstruction of the Besicovitch space $B^2(R,\mathcal{C})$, is provided in the author's paper [165]. In order to obtain the structure of a Banach space, it is necessary to take the space factor of a semi-normed linear space, which naturally appears when we deal with series. The original definition of Besicovitch uses the semi-distance

$$\left\{ \lim_{\ell \to \infty} (2\ell)^{-1} \int_{-\ell}^{\ell} |f(t) - g(t)|^2 \, dt \right\}^{\frac{1}{2}}, \tag{A8}$$

which generates the semi-norm

$$\left\{ \lim_{\ell \to \infty} (2\ell)^{-1} \int_{-\ell}^{\ell} |f(t)|^2 \, dt \right\}^{\frac{1}{2}}. \tag{A9}$$

The fact that the semi-norm can be zero, even though the function inside the norm is not identically zero, leads to the necessity of constructing the factor space.

For example, the zero element of the space $B^2(R,\mathcal{C})$ is the class of locally integrable functions for which the semi-norm, given on the left-hand side of (A9), is zero. One can easily construct such functions (e.g., $f(t) = e^{-|t|}, t \in R$).

With regard to the spaces $AP_1(R,\mathcal{C})$ and $AP_2(R,\mathcal{C}) \equiv B^2(R,\mathcal{C})$, we notice that the convergence in $AP_1(R,\mathcal{C})$ is the uniform convergence on R, while the convergence in $B^2(R,\mathcal{C})$ implies the convergence in the space $L^2_{\text{loc}}(R,\mathcal{C})$, which means the L^2- convergence on any bounded interval of R.

This property is guaranteeing that to each series in $AP_2(R,\mathcal{C})$, there *corresponds a function defined on* R, with complex values, which is an L^2- function on each bounded interval. If one starts with series (A2), and denote by $f(t)$ the function which corresponds to it, then the following relationship holds:

$$\lim_{\ell \to \infty} (2\ell)^{-1} \int_{-\ell}^{\ell} \left| f(t) - \sum_{k=1}^{n} a_k e^{i\lambda_k(t)} \right|^2 dt \leq \sum_{k=n+1}^{\infty} |a_k|^2. \tag{A10}$$

On behalf of (A7), that is, $\phi(u) = u^2$ in (A5), we see that the series (A2) converges in the sense of $AP_2(R,\mathcal{C})$. The limit is not unique, due to the fact that in (A9) we deal with a semi-norm. Of course, we have to keep in mind that the connection between f and $\{a_k; k \geq 1\}$ is given by (A4).

An important aspect, to be mentioned with regard to the spaces $AP_r(R,\mathcal{C})$, with $1 < r < 2$, consists in the inclusion $AP_r(R,\mathcal{C}) \subset AP_2(R,\mathcal{C})$, $1 < r < 2$. The inclusion is not topological, so AP_r are not subspaces of AP_2. But the existence of a function, associated to a series satisfying (A5) for $\phi(u) = u^r$, $1 \leq r < 2$, is assured. This function will be surely in $L^2_{\text{loc}}(R,\mathcal{C})$, and there may be several such functions. On the other hand, in the space AP_r, with the norm of Minkowski

$$|f|_r = \left(\sum_{k=1}^{\infty} |a_k|^r \right)^{\frac{1}{r}}, \quad 1 \leq r < 2, \tag{A11}$$

the correspondence series functions are single-valued:

$$|f - g|_r = \left(\sum_{k=1}^{\infty} |a_k - b_k|^r \right)^{\frac{1}{r}}, \quad 1 \leq r < 2. \tag{A12}$$

The space of (Bohr) almost periodic functions, denoted by $AP(R,\mathcal{C})$, can also be characterized simply if we answer the following question: Under what conditions does (A1) represent an almost periodic function (Bohr)?

The answer is rather simple, if we are acquainted with the classical theory of almost periodic functions, the necessary and sufficient condition for a series of the form (A2) to be the Fourier series of a function in $AP(R,\mathcal{C})$ is its summability by a linear method (Cesaro, Fejér, Bochner), with respect to the *uniform convergence* on R.

How we can reconstruct the theory of almost periodic functions (Bohr) if we start from the definition resulting from the earlier statement is shown in our paper [165]. Of course, some familiarity with the Bohr's theory is very helpful.

A remark concerning the situation of the space $AP(R,\mathcal{C})$, with respect to the spaces belonging to the scale $\{AP_r : 1 \leq r \leq 2\}$, is the fact that AP has nonempty intersection with each of them. At least, the trigonometric polynomials belong to each $AP_r(R,\mathcal{C})$, $1 \leq r \leq 2$. Of course, $AP_1(R,\mathcal{C}) \subset AP(R,\mathcal{C})$. Moreover, we know that $AP(R,\mathcal{C}) \subset AP_2(R,\mathcal{C})$. One can ask the following question: Which functions in $AP(R,\mathcal{C})$ do belong to AP_r?

There are other interesting facts in relationship with the spaces $AP_r(R,\mathcal{C})$. We will consider the definition of a generalized convolution product between an L^1-function and an $AP_r(R,\mathcal{C})$-element. This definition is an immediate extension of a formula, which holds in the usual sense, between a kernel in $L^1(R,\mathcal{C})$ and an arbitrary element of $AP_1(R,\mathcal{C})$

$$(k*f)(t) = \sum_{j=1}^{\infty} a_j \int_R k(t-s)\, e^{i\lambda_j s}\, ds, \tag{A13}$$

where $\{a_j; j \geq 1\}$ are the Fourier coefficients of $f : R \to \mathcal{C}, f \in AP_1(R,\mathcal{C})$ and $\lambda_j{}'s$ are the Fourier exponents of f (actually, any sequence of real numbers).

The right-hand side of (A13) makes sense in case $\{a_j; j \geq 1\} \subset \ell^r$, $1 \leq r \leq 2$, that is, when we deal with an $f \in AP_r(R,\mathcal{C})$, $1 \leq r \leq 2$. Formula (A13) leads to the inequality

$$|h*f|_r \leq |k|_{L^1} \cdot |f|_r, \quad f \in AP_r, \tag{A14}$$

where the index r runs from 1 to 2.

The inequality (A14) reminds us from a similar one, which is related to L^p-spaces. It was used in connection with the investigation of certain functional equations in AP_r-spaces, by Corduneanu [162], Corduneanu and Li [171], Corduneanu and Mahdavi [182]. Equations of the form

$$x(t) + (k*x)(t) = f(t), \quad t \in R, \tag{A15}$$

or

$$\dot{x}(t) + (k * x)(t) = (fx)(t), \quad t \in R, \tag{A16}$$

were investigated and frequency-type criteria of existence of AP_r-solutions provided. There is more to be done in this regard, looking for various properties like stability or asymptotic behavior of solutions.

In concluding this section, which provides a unitary treatment for the almost periodic series/functions, we stress the feature that it is possible to obtain the most usual classes of almost periodic functions, starting from conditions of the form (A5), imposed on the coefficients of trigonometric series of the form (A2).

As we will see in the next sections of Appendix, this procedure, which requires to pass from series to functions, by using adequate conditions on the coefficients (like (A6) or its special case (A1)), one can obtain various types of *almost periodic and oscillatory* functions.

Also, when imposing the summability condition on the series, with respect to a certain procedure and type of convergence, it is possible to redefine the classical almost periodic functions of Bohr, using the series as primary element.

But before we move to more general types of oscillatory function spaces, it is useful to emphasize the case of the space B of Besicovitch, which is richer than the space $B^2(R,\mathcal{C}) = AP_2(R,\mathcal{C})$ and has the remarkable property of *existence* of the Poincaré's mean value (on the whole real axis). The significant role of this *mean value*, used and defined by Poincaré in his famous treatise [454] on Celestial Mechanics, has been demonstrated during the whole history of almost periodicity, helping to prove the fact that any function in B_2 (first one, in this scope, is due to Besicovitch [61]) is characterized by its (generalized) Fourier series.

We take the opportunity to mention that in the treatise of Poincaré, quoted already, one can find the first example of an almost periodic function in the sense of Bohr, represented by the series

$$f(t) = \sum_{k=1}^{\infty} a_k \sin \lambda_k t, \quad t \in R, \tag{A17}$$

which series must be uniformly convergent (in particular, in the space $AP_1(R,\mathcal{C})$). The exponents λ_k, $k \geq 1$, are arbitrary real numbers. This example is not very often mentioned in the literature. Of course, Poincaré did not use the term *almost periodicity*, which appeared 30 years later in Bohr's work.

A.3 CONSTRUCTION OF ANOTHER CLASSICAL SPACE

This section is a sample for constructing spaces of almost periodic functions, starting from a certain class of trigonometric series of the form (A2). The content of this section is a translation from, French into English, of our paper [165]. The space we have in mind is the space $B(R,\mathcal{C})$ of Besicovitch, the largest among the spaces of almost periodic functions with deep representation in the literature.

The construction we have used in Corduneanu [165] is based on the following scheme: first, we start with the classical space of Bohr, $AP(R,\mathcal{C})$, whose construction consists in choosing from the set of all trigonometric series, of the form (A2), those which are summable, with respect to uniform convergence on R, by the method of Cesaró–Fejér–Bochner (see, e.g., Corduneanu [129]). The linear space of such series, which consists of series summable uniformly on R to functions in $BC(R,\mathcal{C})$, denoted by $AP(R,\mathcal{C})$, is a closed subspace of $BC(R,\mathcal{C})$, hence a Banach space.

But besides the supremum norm, used in constructing $AP(R,\mathcal{C})$, one can define another norm on the elements of $AP(R,\mathcal{C})$. Namely, the Poincaré's norm

$$\lim_{\ell\to\infty}(2\ell)^{-1}\int_{-\ell}^{\ell}|f(t)|\,dt = M(|f|). \tag{A18}$$

Let us denote by $AP_M(R,\mathcal{C})$ the same elements of $AP(R,\mathcal{C})$, with the norm (A18), which constitutes a normed space. It turns out that $AP_M(R,\mathcal{C})$ is not a complete normed space, hence not a Banach space.

Indeed, since

$$M\{|f|\} = \lim_{\ell\to\infty}(2\ell)^{-1}\int_{-\ell}^{\ell}|f(t)|\,dt \le |f|_{BC}, \tag{A19}$$

for any $f\in AP(R,\mathcal{C})$, there follows that the topology of $AP_M(R,\mathcal{C})$ is weaker, or at most equivalent to the topology of $AP(R,\mathcal{C})$. If we assume now that the topology of $AP_M(R,\mathcal{C})$ is equivalent to that of $AP(R,\mathcal{C})$, then inequality for the norms, of the form

$$|f|_{BC}\le CM\{|f|\},\quad C>0, \tag{A20}$$

should be valid for some C, and some $f\in AP(R,\mathcal{C})$. But (A20) is impossible, for the following reason: the identity map from $AP(R,\mathcal{C})$ is the same as in $AP_M(R,\mathcal{C})$. Let us also notice that (A20) is the consequence of Banach theorem about the continuity of the inverse linear operator.

We will now construct an example which shows that (A20) cannot hold, for any $C>0$.

Indeed, let us consider the sequence of periodic functions $\{f_n(t); n \geq 2\}$, defined by $f_n(t+1) = f_n(t)$, $t \in [0,1)$, where

$$f_n(t) = \begin{cases} 1 - nt, & 0 \leq t < n^{-1}, \\ 0, & n^{-1} \leq t < 1 - n^{-1}, \\ 1 - n + nt, & 1 - n^{-1} \leq t < 1. \end{cases} \tag{A21}$$

One easily finds from (A21) that $M\{f_n\} = n^{-1}$, $n \geq 1$, while $\sup f_n = 1$ on R_+ (we can extend f_n from R_+ to R, also by periodicity). Therefore, (A20) should hold, which leads to the inequality

$$1 \leq Cn^{-1}, \quad n \geq 1. \tag{A22}$$

This contradiction proves that the topology of $AP_M(R,\mathcal{C})$ is, generally, weaker than the topology of $AP(R,\mathcal{C})$.

Therefore, it makes sense to consider the completion of $AP_M(R,\mathcal{C})$, which will lead to a space (finally, Banach), $B(R,\mathcal{C})$.

What can we say about the completed space $B(R,\mathcal{C})$, relying on the basic results of Functional Analysis? Likely, many properties of $B(R,\mathcal{C})$ can be derived starting from the properties of $AP(R,\mathcal{C})$ or $AP_M(R,\mathcal{C})$.

We shall dwell here only with the existence of the mean value for the elements of $B(R,\mathcal{C})$, a property which has many applications and, first of all, the construction of the Fourier series attached to an element/function of $B(R,\mathcal{C})$.

We recall the fact that the Fourier series of a function from the Bohr space $AP(R,\mathcal{C})$ can be constructed in a rather elementary way. The basic properties of the mean value $M\{f\}$, with $f \in AP(R,\mathcal{C})$, are the following:

1) $f \to M\{f\}$ is a linear continuous map from $AP(R,\mathcal{C})$ into $\mathcal{C}$.
2) $|M\{f\}| \leq M\{|f|\}, f \in AP(R,\mathcal{C})$, which proves continuity.
3) $f(t) \geq 0$, $t \in R$, implies $M\{f\} \geq 0$ and $M\{f\} = 0$ only in case $f \equiv 0$;
4) $M\{\bar{f}\} = \overline{M\{f\}}$.

One can now apply the Hahn–Banach theorem for the extension of functionals, from $AP(R,\mathcal{C})$ to $B(R,\mathcal{C})$.

The functional to be extended from the space $AP(R,\mathcal{C})$ to $B(R,\mathcal{C})$ is the one representing the mean value, $f \to M\{f\}$, and we denote its extension by $\tilde{M}\{f\}$. For motivation of the existence, see Yosida [541] or another treatise on Functional Analysis, dealing with this important result. The extended functional $\tilde{M}\{f\}$ satisfies the conditions $\tilde{M}\{f\} = M\{f\}$ for $f \in AP(R,\mathcal{C})$, while $\|\tilde{M}\| = \|M\|$. Since the elements of the space $B(R,\mathcal{C})$ are locally integrable functions on R, the expression of $\tilde{M}$ remains the same as for M, that is, the formula (A18) above.

Taking into account that $AP(R,\mathcal{C})$ is dense in $B(R,\mathcal{C})$, one can easily derive for $\widetilde{M}$ the properties listed earlier.

An important remark is connected to the null space in $B(R,\mathcal{C})$, which consists of those functions satisfying

$$\lim_{\ell\to\infty} (2\ell)^{-1} \int_{-\ell}^{\ell} |f(t)|\, dt = 0. \tag{A23}$$

If we denote by N_0 the null space of $B(R,\mathcal{C})$, it is obvious that N_0 contains functions from R to $\mathcal{C}$, not everywhere zero. The last step to obtain a Banach space, which we continue to denote by $B(R,\mathcal{C})$, consists in taking the factor space of $B(R,\mathcal{C})$, as defined earlier, with respect to the null space $N_0 = N_0(R,\mathcal{C})$, say B/N_0. It is well known that each element of B/N_0 is a class of equivalence, according to the relationship $f_1 \simeq f_2$,

$$\lim_{\ell\to\infty} (2\ell)^{-1} \int_{-\ell}^{\ell} |f_1(t) - f_2(t)|\, dt = 0. \tag{A24}$$

Also, it is well known that each element of the factor space is completely characterized by any single function belonging to it.

In concluding this section, we notice that a similar discussion/construction can be achieved to obtain the space $B^2(R,\mathcal{C}) = AP_2(R,\mathcal{C})$, substituting to the norm $f \to M\{|f|\}$ on $AP(R,\mathcal{C})$, the norm

$$f \to (M\{|f|^2\})^{\frac{1}{2}}. \tag{A25}$$

A.4 CONSTRUCTING SPACES OF OSCILLATORY FUNCTIONS: EXAMPLES AND METHODS

With the third stage of Fourier analysis came the generalized trigonometric series of the form (A1). The exponents are functions, generally nonlinear, instead of the linear products $\lambda_k t$, $k \geq 1$, $t \in R$. One can expect more complexity from this point of view, but it is already known that such series can be organized as Banach spaces, which implies the fact that their corresponding functions can also be organized as Banach spaces.

The literature (and we mean the mathematical one, even though this type of series appeared first in the engineering literature) in book format, mentions Osipov's and Zhang's (see Bibliography). If we consider only the mathematical literature, we can say that the third stage started in the last decade of the twentieth century, when spaces of oscillatory functions were constructed and a theory was begun.

With regard to the Bohr–Fresnel (B–F) almost periodic functions, as Osipov called his class/space of oscillating functions, there is a representation result (due to Osipov), which can be stated as follows:

Let $f: R \to C$ be a B–F almost periodic function. Then, there exists an almost periodic function of two real variables, in the sense of Bohr, say $F: R \times R \to C$, such that $f(t) = F(t, t^2)$.

In other words, the B–F almost periodic functions, can be represented in a simple manner by means of a Bohr almost periodic function, from R^2 into C.

But when we want to construct the Fourier series associated with $f: R \to C$, assuming f to be B–F, one obtains a generalized Fourier series of the form (A1). Namely, as shown in Osipov's book [432], one constructs the generalized Fourier series of the form

$$f \simeq \sum c(a,k)\, e^{(iat^2 + 2i\lambda_k t)}, \tag{A26}$$

where a and $\lambda_k{}'s$ denote real numbers, while $c(a,k)$ are complex. The series (A18) is, indeed, of the form (A1), the functional exponents being quadratic polynomials in t.

Several basic properties encountered for the Bohr almost periodic functions remain valid for the B–F almost periodic functions. We illustrate only a few of them.

I. The Poincaré *mean value* exists for any B–F almost periodic function f, namely

$$M(f : a, \lambda) = \lim_{\ell \to \infty} (2\ell)^{-1} \int_{-\ell}^{\ell} f(t)\, e^{(iat^2 + 2i\lambda_k t)}\, dt, \tag{A27}$$

for each couple $(a, \lambda) \in R \times R$. But it is nonzero only for a countable set of $\lambda' s$, say $\{\lambda_k;\, k \geq 1\} \subset R$.

II. The Fourier generalized series for an almost periodic function, of B–F type, converges in the sense of quadratic mean to the generating function. The coefficients are given by the following formula:

$$c(a,k) = M\{f(t)\, e^{(-iat^2 - 2i\lambda_k t)}\}, \quad k \geq 1. \tag{A28}$$

III. Each B–F almost periodic function can be uniformly approximated on R, by means of generalized trigonometric polynomials of the form

$$\sum c(a,k)\, e^{(iat^2 + 2i\lambda_k t)}, \tag{A29}$$

with a and k running on finite sets of real values and $c(a,k)$ taking complex values.

IV. If f is B–F almost periodic, then $M\{|f|^2\} = 0$ only when $f \equiv 0$.

Remark A.1 *Since the functional exponents are quadratic polynomials for the B–F almost periodic functions, it is a valid question asking what kind of such polynomials, or even more general functions can be used to construct other spaces of oscillatory functions?*

This question was answered by Zhang and some of his students and collaborators [552, 553, 557, 558] who, besides the space of *pseudo-almost periodic* functions, denoted by $PAP(R,\mathcal{C})$, also constructed the space of *strong limit power* functions, denoted $\mathcal{SLP}(R,\mathcal{C})$, as well as several other spaces of oscillatory functions.

The space of pseudo-almost periodic functions, $PAP(R,\mathcal{C})$, consists of elements/functions which can be represented (uniquely) in the form $f = g + \phi$, with $f \in AP(R,\mathcal{C})$ and ϕ an *ergodic* perturbation, that is, a continuous and bounded function on R, such that its Poincaré's mean value is zero.

We will now dwell on the case of the space $\mathcal{SLP}(R,\mathcal{C})$, which was constructed and investigated in several of the quoted papers under the name of Zhang et al. The expository paper [556] presents the basic definition of the space and properties (like, for instance, compactness criteria). This was achieved by generalizing a result due to Kronecker (found in several books on almost periodic functions).

Let us mention that by limit power function, one understands a map $f : R \to \mathcal{C}$, locally integrable and such that

$$\lim_{\ell\to\infty}(2\ell)^{-1}\int_{-\ell}^{\ell}|f(t)|^2\,dt, \tag{A30}$$

exists as a finite number. The set of maps, satisfying the condition (A.30), appears to be too wide for application purposes. A restricted class of functions within the set of those with property (A30), possesses more properties that make it closer to the space $AP(R,\mathcal{C})$.

According to Zhang [556], the definition of the concept of a *strong limit power* function/space is as follows:

Let $Q(R,R)$ be the set of functions of the form

$$q(t) = \begin{cases} \sum_{j=1}^{m}\lambda_j t^{\alpha_j}, & t \geq 0, \\ -\sum_{j=1}^{m}\lambda_j(-t)^{\alpha_j}, & t < 0, \end{cases} \tag{A31}$$

where $m \geq 1$, $\lambda_j \in R, j = 1,2,\ldots,m$, $\alpha_1 > \alpha_2 > \cdots > \alpha_m > 0$. One sees that $Q(R,R)$ consists of odd functions and is a group with respect to addition.

A generalized trigonometric polynomial is a function of the form

$$P(t) = \sum_{k=1}^{n} c_k e^{i q_k(t)}, \tag{A32}$$

with $c_k \in \mathcal{C}$, $q_k \in \mathcal{Q}(R,R)$, $k = 1,2,\ldots,n$.

The space of *strong limit power functions*, $\mathcal{SLP}(R,\mathcal{C})$, is the closure, with respect to uniform convergence on R, that is, in $BC(R,\mathcal{C})$, of the set of polynomials of the form (A32).

When substituting the usual trigonometric polynomials with those of the form (A32), we get *generalized* oscillatory functions, as limits in $BC(R,\mathcal{C})$, and generalized trigonometric or Fourier series of the form

$$f(t) = \sum_{k=1}^{\infty} c_k e^{i q_k(t)}. \tag{A33}$$

It turns out that, for every $f \in \mathcal{SLP}(R,\mathcal{C})$ the limit

$$M\{f(t)\,e^{-iq(t)}\} = \lim_{\ell\to\infty} \int_{-\ell}^{\ell} f(t)\,e^{-iq(t)}\,dt$$

exists and, moreover provides the formula for the Fourier coefficients in the series (A33):

$$c_k = \lim_{\ell\to\infty} (2\ell)^{-1} \int_{-\ell}^{\ell} f(t)\,e^{-iq_k(t)}\,dt. \tag{A34}$$

The analogy to the case of the space $AP(R,\mathcal{C})$ can be continued, leading to similar results. For instance, the Parseval's equality holds true:

$$\sum_{k=1}^{\infty} |c_k|^2 = M\{|f|^2\} = \lim_{\ell\to\infty} (2\ell)^{-1} \int_{-\ell}^{\ell} |f(t)|^2\,dt. \tag{A35}$$

Also, if the Fourier series is uniformly convergent on R, then it is convergent to its generating function.

Defining a *generalized translate* of a polynomial of the form (A32), by the formula

$$R_s^{\alpha}\,P(t) = \sum_{k=1}^{n} c_k e^{i(q_k(t)+q_k(s))}, \tag{A36}$$

for $s \in R$ and all $t \in R$, one easily finds out that R_s^{α} is a linear operator (on the set of polynomials like $P(t)$), continuous and with norm $\|R_s^{\alpha}\| \leq 1$. The concept of generalized operator of translation allows to establish other properties, similar to those encountered in the classical theory.

For instance, any $f \in \mathcal{SLP}(R,\mathcal{C})$ is normal, which means that the set of the generalized translates $\{R_s^\alpha f : s \in R\}$ is relatively compact in $BC(R,\mathcal{C})$, which reminds us of Bochner's property for Bohr almost periodic functions.

An important result in Zhang [554] is the following profound equivalence: In the space $\mathcal{SLP}(R,\mathcal{C})$, the following three statements are equivalent:

1) The uniqueness theorem: distinct Fourier series of functions in $\mathcal{SLP}(R,\mathcal{C})$, correspond to different generating functions.
2) Parseval's equality holds true for any Fourier series of a function in $\mathcal{SLP}(R,\mathcal{C})$:

$$\sum_{k=1}^{\infty} |c_k|^2 = M\{|f|^2\}, \quad f \in \mathcal{SLP}(R,\mathcal{C}).$$

3) Cauchy's multiplication rule holds true:
if

$$f \sim \sum_{k=1}^{\infty} c_k\, e^{i\, q_k(t)} \quad \text{and} \quad g \sim \sum_{k=1}^{\infty} d_k\, e^{i\, r_k(t)},$$

then

$$f g \sim \sum_{k=1}^{\infty} h_k\, e^{i w_k(t)},$$

where

$$h_k = \sum c_m d_n, \quad \textit{with } q_m + r_n = w_k.$$

Another result from classical theory, transposed to the generalized case, states:

Assume all generalized Fourier coefficients of a function $f \in \mathcal{SLP}(R,\mathcal{C})$ are positive. Then the generalized Fourier series converges uniformly on R to f.

In concluding this section, we can state that generalized Fourier series attached to $\mathcal{SLP}(R,\mathcal{C})$ functions enjoy similar properties to the classical Bohr model. This fact implies the necessity of finding new sets of generalized Fourier exponents that can be used in the construction of new spaces of oscillatory functions, capable to serve in describing, mathematically, phenomena of higher and higher complexity.

A.5 CONSTRUCTION OF ANOTHER SPACE OF OSCILLATORY FUNCTIONS

From the construction of the space $\mathcal{SLP}(R,\mathcal{C})$, we can see that a certain constructive approach was adopted. Namely, to produce first some generalized "trigonometric" polynomials, of the form (A32), and to define the oscillatory functions space as the closure of the set of those polynomials, with respect to the uniform convergence on R (i.e., in the space $BC(R,\mathcal{C})$). This is a rather familiar procedure in modern analysis and, in the case of the space $AP(R,\mathcal{C})$, it was first used by Bogoliubov in the 1930s, who started the theory from the property/definition of Bohr and constructed the approximating trigonometric polynomials (thus, the equivalence was directly established).

In this way, Zhang obtained the space $\mathcal{SLP}(R,\mathcal{C})$, which consists of continuous elements, creating the theory: a tool that provides help in getting new spaces of oscillatory functions. It is adequate to notice that the new space covers the needs expressed in the work of some beneficiaries of the theory (see the references in Zhang's works quoted earlier in this section).

But continuity of the functions describing various processes, encountered in various applications of Fourier analysis, is not always required. For instance, measurable functions/solutions can be useful for the application's needs, if they can be satisfactorily approximated. One knows that the step functions constitute a very good class of approximations for the general measurable functions. To illustrate this aspect, we send the readers to the Lang's book [320], with pertinent treatment of the matter.

In this section we would like to provide the construction of the space we denote by $B_\lambda(R,\mathcal{C})$, which belongs to the category of Besicovitch-type function spaces of oscillatory functions. The index λ stands for the class of $\lambda_k{}'s$ accepted in the representation (A33) of a generalized Fourier series. The case when λ designates the set of linear functions $\{\lambda_k t; k \geq 1, t \in R\}$ was discussed earlier, obtaining the classical Besicovitch space $B = B^1(R,\mathcal{C})$ of almost periodic functions.

We choose as set of generalized Fourier exponents, the set $\mathcal{Q}(R,R)$ we defined already, consisting of the set of generalized polynomials (A31) and (A32) forming the Gelfand algebra.

The corresponding generalized Fourier series is of the form (A33), with $\lambda_k(t) = P_k(t) \in \mathcal{Q}(R,R)$.

Using again the method of completion of the space $\mathcal{SLP}(R,\mathcal{C})$, with respect to the norm induced by Poincaré's mean value, $f \to M\{|f|\}$, we will obtain a Banach space of oscillatory functions. In the work of Zhang [554], there is a construction of this space, based on a different procedure than that we use in the following.

First, let us notice the fact that, for the space $\mathcal{SLP}(R,\mathcal{C})$, we can give a characterization, similar to that for the space $AP(R,\mathcal{C})$, within the class of oscillatory functions with generalized Fourier series of the form (A1). Namely, the following statement is a consequence of the properties of the space $\mathcal{SLP}(R,\mathcal{C})$, as they appear in the Zhang's paper quoted before.

The space $\mathcal{SLP}(R,\mathcal{C})$ corresponds to all generalized Fourier series of the form (A33), with exponents from $Q(R,R)$, which are summable by the method of Cesaró–Fejér–Bochner.

Starting from this fact and relying on Zhang's paper [554], among other things on the properties of the mean value, which are the same as in the case of the space $AP(R,\mathcal{C})$. We remind them of the following:

1) $M\{f\}$ is a linear continuous map on $\mathcal{SLP}(R,\mathcal{C})$.
2) $M\{\bar{f}\} = \overline{M\{f\}}$.
3) $|M\{f\}| \leq M\{|f|\}$.
4) $M\{f\} \geq 0$ if $f(t) \geq 0$, with $M\{f\} = 0$ only for $f(t) \equiv 0$.

We consider now the completion of the Banach space $\mathcal{SLP}(R,\mathcal{C})$ with respect to the norm $f \to M\{|f|\}$. This is a unique Banach space, which will be denoted by $B_\lambda(R,\mathcal{C})$, where λ stands for the set of all $\lambda_k(t)'s$, the generalized functional exponents in (A1).

We extend, according to Hahn–Banach theorem, the functional $f \to M\{f\}$, from $\mathcal{SLP}(R,\mathcal{C})$ to B_λ. Denoting the extended functional by $\widetilde{M}\{f\}$ we shall have $\widetilde{M}\{f\} = M\{f\}$ for $f \in \mathcal{SLP}(R,\mathcal{C})$ and $\|\widetilde{M}\| = \|M\|$, on $B_\lambda(R,\mathcal{C})$, respectively $\mathcal{SLP}(R,\mathcal{C})$.

The extension of the mean value functional, to the space $B_\lambda(R,\mathcal{C})$, allows us to construct the Fourier series of the functions in $B_\lambda(R,\mathcal{C})$.

To summarize the process of constructing the Besicovitch-type space of oscillatory functions, we start with the Zhang's space $\mathcal{SLP}(R,\mathcal{C})$, consisting of those series of the form (A1), with generalized exponents from the algebra $Q(R,R)$, which are summable with respect to the uniform convergence on R (i.e., in the supremum norm). In this space, one defines the functional (mean value) $f \to M\{f\}$, by the same procedure as in case of the space $AP(R,\mathcal{C})$. The map $f \to M\{|f|\}$ is a norm on $\mathcal{SLP}(R,\mathcal{C})$, leading to a weaker topology than the one induced by the supremum norm (generally speaking, we have seen the example of this type, in case of the AP-space). Depending on the set of generalized Fourier exponents we may ask, whether or not, this situation arises in the case of arbitrary sets of generalized Fourier exponents, or there are some sets of generalized exponents when the new norm leads to a complete space (this will mean, a topology equivalent to the original one for the space $\mathcal{SLP}(R,\mathcal{C})$). This problem must be clarified in the future.

Therefore, the last step, in getting the Banach space $B_\lambda(R,\mathcal{C})$, $\lambda = Q(R,R)$, is the completion of the Zhang's space $\mathcal{SLP}(R,\mathcal{C})$. In terms of the functions naturally belonging to the space $B_\lambda(R,\mathcal{C})$, we can say that they are the result of summability of the series (A1), with respect to the semi-norm of Poincaré $f \to \lim_{\ell\to\infty}(2\ell)^{-1}\int_{-\ell}^{\ell}|f(t)|\,dt$, for those $f \in L^1_{\mathrm{loc}}(R,\mathcal{C})$ for which the limit exists (as a finite number). Of course, this semi norm becomes a norm in the factor space $B_\lambda(R,\mathcal{C})$, because the null space in $\mathcal{SLP}(R,\mathcal{C})$ consists only of zero; hence, it is a closed subspace of $\mathcal{SLP}(R,\mathcal{C})$.

As in the classical case of almost periodic function spaces, the space $B_\lambda(R,\mathcal{C})$ is the richest space in Besicovitch-type category. For instance, compared with the most often encountered Besicovitch space $B^2_\lambda(R,\mathcal{C})$, which appears in Zhang [554] with its entire construction, the inclusion $B^2_\lambda(R,\mathcal{C})$ in $B_\lambda(R,\mathcal{C})$, for $\lambda = Q(R,R)$, follows from the integral inequality

$$(2\ell)^{-1}\int_{-\ell}^{\ell}|f(t)|\,dt \le \left[(2\ell)^{-1}\int_{-\ell}^{\ell}|f(t)|^2\,dt\right]^{\frac{1}{2}},$$

obtained by applying the well known Cauchy inequality $|\int_a^b fg|^2 \le \int_a^b f^2 \int_a^b g^2$.

Unlike other spaces of almost periodic functions, or spaces of oscillatory functions, the spaces $B(R,\mathcal{C})$, not to mention yet $B_\lambda(R,\mathcal{C})$, have not been involved in many applications. Perhaps, the most frequently encountered application is related to the introduction of pseudo-almost periodic functions, see Zhang [553], in which case the perturbations of the almost periodic component (of a pseudo-almost periodic function) must also belong to the space $B(R,\mathcal{C})$. See also Diagana [193].

In concluding this section, we remark that, an important tool in getting new spaces of oscillatory functions, is the possession of the set λ, consisting of generalized Fourier exponents, with the property of orthogonality. We attempt to discuss this problem, still in its infancy. We do not claim we have gotten a decisive answer to the problem of "producing" systems/sets of generalized Fourier exponents, but we make just an attempt to further a bit the search of those functional exponents.

A.6 SEARCHING FUNCTIONAL EXPONENTS FOR GENERALIZED FOURIER SERIES

We notice that, except for the pseudo-almost periodic functions, which are largely represented in the mathematical literature, only sporadic contributions have been made to advance and apply the theory started by Osipov [432] and Zhang [552, 554, 556]. More than two decades have elapsed since the birth of the *third stage* in Fourier analysis, and only a small number of papers has been dedicated to this new and promising field.

The main problem we discuss in this section is how can we proceed to produce systems of functions, even a sequence like $\{\lambda_k(t); t \in R, k \geq 1\}$, which consists of maps from R into R, and satisfies the condition

$$\lim_{\ell \to \infty} (2\ell)^{-1} \int_{-\ell}^{\ell} e^{i[\lambda_k(t)-\lambda_j(t)]}\, dt = \begin{cases} 0, & k \neq j, \\ 1, & k = j. \end{cases} \tag{A37}$$

Condition (A37) will be satisfied by a sequence of potential generalized exponents, if we can get distinct solutions of the following functional equation/relation:

$$\lim_{\ell \to \infty} (2\ell)^{-1} \int_{-\ell}^{\ell} e^{i\lambda(t)}\, dt = \begin{cases} 0, & \lambda(t) \not\equiv 0, \\ 1, & \lambda(t) \equiv 0. \end{cases} \tag{A38}$$

Let us dwell now on relation (A38), assuming that $\lambda(t), t \in R$, the unknown quantity there, is the restriction to R of an entire function $\lambda(z)$, $z = t + is \in C$, $t, s \in R$.

We will now apply Cauchy's integral theorem, to the function $e^{i\lambda(z)}$, along a closed contour, which consists of the semicircle in the semi-plane $s \geq 0$, with the segment $[-\ell, \ell]$, $\ell > 0$ as diameter. Let us denote this contour by C_ℓ for any $\ell > 0$. The result is the equation

$$\int_{-\ell}^{\ell} e^{i\lambda(t)}\, dt + \int_{S_\ell} e^{i\lambda(z)}\, dz = 0, \tag{A39}$$

where S_ℓ is the semicircle on which z runs from $(\ell, 0)$ to $(-\ell, 0)$.

If one denotes by $\Lambda(z)$ a primitive of $e^{i\lambda(z)}$, which is also an entire function, equation (A39) gives

$$\int_{-\ell}^{\ell} e^{i\lambda(t)}\, dt = \Lambda(-\ell) - \Lambda(\ell). \tag{A40}$$

One can, therefore, rephrase the basic equation/relation (A38) as follows:

$$\ell^{-1}[\Lambda(\ell) - \Lambda(-\ell)] = o(1), \quad \text{as} \quad \ell \to \infty. \tag{A41}$$

Relation (A41) is certainly a source for finding entire functions $e^{i\lambda(z)} = \Lambda'(z)$, such that (A38) is valid.

Obviously, (A40) leads to the following relationship:

$$\lim_{\ell \to \infty} (2\ell)^{-1} \int_{-\ell}^{\ell} e^{i\lambda(t)}\, dt = \lim_{\ell \to \infty} (2\ell)^{-1} [\Lambda(-\ell) - \Lambda(\ell)]. \tag{A42}$$

More precisely, one has to estimate the limit on the right-hand side of (A42), with $\Lambda'(z) = e^{i\lambda(z)}$, in order to satisfy equation (A38).

When $\lambda(t) \equiv 0$, $t \in R$, it is obvious that the second part of (A38) is satisfied.

When $\lambda(t) \not\equiv 0$, $\lambda(z) = \lambda z$, $\lambda \in R$, $\lambda \neq 0$, $z \in \mathcal{C}$, we obtain from (A42)

$$\lim_{\ell \to \infty} (2\ell)^{-1} \int_{-\ell}^{\ell} e^{i\lambda(t)}\, dt = (i\lambda)^{-1} \lim_{\ell \to \infty} (2\ell)^{-1} [e^{i\lambda\ell} - e^{-i\lambda\ell}],$$

and since the bracket is bounded in ℓ, $\ell \in R$, there results the first part in (A38).

In other words, Fourier series of the form (A2), that is, with linear exponents

$$\sum_{k=1}^{\infty} a_k e^{i\lambda_k t}, \quad t \in R, \quad \lambda_k \in R, \quad k \geq 1,$$

are candidates for oscillatory functions. As we know, these series correspond to various classes of almost periodic functions. For instance, if the series is summable by the method of Cesaró–Fejér–Bochner, it will represent a Bohr almost periodic function. Also, if $\sum_{k=1}^{\infty} |a_k|^r < \infty$, with $1 \leq r \leq 2$, the series in (A2) will correspond to an $AP_r(R, \mathcal{C})$ almost periodic function. For $r = 2$, we obtain the Besicovitch function space $B^2(R, \mathcal{C}) = AP_2(R, \mathcal{C})$.

The case $p = 1$ is an open problem, that is, the Besicovitch space $B(R, \mathcal{C})$. We do not have yet a characterization in terms of the Fourier coefficients series only. We know only $|\lambda_k| \leq M\{|f|\}$.

Let us also notice that the periodic case, corresponding to $\lambda = (0, \lambda_1, \lambda_2, \ldots, \lambda_k, \ldots)$, occurs only in case $\lambda_k = \pm k\omega$, $k \geq 0$, $\omega > 0$.

All the cases considered here are covered by the fact that one chooses in (A42) $\lambda(z) = \lambda z$, $\lambda \in R$.

Without trying to get the general solution for the equation (A42), we shall continue its investigation, in order to establish other connections related to the third stage of Fourier analysis, with particular regard to the relationship between series and attached function (as the sum, according to a certain norm or semi-norm).

One cannot be, but amazed, by the fact that the simplest solutions of equation (A38), lead to the findings of our forefathers, in constructing the first two stages of Fourier analysis (covering the cases of periodicity and almost periodicity). Of course, this happened during a period encompassing more than two centuries in the history of the Fourier analysis.

We invite the readers to look for other analytic solutions to (A41).

We will consider a special choice for $\lambda(t)$, namely, assuming

$$\lambda(-t) = -\lambda(t), \quad t \in R, \tag{A43}$$

that is, we deal with possible generalized Fourier exponents among the class of odd functions on R. This is a category that is richer than the one considered by Zhang [553] in constructing the space $\mathcal{SLP}(R,\mathcal{C})$. Indeed, $Q(R,R)$ contains only odd functions (by construction).

We estimate the first term/integral which appears above (A38) as follows:

$$\int_{-\ell}^{\ell} e^{i\lambda(t)}\,dt = \int_{-\ell}^{0} e^{i\lambda(t)}\,dt + \int_{0}^{\ell} e^{i\lambda(t)}\,dt = 2\int_{0}^{\ell} \cos\lambda(t)\,dt,$$

due to (A43). Therefore, to satisfy the first part of condition (A38), one needs to impose on $\lambda(t)$ the following restriction:

$$\int_{0}^{\ell} \cos\lambda(t)\,dt = o(\ell), \quad \ell \to \infty. \tag{A44}$$

Only those odd functions $\lambda : R \to R$, at least locally integrable, can be taken as candidates for generalized Fourier exponents.

Equation/Relation (A44) has nontrivial solutions. For instance, (A44) is verified for any linear function $\lambda(t) = \mu t$, $\mu \in R$, in which case we get the results obtained before, when choosing $\lambda(z) = \lambda z$, $\lambda \in R$, which led to the class of almost periodic functions (in various senses).

Can we get new solutions to equation/relation (A44)? The answer is positive and one good example relies on the Fresnel integrals. It is well known that

$$\int_{0}^{\infty} \cos t^2\,dt = \frac{1}{2}\sqrt{\frac{\pi}{2}}, \quad \int_{0}^{\infty} \sin t^2\,dt = \frac{1}{2}\sqrt{\frac{\pi}{2}}. \tag{A45}$$

Assuming that (A45) holds, that is, $\lambda(t) = t^2$, $t \in R$, there follows that t^2 is one of the admissible choices as a generalized Fourier functional exponent, which takes us to the Osipov's kind of almost periodicity. Indeed, let us first notice that by substituting $t = \alpha\tau$, say $\alpha > 0$, then (A45) leads to the following relationships:

$$\alpha\int_{0}^{\infty} \cos(\alpha\tau)^2\,d\tau = \frac{1}{2}\sqrt{\frac{\pi}{2}} \quad \text{or} \quad \int_{0}^{\infty} \cos(\alpha\tau)^2\,d\tau = \frac{1}{2\alpha}\sqrt{\frac{\pi}{2}}.$$

This means that αt^2, $\alpha > 0$, is also an acceptable solution to (A44).

It is a simple fact that a system of the form $\{\phi(t) + \lambda_k t; k \geq 1\}$, with $\phi(t)$ a function $R \to R$ possessing a (finite) mean value in Poincaré's sense, where $\{\lambda_k t; k \geq 1\}$ is already orthogonal, because $\lambda_k \neq \lambda_j$ for $k \neq j$, is also orthogonal. One has to take into account the obvious relationship $(\phi + \lambda_k t) - (\phi + \lambda_j t) = (\lambda_k - \lambda_j)t$ (see (A37)).

If the Fourier series attached to various types of almost periodic functions has the form $\sum_{k=1}^{\infty} a_k e^{i\lambda_k t}$, as seen earlier (A3), the generalized Fourier series for B–F type of almost periodic functions (Osipov) has the form shown by (A26), with α running on a finite set of reals and $k \geq 1$.

We have mentioned already some properties of the B–F almost periodic functions, showing their similarity with the classical model of Bohr almost periodic functions. In the quoted book by Osipov, we find the statement according to which, Wiener dealt with the so-called (by Osipov) Wiener's waves, when considering the functions $e^{(i\alpha t^2 + i\beta t)}$, in his investigations within harmonic analysis.

In concluding the reference to the B–F almost periodic functions, we can state that Osipov produced the first full-fledged theory related to the third stage of Fourier analysis in his investigation of this concept and the attached generalized Fourier series.

The simplest definition of the space of B–F almost periodic function is the one based on the property of uniform approximation on R, by generalized trigonometric polynomials of the form

$$T(t) = \sum_{(\alpha,\lambda)} c_{(\alpha,\lambda)} e^{(i\alpha t^2 + 2\lambda_k t)}, \tag{A46}$$

where (α, λ) runs on a finite set of points in the real plane. In other words, the space is the closure of the set of all polynomials of the form (A46), with respect to the uniform convergence on R. Let us note that the polynomials (A46) are obtained by truncating the generalized Fourier series of the form (A26).

A remark is necessary to explain the use of t^2, which is not an odd function. Actually, we are dealing (on R) with the odd function obtained by extending to R_- the function t^2 by letting $\lambda(t) = -t^2$, for $t < 0$.

We will now consider the choice $\lambda(t) = \sin \lambda t$, $\lambda \in R_+$, to get a new type of generalized Fourier series. These series, pertaining to the third stage of Fourier Analysis, are of the form

$$\sum_{k=1}^{\infty} a_k e^{i \sin \lambda_k t}, \tag{A47}$$

with $a_k \in \mathcal{C}$, $\lambda_k \in R_+$, $k \geq 1$. Depending on the conditions imposed on $\{a_k; k \geq 1\}$, the series of the form (A47) may have as sum an oscillatory function belonging to the Bohr space of almost periodic functions. Of course, this should not be considered as a curiosity, due to the fact that periodic and almost periodic function spaces are included in the largest class of oscillatory function spaces.

Indeed, if we assume $\{a_k; k \geq 1\} \subset \ell^1(N,\mathcal{C})$, we secure the uniform convergence on R, of the series (A47). Since each term of this series is a periodic function, there results that its sum is a Bohr almost periodic function, that is, belongs to $AP(R,\mathcal{C})$. Is the space of the sum functions a proper subspace of $AP(R,\mathcal{C})$? We invite the readers to consider this problem.

If one accepts the condition $\{a_k; k \geq 1\} \subset \ell^2(N,\mathcal{C})$, then it is normal to expect sum functions that are Besicovitch almost periodic.

Of course, the question could be extended from $AP(R,\mathcal{C})$ and $AP_2(R,\mathcal{C})$ to any $AP_r(R,\mathcal{C})$, with $1 < r < 2$. We leave readers the task of carrying out the details.

Before concluding this section, let us comment on formula (A39), obtained under the assumption that $\lambda(t)$ is the restriction, to the real axis R, of an entire function $\lambda(z)$, $z \in \mathcal{C}$. This condition, quite restrictive, can be extended, assuming $\lambda(z)$ a meromorphic function on $\mathcal{C}$. Instead of the Taylor series for $\lambda(z)$, we have to work with the Laurent series, including the terms defining the residues. Under the extra assumption that the function $\lambda(z)$ has a finite number of poles (the Laurent series has a finite number of terms with negative exponents), formula (A39) should be substituted with

$$\int_{-\ell}^{\ell} e^{i\lambda(t)}\,dt + \int_{S_\ell} e^{i\lambda(z)}\,ds = 2\pi i \sum Res[e^{i\lambda(z)}],$$

the $\sum$ being extended to all residues of $\lambda(z)$, with ℓ sufficiently large, such that all poles are in the semi-disk, bounded by the segment $(-\ell,\ell)$ and S_ℓ. Of course, no poles admitted on R.

So far, we do not know any attempt in using formula (A41), in finding generalized Fourier exponents for series/functions related to the third stage of Fourier analysis.

In concluding Section A.6, we return to the case which was used by Zhang to prove the relation (A38), when $\lambda(t) = \sum_{j=1}^{m} \lambda_j t^{\alpha_j}$ for $t \geq 0$, with $\lambda_j \in R$ and α_k's such that $\alpha_1 > \alpha_2 > \cdots > \alpha_m > 0$, with antisymmetry with respect to the origin. Zhang has proven directly the validity of (A38), in this case, and we send the readers to his paper Zhang [554]. This choice led to the space $\mathcal{SLP}(R,\mathcal{C})$, which we described in Section A.4.

We will conclude this section with a brief list of the steps, one has to undertake, in order to construct new spaces of oscillatory functions, starting from a class of formal generalized Fourier series, as they appear after we get a set of generalized Fourier exponents. In addition to the spaces we presented earlier (Osipov, Zhang), we found other classes of generalized Fourier exponents, using the method sketched before. Namely, using Equation/Relation (A41), we instantly obtained the classical cases of almost periodic functions, which were known for a long time: the first and second stage of Fourier Analysis. Using

(A44), we re-obtained the generalized Fourier exponents for the B–F almost periodic functions (Osipov). In his book [432], Osipov started from Fourier–Fresnel transform and the space of maximal ideals of the algebra $AP(R,\mathcal{C})$ of Bohr almost periodic functions.

In the last example, we treated before, by means of relation/equation (A44), the case of $\lambda(t) = \sin \lambda_0 t$, $\lambda : R_+ \to R$, $\lambda_0 \in R_+$, led to generalized series of the form (A47), representing, under conditions $\{a_k; k \geq 1\} \subset \ell^1(R,\mathcal{C})$, a function in $AP(R,\mathcal{C})$. The topology may be different than in AP.

A challenge to our readers: find $\lambda(t)$ producing spaces of oscillatory functions, which are not almost periodic function spaces, but in the third stage of Fourier analysis.

The last part of this section contains a scheme, to be followed in constructing new spaces of oscillatory functions, after a set of generalized Fourier exponents is found. Of course, the practice in various applications should suggest those types of exponents, in accordance with their needs.

I. First, a set (at least countable) of functions of the form $e^{i\lambda(t)}$ should be identified, such that (A34) is valid. This set will contain the generalized Fourier exponents. We know that such exponents exist, for instance, the linear function λt, $\lambda \in R$ and $t \in R$, which leads to different spaces of almost periodic functions (depending on the choice of convergence, we look for series like $\sum_{k=1}^{\infty} a_k e^{i\lambda_k(t)}$). We also indicated earlier, in this section, some acceptable $\lambda_k(t)'s$ that leads to series pertaining to the third stage of Fourier analysis. It is proven (see Besicovitch [61] or Corduneanu [156]) that any function in $B(R,\mathcal{C})$ possesses a (finite) Poincaré's mean value. From such a set of functions, of the form $e^{i\lambda(t)}$, one can choose sequences with distinct generalized Fourier exponents.

II. With the set of formal generalized trigonometric series, constructed by means of such set of candidate exponents, organized as a linear space over $\mathcal{C}$, one needs to introduce a convergence/topology. This is usually achieved by choosing a norm or a semi-norm on the space of generalized trigonometric series, such as $\{a_k; k \geq 1\} \in \ell^1(N,\mathcal{C})$ or $\ell^2(N,\mathcal{C})$. Also, it is possible to choose a certain property of the series belonging to the same space of oscillatory functions. For instance, in order to define the Bohr space $AP(R,\mathcal{C})$, we must retain from the set of all generalized trigonometric series only those which are summable according to the linear Cesaró–Fejér–Bochner procedure, see Corduneanu [165] for details.

III. When we chose a semi-norm to define convergence, it becomes necessary to take the factor space, determined by means of this semi-norm, namely, the classes of equivalence being dictated by the fact that the formal series can converge (because of the non-uniqueness of the limit) to any of the functions in the same class of equivalence. Taking the factor space of the semi-normed linear space of generalized trigonometric series, one obtains a Banach space, which is the space of oscillatory functions we wanted to construct. See also Corduneanu [165].

Remark A.2 *In both the cases investigated by Osipov [432] and Zhang [554, 556], in view of constructing new spaces of oscillatory functions, the result is a space of continuous functions.*

We know, there exist spaces of periodic or almost periodic functions whose elements are not necessarily continuous. The functions belonging to such spaces (i.e., the sums of the trigonometric series, in the sense of the topology induced by norm or semi-norm), when discontinuous, but not compatible with the reality, can be approximated, in various senses, by means of continuous ones.

Let us point out that all the steps, mentioned earlier, have been taken in constructing the new spaces of oscillatory functions. This is because we want to illustrate the fact that the theory of spaces consisting of oscillatory functions, including the almost periodic case, can be started taking the trigonometric series, classical or generalized, as the point of departure (as primary element, instead of the trigonometric polynomials, in each stage).

A.7 SOME COMPACTNESS PROBLEMS

The compactness criteria play a significant role in connection with the applications to functional equations. They are rather complicated and, often, not easy to apply in existence or approximations of the exact solutions. One possible reason for this situation is, likely, the fact that the spaces of almost periodic functions are richer than most of the spaces encountered in modern analysis. For instance, classical spaces, like $L^2(a,b)$ or $C(a,b)$, have a countable subset which is dense in that space. This feature is not present in the case of almost periodic function spaces.

It appears, as a reasonable alternative, to the problem of finding handy compactness criteria in the spaces of oscillatory functions, to limit the search to some "subspaces," which may possess a countable basis. Such "subspaces" appear, usually, when searching for solutions of a functional equation, which involves only a finite number of oscillatory functions, each being determined by a set, at most countable, of Fourier exponents. The set of all Fourier exponents thus involved is also countable, a feature that allows us to look for a series solution within the "subspace" (I am tempted to call it a fiber) generated by the series with exactly the generalized exponents produced by the given equation. This approach may not be easy to pursue in all cases. We illustrated the case of the Zhang's space $\mathcal{SLP}(R,\mathcal{C})$.

We will rely on the concept of generalized translate, defined in Section A.4. The following compactness criterion in the whole space $\mathcal{SLP}(R,\mathcal{C})$ is given by Zhang [556].

Let $M \subset \mathcal{SLP}(R,\mathcal{C})$, then M is relatively compact, if and only if the following properties are satisfied:

$a)$ For each $t \in R$, the set $\{f(t); f \in M\}$ is bounded.

$b)$ The set M consists of equicontinuous functions, that is, to each $\epsilon > 0$, there corresponds $\delta > 0$, such that $|R_t^\alpha f - R_s^\alpha f| < \epsilon$, whenever $|t-s| < \delta$, for all $f \in M$.

$c)$ To each $\epsilon > 0$, the functions in M have common ϵ-translation numbers, that is, there exists a finite number of $a_k \in R$, $k = 1,\dots,n$, such that for each $s \in R$, one can find $a_s \in (a_1,\dots,a_n)$, with the property

$$|R_s^\alpha R_{\alpha_s}^\alpha f - f| < \epsilon, \quad f \in M.$$

At the time this Appendix was written, we do not know another case of space of oscillatory functions for which a criterion of compactness is known. Also, we do not have yet examples which show that a simpler type of criterion of compactness can be obtained, if we restrict our considerations to a "subspace," for instance, one generated by a countable set of generalized Fourier exponents.

We invite the readers to examine the above mentioned suggestions and start with the classical Bohr space $AP(R,\mathcal{C})$. Take into account the relationships based on the Parseval formula $M\{|f|^2\} = \sum_{k=1}^{\infty} |a_k|^2$, which establishes a correspondence between subsets of $AP(R,\mathcal{C})$ and those of $\ell^2(N,\mathcal{C})$:

$$M\{|f(t) - g(t)|^2\} = \sum_{k=1}^{\infty} |a_k - b_k|^2,$$

$$M\{|f(t+\tau) - f(t)|^2\} = \sum_{k=1}^{\infty} |a_k e^{i\lambda_k \tau} - a_k|^2,$$

$$f(t) \sim \sum_{k=1}^{n} a_k e^{i\lambda_k t}, \quad g(t) \sim \sum_{k=1}^{\infty} b_k e^{i\lambda_k t}.$$

The development of the third stage in Fourier analysis appears to be something to be achieved in the future.

BIBLIOGRAPHY

1. S. Abbas and M. Benchohra. Fractional order Riemann-Liouville integral inclusions with two independent variables and multiple time delay. *Opuscula Mathematica*, 33:209–222, 2013.
2. S. Abbas, M. Benchohra, and G. M. N'Guerekata. *Topics in Fractional Differential Equations*. Springer, New York, 2012.
3. U. Abbas and V. Lupulescu. Set functional differential equations. *Communications on Applied Nonlinear Analysis*, 18(1):97–110, 2011.
4. R. A. Aftabizadeh. Bounded solutions for some gradient type systems. *Libertas Mathematica*, 2:121–130, 1982.
5. R. P. Agarwal, M. Bohner, A. Domoshnitsky, and Y. Goltser. Floquet theory and stability of nonlinear integro-differential equations. *Acta Mathematica Hungarica*, 109(4):305–330, 2005.
6. R. P. Agarwal, B. De Andrade, and C. Cuevas. On type of periodicity and ergodicity to a class of integral equations with infinite delay. *Nonlinear and Convex Analysis*, 11:309–333, 2010.
7. R. P. Agarwal, A. Domoshnitsky, and Y. Goltser. Stability of partial functional integro-differential equations. *Dynamical and Control Systems*, 12:1–31, 2006.
8. R. P. Agarwal and L. Górniewicz. Aronszajn type results for Volterra equations and inclusions. *Topological Methods in Nonlinear Analysis*, 23:149–159, 2004.

Functional Differential Equations: Advances and Applications, First Edition.
Constantin Corduneanu, Yizeng Li and Mehran Mahdavi

9. R. P. Agarwal, S. K. Ntouyas, B. Ahmad, and M. S. Alhothuali. Existence of solutions for integro-differential equations of fractional order with nonlocal three-point fractional boundary conditions. *Advances in Difference Equations*, 2013:128, 2013.
10. R. P. Agarwal, D. O'Regan, and P. J. Y. Wong. *Positive Solutions of Differential, Difference and Integral Equations*. Kluwer Academic, Dordrecht, 1999.
11. B. Ahmad and S. Sivasundaram. Some stability results for set integro-differential equations. *Mathematical Inequalities and Applications*, 10:597–605, 2007.
12. H. M. Ahmed. Fractional neutral evolution equations with nonlocal conditions. *Advances in Difference Equations*, 2013:117, 2013.
13. AitDads, ElHadi, P. Cieutat, and K. Ezzinbi. The existence of pseudo-almost periodic solutions for some nonlinear differential equations in Banach spaces. *Nonlinear Analysis: Theory, Methods & Applications*, 69(4):1325–1342, 2008.
14. S. Aizicovici and V. Staicu. Multivalued evolution equations with nonlocal initial conditions in Banach spaces. *Nonlinear Differential Equations and Applications*, 14:361–376, 2007.
15. O. Akinyele. On partial stability and boundedness of degree k. *Rendiconti Accad. Naz. Lincei (Scienze Fisico-Matematiche)*, 65(6):259–264, 1978.
16. O. Akinyele. Conditional asymptotic invariant sets and perturbed systems. *Rendiconti Accademic Nazionale dei Lincei (Scienze Fisico-Matematiche)*, 67(4):214–220, 1979.
17. O. Akinyele. On non-uniform partial stability and perturbation for delay systems. *Rendiconti Accademic Nazionale dei Lincei (Scienze Fisico-Matematiche)*, 67(1–2):39–44, 1979.
18. O. Akinyele. On partial boundedness of differential equations with time delay. *Rivista Matematica Universita di Parma*, 7(4):9–21, 1981.
19. G. Alexits and I. Földes. *Convergence Problems of Orthogonal Series*. Pergamon Press, New York, 1961.
20. H. Amann and J. Escher. *Analysis III*. Springer, Berlin, 2009.
21. L. Amerio and G. Prouse. *Almost-periodic Functions and Functional Equations*. Van Nostrand Reinhold, New York, 1971.
22. L. Y. Anapolsky and V. M. Matrosov. *Problems in Qualitative Theory of Differential Equations*. Siberian Division, Nauka, 1988.
23. J. Andres and A. M. Bersani. Almost-periodicity problem as a fixed-point problem for evolution inclusions. *Topological Methods in Nonlinear Analysis*, 18:337–350, 2001.
24. J. Andres, A. M. Bersani, and R. F. Grande. Hierarchy of almost-periodic functions spaces. *Rendicente Matematica, Senesvir*, 26:121–188, 2008.
25. H. A. Antosiewicz. An inequality for approximate solutions of ordinary differential equations. *Mathematische Zeitschrift*, 78(1):44–52, 1962.

26. M. Ashordia, Sh. Akhalaia, and N. Kekelia. On necessary and sufficient conditions for stability of linear systems of generalized ordinary differential equations. *Memoirs on Differential Equations and Mathematical Physics*, 46:115–128, 2009.
27. F. V. Atkinson and J. R. Haddock. On determining phase spaces for functional differential equations. *Funkcialaj Ekvacioj*, 31:331–347, 1988.
28. J. P. Aubin. *Viability Theory. Systems & Control: Foundations & Applications.* Birkhäuser, Boston, 1991.
29. J. P. Aubin and A. Cellina. *Differential Inclusions: Set-Valued Maps and Viability Theory*. Springer-Verlag, New York, 1984.
30. A. Avantaggiati, G. Bruno, and R. Iannacci. The Hausdorff-Young theorem for almost periodic functions and some applications. *Nonlinear Analysis: Theory, Methods & Applications*, 25(1):61–87, 1995.
31. C. Avramescu. Asymptotic behavior of solutions of nonlinear differential equations and generalized guiding functions. *Electronic Journal of Qualitative Theory of Differential Equations*, (13):1–9, 2003.
32. N. V. Azbelev. Recent trends in the theory of nonlinear functional differential equations. In *Proceedings of the first world congress of nonlinear analysts*, pages 1807–1814, Tampa, FL, 1996.
33. N. V. Azbelev, V. P. Maksimov, and L. F. Rakhmatullina. *Introduction to the Theory of Functional Differential Equations: Methods and Applications.* Hindawi Publishing Corporation, New York, 2007.
34. N. V. Azbelev and P. M. Simonov. *Stability of Differential Equations with Aftereffect*. Taylor & Francis, London, 2003.
35. C. T. H. Baker, E. O. Agyingi, E. I. Parmuzin, F. A. Rihan, and Yihong Song. Sense from sensitivity and variation of parameters. *Applied Numerical Mathematics*, 56(3–4):397–412, 2006.
36. C. T. H. Baker, G. Bocharov, E. I. Parmuzin, and F. A. Rihan. Some aspects of causal and neutral equations used in modelling. *Journal of Computational and Applied Mathematics*, 229(2):335–349, 2009.
37. R. Balan. An extension of Barbashin-Krasovski-Lasalle theorem to a class of nonautonomous systems. *Princeton University: Program in Applied and Computational Mathematics*, pages 1–12, 1995.
38. D. Baleanu and G.-Ch. Wu. New applications of the variational iteration method—from differential equations to q-fractional difference equations. *Advances in Difference Equations*, 2013:21, 2013.
39. S. Banach. *Théorie des opérations linéaires, 1932: Theory of Linear Operations.* North-Holland, New York, 1987. English Translation.
40. J. Banaś and I. J. Cabrera. On existence and asymptotic behavior of solutions of a functional integral equation. *Nonlinear Analysis: Theory, Methods & Applications*, 66(10):2246–2254, 2007.
41. G. Bantaș. On the asymptotic behavior in the theory of Volterra integro-functional equations. *Periodica Mathematica Hungarica*, 5:323–332, 1974.

42. I. Barbălat. Systēmes d'équations differentielles d'oscillations nonlineaires. *Revue Roumaine de Mathématiques Pures et Appliquées*, IV:267–270, 1959.
43. E. A. Barbashin. *Liapunov Functions*. Nauka, Moscow, 1970. Russian.
44. V. Barbu. *Nonlinear Semigroups and Differential Equations in Banach Spaces*. Noordhoff, Leyden, 1976.
45. V. Barbu. *Nonlinear Differential Equations of Monotone Types in Banach Spaces*. Springer, New York, 2010.
46. V. Barbu and Th. Precupanu. *Convexity and Optimization in Banach Spaces*. Sijthoff & Noordhoff, Bucarest, 1978.
47. N. K. Bary. *A Treatise on Trigonometric Series*. Pergamon Press, New York, 1964.
48. B. Basit. Harmonic analysis and asymptotic behavior of solutions to the abstract Cauchy problem. *Semigroup Forum*, 54(1):58–74, 1997.
49. B. Basit and H. Günzler. Generalized Esclangon-Landau results and applications to linear difference-differential systems in Banach spaces. *Difference Equations and Applications*, 10(11):1005–1023, 2004.
50. B. Batko and J. Brzdek. A fixed point theorem and the Hyers-Ulam stability in Riesz spaces. *Advances in Difference Equations*, 2013:138, 2013.
51. H. Baudrand, M. Titaouine, N. Raveu, and G. Fontgland. Electromagnetic modeling of planar almost periodic structures. In *IEEE: International Microwave and Optoelectronics Conference*, pages 427–431, Belem, Brazil, 2009.
52. D. M. Bedivan and D. O'Regan. Fixed point sets for abstract Volterra operators on Fréchet spaces. *Applicable Analysis*, 76(1–2):131–152, 2000.
53. R. Bellman. *Stability Theory of Differential Equations*. Courier Dover Publications, New York, 2013.
54. M. Benchohra, J. Henderson, and F. Ouaar. Bounded solutions to an initial value problem for fractional differential equations on the half-line. *Pan American Mathematical Journal*, 21(2):35–44, 2011.
55. M. Benchohra and S. K. Ntouyas. Existence results on infinite intervals for neutral functional differential and integrodifferential inclusions in Banach spaces. *Georgian Mathematical Journal*, 7(4):609–625, 2000.
56. L. Berezansky. The positiveness of Cauchy functions and stability of linear differential equations with aftereffect. *Differential Equations*, 26:1440–1500, 1990. Russian.
57. L. Berezansky and E. Braverman. Boundedness and stability of impulsively perturbed systems in a Banach space. *International Journal of Theoretical Physics*, 33(10):2075–2090, 1994.
58. L. Berezansky and E. Braverman. On integrable solutions of impulsive delay differential equations. *eprint arXiv: funct-an/9402001*, pages 1–23, 1994.
59. L. Berezansky, E. Braverman, and A. Domoshnitsky. First order functional differential equations: nonoscillation and positivity of Green's functions. *Functional Differential Equations*, 15:57–94, 2008.

60. S. R. Bernfeld, C. Corduneanu, and A. O. Ignatyev. On the stability of invariant sets of functional differential equations. *Nonlinear Analysis: Theory, Methods & Applications*, 55(6):641–656, 2003.
61. A. S. Besicovitch. *Almost Periodic Functions*. Cambridge University Press, London and New York, 1932.
62. A. S. Besicovitch and H. Bohr. Almost periodicity and general trigonometric series. *Acta Mathematica*, 57(1):203–292, 1931.
63. N. P. Bhatia and O. Hájek. *Local Semi-Dynamical Systems: Lecture Notes in Mathematics 90*. Springer-Verlag, New York, 1969.
64. N. P. Bhatia and G. P. Szegö. *Stability Theory of Dynamical Systems*. Springer-Verlag, Berlin, 1970.
65. A. Bielecki. Une remarque sus la méthode de Banach-Cacciopoli-Tikhonov daus la theory des equation differentielles ordinaires. *Bulletin de l'Academic Polonaise des Sciences*, IV:261–264, 1956.
66. G. D. Birkhoff. *Dynamical Systems*. American Mathematical Society, Providence, 1927.
67. J. Blot and D. Pennequin. Existence and structure results on almost periodic solutions of difference equations. *Difference Equations and Applications*, 7(3):383–402, 2001.
68. S. Bochner. Über gewisse differential-und allgemeinere gleichungen deren lösungen fast-periodisch Sind.I. Teil. Der Existenzsatz. *Mathematische Annalen*, 102:489–504, 1929.
69. S. Bochner. Über gewisse differential-und allgemeinere gleichungen deren lösungen fast-periodisch Sind.II. Teil. Der Beschränktheitssatz. *Mathematische Annalen*, 103:588–597, 1929.
70. S. Bochner. Beitrag zur absoluten konvergenz fastperiodischer funktionen. *Jahresbericht der Deutschen Mathematiker-Vereinigung*, 39:52–54, 1930.
71. S. Bochner. Über gewisse differential-und allgemeinere gleichungen deren lösungen fast-periodisch Sind.III. Teil. Systeme von Gleichungen. *Mathematische Annalen*, 104:579–587, 1931.
72. H. Bohr. *Fastperiodische Funktionen*. Springer-Verlag, Berlin, 1932.
73. H. Bohr and E. Folner. On some types of functional spaces. *Acta Mathematica*, 76(1–2):31–155, 1944.
74. O. Boruwka. Uber verallgemeinerung der eindeutigkeitssatze fur integralle der differentialgleichung $y' = f(x,y)$. *Acta Universitatis Comenianae*, IV–VI: 155–167, 1956.
75. F. Brauer. Some results on uniqueness and successive approximations. *Canadian Journal of Mathematics*, 11:527–533, 1959.
76. F. Brauer and S. Sternberg. Local uniqueness, existence in the large, and the convergence of successive approximations. *American Journal of Mathematics*, 80(2):421–430, 1958.

77. H. Brézis. On a characterization of flow-invariant sets. *Communications on Pure and Applied Mathematics*, 23(2):261–263, 1970.
78. L. Brugnano and D. Trigiante. *Solving ODEs by Linear Multistep Initial and Boundary Value Methods*. Gordon & Breach, Amsterdam, 1998.
79. C. Burnap and M. A. Kazemi. Optimal control of a system governed by nonlinear Volterra integral equations with delay. *IMA Journal of Mathematical Control and Information*, 16(1):73–89, 1999.
80. T. A. Burton. *Volterra Integral and Differential Equations*. Academic Press, New York, 1983.
81. T. A. Burton. *Stability and Periodic Solutions of Ordinary and Functional Differential Equations*. Academic Press, Orlando, 1985.
82. T. A. Burton. Integral equations, implicit functions, and fixed points. In *Proceedings of the American Mathematical Society*, volume 124, pages 2383–2390, 1996.
83. T. A. Burton. *Stability by Fixed Point Theory for Functional Differential Equations*. Dover Publications, New York, 2006.
84. T. A. Burton. *Liapunov Functionals for Integral Equations*. Trafford Publishing, Victoria, 2008.
85. T. A. Burton and L. Hatvani. Stability theorems for nonautonomous functional differential equations by Liapunov functionals. *Tohoku Mathematical Journal*, 41:65–104, 1989.
86. R. Campanini. Sulla prolungabilita e sulla limitatezza parziale delle soluzioni di sistemi di equazioni differenziali. *Bolletino U M I, Analisi Funzionale e Applicazioni*, VI(III):75–84, 1984.
87. G. Cantarelli. On the stability of the equilibrium of mechanical systems. *Periodica Mathematica Hungarica*, 44(2):157–167, 2002.
88. M. Caputo. *Elasticità e Dissipazione*. Zanichelli, Bologna, 1969.
89. O. Carja, M. Necula, and I. I. Vrabie. Local invariance via comparison functions. *Electronic Journal of Differential Equations*, 2004(50):1–14, 2004.
90. O. Carja, M. Necula, and I. I. Vrabie. *Viability, Invariance and Applications*. Elsevier, North-Holland, Amsterdam, 2007. Mathematics Studies 207.
91. O. Carja, M. Necula, and I. I. Vrabie. Private communication, 2013.
92. H. Cassago, Jr. and C. Corduneanu. The ultimate behavior for certain nonlinear integro-differential equations. *Journal of Integral Equations and Applications*, 9:113–124, 1985.
93. L. P. Castro and R. C. Guevara. Hyers-Ulam-Rassias stability of Volterra integral equations with delay within weighted spaces. *Libertas Mathematica*, 32:1–14, 2012.
94. R. Ceppitelli and L. Faina. Differential equations with hereditary structure induced by a Volterra type property. *Fields Institute Communications*, 29:73–91, 2001.

95. A. Cernea. Continuous selections of solutions sets of fractional differential inclusions involving Caputo's fractional derivative. *Revue Roumaine de Mathématiques Pures et Appliquées*, 55:121–129, 2010.
96. A. Cernea. Continuous version of Filippov's theorem for fractional differential inclusions. *Nonlinear Analysis: Theory, Methods & Applications*, 72(1):204–208, 2010.
97. L. Cesari. *Asymptotic Behavior and Stability Problems in Ordinary Differential Equations*. Springer-Verlag, Berlin, second edition, 1963.
98. V. Cesarino. Spectral properties of weakly asymptotic almost periodic semigroups in the sense of Stepanov. *Rendiconti Accademic Nazionale dei Lincei*, 8:167–181, 1997.
99. D. N. Cheban. *Asymptotically Almost Periodic Solutions of Differential Equations*. Hindawi Downloads, Hindawi.com, 2009.
100. D. N. Cheban. Private communication, 2012.
101. D. N. Cheban and C. Mammana. Invariant manifolds and almost automorphic solutions of second-order monotone equations. In *Proceedings of Equadiff 11*, pages 141–147, Bratislava, Czach Republic, 2005.
102. Y. Q. Chen and Y. J. Cho, editors. *Nonlinear Operator Theory in Abstract Spaces and Applications*. Nova Science Publishers, New York, 2004.
103. N. G. Chetayev, editor. *The Stability of Motion*. Pergamon, London, 1961.
104. K. S. Chu and M. Pinto. Oscillatory and periodic solutions in alternatively advanced and delayed differential equations. *Carpathian Journal of Mathematics*, 29:149–158, 2013.
105. E. N. Chukwu, editor. *Stability and Time-Optimal Control of Hereditary Systems*, volume 188 of *Mathematics in Science and Engineering*. Academic Press, Boston, 1992.
106. E. A. Coddington and N. Levinson. *Theory of Ordinary Differential Equations*. McGraw-Hill, New York, 1955.
107. B. D. Coleman and V. J. Mizel. On the stability of solutions of functional-differential equations. *Archive for Rational Mechanics and Analysis*, 30(3):173–196, 1968.
108. L. Collatz. *Functional Analysis and Numerical Mathematics*. Academic Press, New York, 1966.
109. R. M. Colombo, A. Fryszkowski, T. Rzezuchowski, and V. Staicu. Continuous selections of solutions sets of Lipschitzean differential inclusions. *Funkcialaj Ekvacioj*, 34:321–330, 1991.
110. R. Conti. Sulla prolungabilità delle soluzioni di un sistema di equazioni differenziali ordinarie. *Bollettino dell'Unione Mathematica Italiana*, 11(3):510–514, 1956.
111. J. B. Conway. *A Course in Functional Analysis*. Springer, New York, 1990.
112. A. Corduneanu. Exponential stability for a delay-differential equation. *Analele Stiintifice Universitatii "Al I. Cuza"*, XLIV:497–504, 1998. Supplement.

113. A. Corduneanu. A linear second order delay-differential equation. *Libertas Mathematica*, XVIII:59–70, 1998.
114. C. Corduneanu. Application of differential inequalities to stability theory. *Annals of the Alexandru Ioan Cuza University–Mathematics*, 6:47–58, 1960. Russian.
115. C. Corduneanu. Complement to the paper: Application of differential inequalities to stability theory. *Annals of the Alexandru Ioan Cuza University–Mathematics*, 7:247–252, 1961. Russian.
116. C. Corduneanu. Sur la stabilite partielle. *Revue Roumaine de Mathematiques Pures et Appliquees*, 9:229–236, 1964.
117. C. Corduneanu. Sur certaines équations fonctionnelles de Volterra. *Funkcialaj Ekvacioj*, 9:119–127, 1966.
118. C. Corduneanu. Quelques problèms qualitatifs de la théorie des équations intégro-différentielles. *Colloquium Mathematicum*, 18:77–87, 1967.
119. C. Corduneanu. Some problems concerning partial stability (*). In *Symposia Mathematica*, volume 6, pages 141–154. Academic Press, New York, 1971.
120. C. Corduneanu. *Integral Equations and Stability of Feedback Systems*. Academic Press, New York, 1973.
121. C. Corduneanu. On partial stability for delay systems. *Annales Polonici Mathematici*, 29:357–362, 1975.
122. C. Corduneanu. Asymptotic behavior for some systems with infinite delay. In *VIII Internationale Konferenz Uber Nichthlineare Schwingungen Akademie Verlag, Band I, Berlin*, pages 155–160, 1977.
123. C. Corduneanu. *Principles of Differential and Integral Equations*. Chelsea Publishing Company, New York, 1977. Distributed by AMS.
124. C. Corduneanu. Almost periodic solutions for infinite delay systems. *North-Holland Mathematics Studies*, 55:99–106, 1981.
125. C. Corduneanu. Bielecki's method in the theory of integral equations. *Annales Universitatis 'Mariae Curie Sklodowska'*, 38(6):23–40, 1984.
126. C. Corduneanu. Ultimate behavior of solutions to some nonlinear integrodifferential equations. *Libertas Mathematica*, 4:61–72, 1984.
127. C. Corduneanu. An existence theorem for functional equations of Volterra type. *Libertas Mathematica*, 6:117–124, 1986.
128. C. Corduneanu. Two qualitative inequalities. *Journal of Differential Equations*, 64(1):16–25, 1986.
129. C. Corduneanu. *Almost Periodic Functions*. Chelsea, New York, second edition, 1989. Distributed by AMS and Oxford University Press.
130. C. Corduneanu. Integral representation of solutions of linear abstract Volterra functional differential equations. *Libertas Mathematica*, 9:133–146, 1989.
131. C. Corduneanu. Perturbation of linear abstract Volterra equations. *Integral Equations and Applications*, 2(3):393–401, 1990.
132. C. Corduneanu. Second order functional-differential equations involving abstract Volterra operators. *Libertas Mathematica*, 10:87–93, 1990.

133. C. Corduneanu. Some results on the ultimate behavior of solutions of Volterra functional-differential equations. *Annales Polonici Mathematici*, LI:129–135, 1990.
134. C. Corduneanu. Equations involving abstract Volterra operators. In J. Kato and T. Yoshizawa, editors, *Functional Differential Equations*, pages 55–66. World Scientific, Singapore, 1991.
135. C. Corduneanu. *Integral Equations and Applications*. Cambridge University Press, Cambridege, 1991.
136. C. Corduneanu. Functional differential equations with abstract Volterra operators and their control. In *International Meeting on Ordinary Differential Equations and Their Applications*, pages 61–81, Firenze, Italy, 1993.
137. C. Corduneanu. Equations with abstract Volterra operators as modelling tools in science and engineering. In C. Constanda, editor, *Pitman Research Notes in Mathematics Series: Integral Methods in Science and Engineering*, volume 3, pages 146–154. Longeman Scientific and Technical, London, 1994.
138. C. Corduneanu. Stability problems for Volterra functional differential equations. In X. Liu and D. Siegel, editors, *Comparison Methods and Stability Theory*, volume 162, pages 87–99. Marcel Dekker, New York, 1994.
139. C. Corduneanu. Discrete qualitative inequalities and applications. *Nonlinear Analysis: Theory, Methods & Applications*, 25:933–939, 1995.
140. C. Corduneanu. Some new trends in Lyapunov's direct method. In V. Lakshmikantham, editor, *Proceedings of the First World Congress of Nonlinear Analysts*, volume II, pages 1295–1302. Berlin and New York, 1996. Walter De Gruyter.
141. C. Corduneanu. Almost periodic solutions to differential equations in abstract spaces. *Revue Roumaine de Mathematiques Pures et Appliquees*, XLII:753–758, 1997.
142. C. Corduneanu. Neutral functional differential equations with abstract Volterra operators. In *Advances in Nonlinear Dynamics*, volume 5, pages 229–235. Gordon and Breach, Amsterdam, 1997.
143. C. Corduneanu. Neutral functional equations of Volterra type. *Functional Differential Equations*, 4:265–270, 1997.
144. C. Corduneanu. Abstract Volterra equations: A survey. *Mathematical and Computer Modelling*, 32(11):1503–1528, 2000.
145. C. Corduneanu. Existence of solutions for neutral functional differential equations with causal operators. *Journal of Differential Equations*, 168(1):93–101, 2000.
146. C. Corduneanu. Discrete dynamical systems described by neutral equations. In K. Vajravelu, editor, *Differential Equations and Nonlinear Mechanics*, pages 69–74. Kluwer Academic, Dordrecht, 2001.
147. C. Corduneanu. Some existence results for functional equations with causal operators. *Nonlinear Analysis: Theory, Methods & Applications*, 47:709–716, 2001.

148. C. Corduneanu. Absolute stability for neutral differential equations. *European Journal of Control*, 8:209–212, 2002.

149. C. Corduneanu. *Functional Equations with Causal Operators*. Taylor & Francis, London and New York, 2002.

150. C. Corduneanu. Review of some recent results related to discrete-time functional equations. In *American Institute of Physics Conference Proceedings: Computing Anticipatory Systems, CASYS 2001*, volume 627, pages 160–169, Liege, Belgium 2002.

151. C. Corduneanu. Absolute stability for neutral equations with infinite delay. In C. Constanda et al., editor, *Integral Methods in Science and Engineering*, pages 53–60. Birkhäuser, Boston, 2004.

152. C. Corduneanu. A modified LQ-Optimal control problem for causal functional differential equations. *Nonlinear Dynamics and Systems Theory*, 4(2):139–144, 2004.

153. C. Corduneanu. Some remarks on functional equations with advanced-delayed operators. In *American Institute of Physics Conference Proceedings*, volume 718, pages 204–209, Liese, Belgium. 2004.

154. C. Corduneanu. Almost periodic solutions for a class of functional differential equations. *Functional Differential Equations*, 14:223–229, 2007.

155. C. Corduneanu. Almost periodicity in functional equations. In V. Staicu, editor, *Differential Equations, Chaos and Variational Problems: Progress in Nonlinear Differential Equations and Their Applications*, volume 75, pages 157–163. Birkhäuser, Basel, 2007.

156. C. Corduneanu. *Almost Periodic Oscillations and Waves*. Springer, New York, 2009.

157. C. Corduneanu. Some classes of second order functional differential equations. *Nonlinear Analysis: Theory, Methods & Applications*, 71(21):e865–e871, 2009.

158. C. Corduneanu. Some comments on almost periodicity and related topics. *Communications in Mathematical Analysis*, 8(2):5–15, 2010.

159. C. Corduneanu. Almost periodicity in semilinear systems. In C. Constanda and P. J. Harris, editors, *Integral Methods in Science and Engineering*, pages 141–146. Birkhäuser, Boston, 2011.

160. C. Corduneanu. Boundedness of solutions for a second order differential equation with causal operators. *Nonlinear Studies*, 18(2):135–139, 2011.

161. C. Corduneanu. A scale of almost periodic function spaces. *Differential and Integral Equations*, 24(1–2):1–27, 2011.

162. C. Corduneanu. AP_r-almost periodic solutions to functional differential equations with deviated argument. *Functional Differential Equations*, 19:59–69, 2012.

163. C. Corduneanu. Éléments d'une construction axiomatique de la théorie des fonctions presque périodiques. *Libertas Mathematica*, 32:5–18, 2012.

164. C. Corduneanu. Almost periodicity; a new approach. In *VII-th International Congress of Romanian Mathematicians/Editura Academiei*, pages 121–129, Bucharest, 2013.

165. C. Corduneanu. Formal trigonometric series, almost periodicity and oscillatory functions. *Nonlinear Dynamics and Systems Theory*, 13:367–388, 2013.

166. C. Corduneanu and A. O. Ignatyev. Stability of invariant sets of functional differential equations with delay. *Nonlinear Functional Analysis and Applications*, 10(1):11–24, 2005.

167. C. Corduneanu and V. Lakshmikantham. Equations with unbounded delay: a survey. *Nonlinear Analysis: Theory, Methods & Applications*, 4(5):831–877, 1980.

168. C. Corduneanu and Yizeng Li. A global existence result for functional differential equations with causal operators. In C. Constanda et al., editor, *Integral Methods in Science and Engineering*, pages 69–74. Birkhäuser, Boston, 2002.

169. C. Corduneanu and Yizeng Li. On exponential asymptotic stability for functional differential equations with causal operators. In A. A. Martynyuk, editor, *Advances in Stability Theory at the End of 20th Century*, pages 15–23. Taylor & Francis, London, 2003.

170. C. Corduneanu and Yizeng Li. A duality principle in the theory of dynamical systems. *Nonlinear Dynamics and System Theory*, 5:135–140, 2005.

171. C. Corduneanu and Yizeng Li. A neutral-convolution type functional equation. *Libertas Mathematica*, 31:87–92, 2011.

172. C. Corduneanu, Yizeng Li, and M. Mahdavi. New examples for a duality principle in the theory of dynamical systems. In *Proceedings of CASYS'05, American Institute of Physics*, volume 839, pages 340–343, Liege, Belgium 2006.

173. C. Corduneanu and N. Luca. The stability of some feedback systems with delay. *Journal of Mathematical Analysis and Applications*, 51:377–393, 1975.

174. C. Corduneanu and M. Mahdavi. Asymptotic behavior of systems with abstract Volterra operators. In C. Corduneanu, editor, *Qualitative Problems for Differential Equations and Control Theory*, pages 113–120. World Scientific, Singapore, 1995.

175. C. Corduneanu and M. Mahdavi. On neutral functional differential equations with causal operators. In *Proceedings of the Third Workshop of the Inter. Inst. General Systems Science: Systems Science and Its Applications*, pages 43–48, Tianjin People's Publishing House, Tianjin, 1998.

176. C. Corduneanu and M. Mahdavi. On neutral functional differential equations with causal operators, II. In *Chapman & Hall/CRC Research Notes in Mathematics: Integral Methods in Science and Engineering*, pages 102–106. Chapman & Hall/CRC, London, 2000.

177. C. Corduneanu and M. Mahdavi. Neutral functional equations in discrete time. In *Proceedings of the International Conference on Nonlinear Operators, Differential Equations and Applications, Babes-Bolyai University of Cluj-Napoca, Romania*, volume III, pages 33–40, Cluj-Napoca, Romania 2002.

178. C. Corduneanu and M. Mahdavi. A class of second order functional differential equations of neutral type. *Mathematical Reports, Romanian Academy*, 5(55)(4):293–299, 2003.

179. C. Corduneanu and M. Mahdavi. Some function spaces on R. *Libertas Mathematica*, XXVI:79–82, 2006.

180. C. Corduneanu and M. Mahdavi. Neutral functional equations with causal operators on a semi-axis. *Nonlinear Dynamics and Systems Theory*, 8(4):339–348, 2008.

181. C. Corduneanu and M. Mahdavi. Neutral functional equations of the second order. *Functional Differential Equations*, 16(2):263–271, 2009.

182. C. Corduneanu and M. Mahdavi. Existence of AP_r-almost periodic solutions for some classes of functional differential equations. *African Diaspora Journal of of Mathematics*, 15(2):47–55, 2013.

183. R. F. Curtain, H. Logemann, and O. Staffans. Stability results of Popov-Type for infinite-dimensional systems with applications to integral control. *Proceedings of the London Mathematical Society*, volume 86, pages 779–816, 2003.

184. J. M. Cushing. Admissible operators and solutions of perturbed operator equations. *Funkcialaj Ekvacioj*, 19:79–84, 1976.

185. J. M. Cushing. Strongly admissible operators and Banach space solutions of nonlinear equations. *Funkcialaj Ekvacioj*, 20:237–245, 1977.

186. I. I. L. Daletskii and M. M. G. Krein. *Stability of Solutions of Differential Equations in Banach Space*. American Mathematical Society, Providence, 1974. Translation from Russian. Number 43.

187. S. Das. *Functional Fractional Calculus*. Springer-Verlag, Berlin, second edition, 2011.

188. R. Datko. Lyapunov functionals for certain linear delay differential equations in a Hilbert space. *Journal of Mathematical Analysis and Applications*, 76(1):37–57, 1980.

189. C. R. de Oliveira and M. S. Simsen. Almost periodic orbits and stability for quantum time-dependent Hamiltonians. *Reports on Mathematical Physics*, 60(3):349–366, 2007.

190. K. Deimling. *Nonlinear Functional Analysis*. Springer-Verlag, Berlin, 1985.

191. K. Deimling. *Multivalued Differential Equations*. Walter de Gruyter, Berlin, 1992.

192. R. L. Devaney. *An Introduction to Chaotic Dynamical Systems*. Addison-Wesley, Reading, 1989.

193. T. Diagana. *Pseudo Almost Periodic Functions in Banach Spaces*. Nova Science Publishers, Inc., Commack, 2007.

194. T. Diagana and V. Nelson. Existence results for some higher-order evolution equations with operator coefficients. *Applied Mathematics and Computation*, 219(6):2923–2931, 2012.

195. P. Diamond. Theory and applications of fuzzy Volterra integral equations. *IEEE Transactions on Fuzzy Systems*, 10(1):97–102, 2002.

196. K. Diethelm. *The Analysis of Fractional Differential Equations*. Springer, New York, 2010. Lecture Notes in Mathematics.

197. M. Z. Djordjevič. Stability analysis of nonlinear systems by the matrix Liapunov method. In *Proceedings of IMACS-IFAC Symposium, Villeneuve d'Asq*, pages 209–213, Villeneuve d'Asq. France, 1986.

198. A. Domoshnitsky. On stability of nonautonomous integro-differential equations. In *Proceedings of EQUADIFF*, pages 1059–1064, Hasselt, Balgium, 2003.

199. M. Drakhlin and E. Litsyn. Volterra operators: Back to the future. *Journal of Nonlinear and Convex Analysis*, 6(3):375–391, 2005.

200. Z. Drici, F. A. McRae, and J. Vasundhara Devi. Variation of parameters for differential equations with causal maps. *Mathematical Inequalities & Applications*, 12(1):209–215, 2009.

201. R. D. Driver. Existence and stability of solutions of a delay-differential system. *Archive for Rational Mechanics and Analysis*, 10(1):401–426, 1962.

202. R. D. Driver. Existence and continuous dependence of solutions of a neutral functional-differential equation. *Archive for Rational Mechanics and Analysis*, 19(2):149–166, 1965.

203. R. D. Driver. *Ordinary and Delay Differential Equations*, volume 20. Springer-Verlag, New York, 1977.

204. N. Dunford and J. T. Schwartz. *Linear Operators, Part I*. Wiley (Interscience), New York, 1958.

205. A. Dzieliński. Stability of a class of adaptive nonlinear systems. *International Journal of Applied Mathematics and Computer Science*, 15(4):455–462, 2005.

206. S. Elaydi. Asymptotic stability of linear systems with infinite delay. In J. Kato and T. Yoshizawa, editors, *Functional Differential Equations*, pages 67–74. World Scientific, Singapore, 1991.

207. L. E. Eĺsgoĺts and S. B. Norkin. *Introduction to the Theory and Applications of Differential Equations with Deviating Arguments*. Academic Press, New York, 1973.

208. J. Favard. *Leçons sur les fonctions presque-périodiques*. Gauthier-Villars, Paris, 1933.

209. L. G. Fedorenko. Stability of functional differential equations. *Differential Equations*, 21:1031–1037, 1986. Translation from Russian.

210. V. V. Filippov. Axiomatic theory of solution spaces of ordinary differential equations and differential inclusions. *Soviet Mathematics-Doklady*, 31:86–90, 1985.

211. V. V. Filippov. Topological construction of the space of solutions of an ordinary differential equation. *Uspekhi Matematicheskikh Nauk*, 48:103–154, 1993. Russian.

212. A. M. Fink. Almost periodic functions invented for specific purposes. *SIAM Review*, 14(4):572–581, 1972.
213. A. M. Fink. *Almost Periodic Differential Equations*. Springer-Verlag, Berlin, 1974. Lecture Notes in Mathematics, 377.
214. A. Friedman. *Foundations of Modern Analysis*. Courier Dover Publications, New York, 1970.
215. T. Furumochi. Existence of periodic solutions of one-dimensional differential-delay equations. *Tôhoku Mathematical Journal*, 30:13–35, 1978.
216. V. E. Germaidze and N. N. Krasovskii. On stability under constantly acting perturbations. *Prikladnaya Matematicai Mekhanica*, 21:769–774, 1957. Russian.
217. P. Giesel and M. Rasmussen. A note on almost periodic variational equations, 2008. Preprint Number 13, Institut Fur Mathematik Universitat Augsburg.
218. P. Giesl and S. Hafstein. Construction of Lyapunov functions for nonlinear planar systems by linear programming. *Journal of Mathematical Analysis and Applications*, 388(1):463–479, 2012.
219. M. I. Gil̆. L^2-stability of vector equations with nonlinear causal mappings. *Dynamic Systems and Applications*, 17(1):201–220, 2008.
220. M. I. Gil̆. Absolute and input-to-state stabilities of nonautonomous systems with causal mappings. *Dynamic Systems and Applications*, 18(3–4):655–666, 2009.
221. M. I. Gil̆. Positivity of green's functions to Volterra integral and higher order integro-differential equations. *Analysis and Applications*, 7(4):405–418, 2009.
222. M. I. Gil̆. *Stability of Vector Differential Delay Equations*. Birkhäuser-Springer, Basel, 2013.
223. B. Giordano and A. Pankov. On convolution operators in the spaces of almost periodic functions and L^p spaces. *Zeitschrift Fur Analysis Und Ihre Anwendungen*, 19(2):359–368, 2000.
224. I. C. Gohberg and S. Goldberg. *Basic Operator Theory*. Birkhäuser, Boston-Basel, 1981.
225. K. Gopalsamy. *Stability and Oscillations in Delay Differential Equations of Population Dynamics*. Kluwer Academic, Dordrecht, 1992.
226. L. Górniewicz. *Topological Fixed Point Theory of Multivalued Mappings*. Springer, New York, 2006.
227. W. H. Gottschalk and G. A. Hedlund. *Topological Dynamics*, volume 36. American Mathematical Society, Providence, 1955.
228. G. Gripenberg, S. O. Londen, and O. Staffans. *Volterra Integral and Functional Equations*, volume 34. Cambridge University Press, Cambridge, 1990.
229. D. Guo and V. Lakshmikantham. *Nonlinear Problems in Abstract Cones*. Academic Press, New York, 1988.
230. S. A. Gusarenko. A generalization of the concept of a Volterra operator. *Doklady Akademil SSSR*, 295:1046–1049, 1987.
231. S. A. Gusarenko. On linear equations with generalized Volterra operators. *Functional Differential Equations*, 7(1–2):109–120, 2000.

232. S. A. Gusarenko, E. S. Zhukowskii, and V. P. Maksimov. On the theory of functional-differential equations with locally Volterra operators. *Soviet Mathematics Doklady*, 33:368–370, 1986.

233. I. Györi and G. Ladas. *Oscillation Theory of Delay Differential Equations: With Applications*. Clarendon Press, Oxford, 1991.

234. S. F. Hafstein. A constructive converse Lyapunov theorem on asymptotic stability for nonlinear autonomous ordinary differential equations. *Dynamical Systems*, 20:281–299, 2005.

235. W. Hahn. *Stability of Motion*, volume 138. Springer-Verlag, Berlin, 1967.

236. R. Hakl, A. Lomtatidze, and J. Sremr. On non-negative solutions of a periodic type boundary value problem for first order scalar functional differential equations. *Functional Differential Equations*, 11:363–394, 2004.

237. A. Halanay. *Differential Equations: Stability, Oscillations, Time Lags*. Academic Press, New York, 1966.

238. A. Halanay and V. Răsvan. *Applications of Liapunov Methods in Stability*. Kluwer Academic Publishers, Boston, 1993.

239. J. K. Hale. Linear functional differential equations with constant coefficients. *Contributions to Differential Equations*, 2:291–319, 1963.

240. J. K. Hale. *Theory of Functional Differential Equations*, volume 3. Springer-Verlag, New York, 1977.

241. J. K. Hale and J. Kato. Phase space for retarded equations with infinite delay. *Funkcialaj Ekvacioj*, 21:11–41, 1978.

242. J. K. Hale and S. M. V. Lunel. *Introduction to Functional Differential Equations*, volume 99. Springer-Verlag, New York, 1993.

243. J. K. Hale and K. R. Meyer. *A Class of Functional Equations of Neutral Type*. American Mathematical Society, Providence, 1967. No. 76.

244. T. G. Hallam and S. A. Levin, editors. *Mathematical Ecology: An Introduction*. Springer, New York, 2011.

245. Y. Hamaya. On the existence of almost periodic solutions to a nonlinear Volterra difference equation. *Advanced Studies in Pure Mathematics*, 53:59–66, 2009.

246. Y. Hamaya. Stability properties and almost periodic solutions of abstract functional differential equations with infinite delay. *Libertas Mathematica*, XXXI:93–102, 2011.

247. Y. Hamaya and T. Arai. Permanence of an SIR epidemic model with diffusion. *Nonlinear Studies*, 17:69–79, 2010.

248. P. Hartman. *Ordinary Differential Equations*. John Wiley & Sons, Inc., New York, 1964.

249. P. Hartman. On invariant sets and on a theorem of Ważewski. *Proceedings of the American Mathematical Society*, 32:511–520, 1972.

250. P. Hasil and M. Veselý. Oscillation of half-linear differential equations with asymptotically almost periodic coefficients. *Advances in Difference Equations*, 2013(1):1–15, 2013.

251. L. Hatvani. The application of differential inequalities to stability theory. *Vestnik Moskov Universiteta Series I Mathematics and Mechanics*, 30(3):83–89, 1975. Russian.
252. L. Hatvani. Attractivity theorems for nonautonomous systems of differential equations. *Acta Scientiarum Mathematicarum*, 40:271–283, 1978.
253. L. Hatvani. Stability and partial stability of nonautonomous systems of differential equations. *Alkalmazolt Matematikai Lapok*, 5:1–48, 1979.
254. L. Hatvani. On partial asymptotic stability and instability, I. *Acta Scientiarum Mathematicarum*, 45:219–231, 1983.
255. L. Hatvani. On partial asymptotic stability and instability, II. *Acta Scientiarum Mathematicarum*, 46:143–156, 1983.
256. L. Hatvani. On partial asymptotic stability and instability, III. *Acta Scientiarum Mathematicarum*, 49:157–167, 1985.
257. L. Hatvani. On partial asymptotic stability by the method of limiting equations. *Annali di Matematica Pura ed Applicata, IV*, CXXXIX:65–82, 1985.
258. L. Hatvani. On the asymptotic stability of the solutions of functional-differential equations. In *Colloquia Mathematica Societatis János Bolyai, 53: Qualitative Theory of Differential Equations (Szeged, 1988)*, pages 227–238. North-Holland, Amsterdam, 1990.
259. L. Hatvani. On the asymptotic stability by Lyapunov functionals with semidefinite derivatives. *Nonlinear Analysis: Theory, Methods & Applications*, 30(9):4713–4721, 1997.
260. H. R. Henriquez and C. H. Vasquez. Almost periodic solutions of abstract retarded functional differential equations with unbounded delay. *Acta Applicandae Mathematicae*, 57(2):105–132, 1999.
261. D. Hilbert. *Grundzüge einer allgemeinen Theorie der linearen Integralgleichungen*. Teubner, Leipzig, 1912.
262. Y. Hino, S. Murakami, and T. Naito. *Functional Differential Equations with Infinite Delay*. Springer-Verlag, New York, 1991. Lecture Notes in Mathematics, 1473.
263. Y. Hino, T. Naito, N. V. Minh, and J. S. Shin. *Almost Periodic Solutions of Differential Equations in Banach Spaces*. Taylor & Francis, London, 2002.
264. H. Hochstadt. *Integral Equations*. John Wiley & Sons, Inc., New York, 1973.
265. A. Hof and O. Knill. Cellular automata with almost periodic initial conditions. *Nonlinearity*, 8(4):477–491, 1995.
266. V. M. Hokkanen and G. Morosanu. *Functional Methods in Differential Equations*. Chapman & Hall/CRC, Boca Raton, 2002.
267. J. Hong and R. Obaya. Ergodic type solutions of some differential equations. In K. Vajravelu, editor, *Differential Equations and Nonlinear Mechanics*, pages 135–152. Kluwer Academic, Dordrecht, 2001.
268. D. H. Hyers, G. Isac, and T. M. Rassias. *Stability of Functional Equations in Several Variables*. Birkhäuser, Boston, 1998.

269. A. O. Ignatyev. Partial asymptotic stability in probability of stochastic differential equations. *Statistics and Probability Letters*, 79:597–601, 2009.

270. D. Z. Ch. Ji. Translation invariance of weighted pseudo almost periodic functions and related problems. *Journal of Mathematical Analysis and Applications*, 391:350–362, 2012.

271. G. S. Jordan, O. J. Staffans, and R. L. Wheeler. Subspaces of stable and unstable solutions of a functional differential equation in a fading memory space: The critical case. *SIAM Journal on Mathematical Analysis*, 18(5):1323–1340, 1987.

272. B. S. Kalitvin. Comparison method for a stability problem for periodic systems. *Differential Equations*, 23:423–428, 1985. Translation from Russian.

273. T. Kalmar-Nagy. Stability analysis of delay-differential equations by the method of steps and inverse Laplace transform. *Differential Equations and Dynamical Systems*, 17:185–200, 2009.

274. L. V. Kantorovich and G. P. Akilov. *Functional Analysis*. Pergamon Press, Oxford, second edition, 1982.

275. F. Kappel. Laplace-transform methods and linear autonomous functional-differential equations. *Berichte der Mathematisch-Statistischen Sektion im Forschungszentrum Graz*, 64:1–62, 1976.

276. F. Kappel and W. Schappacher, editors. *Evolution Equations and Their Applications*. Pitman Publishing, London, 1992.

277. G. Karakostas. Causal operators and topological dynamics. *Annali di Matematica Pura ed Applicata*, 131(4):1–27, 1982.

278. J. Kato. On Liapunov-Razumikhin type theorems for functional differential equations. *Funkcialaj Ekvacioj*, 16:225–239, 1973.

279. J. Kato. Stability problem in functional differential equations with infinite delay. *Funkcialaj Ekvacioj*, 21:63–80, 1978.

280. J. Kato. Liapunov's second method in functional differential equations. *Tohoku Mathematical Journal*, 32:487–497, 1980.

281. J. Kato, A. A. Martynyuk, and A. A. Shestakov. *Stability of Motion of Nonautonomous Systems: Methods of Limiting Equations*. Gordon & Breach, Amsterdam, 1996.

282. J. Kato and Y. Sekiya. Existence of almost periodic solutions of nonlinear differential equations in a Banach space. *Mathematica Japonica*, 43:563–568, 1996.

283. S. Kato. On the existence of periodic solutions for nonlinear ordinary differential equations in Banach spaces. *Funkcialaj Ekvacioj*, 23:143–149, 1980.

284. W. G. Kelley and A. C. Peterson. *Difference Equations: An Introduction with Applications*. Academic Press, Boston, 1991.

285. A. A. Kilbas, H. M. Srivastava, and J. J. Trujillo. *Theory and Applications of Fractional Differential Equations*. Elsevier, Amsterdam, 2006.

286. A. V. Kim. *Lyapunov's Direct Method in Stability Theory of Delay Systems*. Ural University Press, Ekaterinburg, 1992. Russian.

287. A. V. Kim and A. N. Krasovskii. *Mathematical and Computer Modeling of Systems with Delay*. UGTU-ULJ, Ekaterinburg, 2010. Russian.

288. A. V. Kim and V. G. Pimenov. *i-Smooth Analysis and Numerical Methods of Solving Functional-Differential Equations*. R & C Dynamics Publication, Moscow, 2004. Russian.

289. V. B. Kolmanovskii. Application of differential inequalities for stability of some functional differential equations. *Nonlinear Analysis: Theory, Methods & Applications*, 25(9–10):1017–1028, 1995.

290. V. B. Kolmanovskii. Functional-differential equations. In *Encyclopedia of Mathematics*. Springer:Online, 2001. reference works.

291. V. B. Kolmanovskii and A. Myshkis. *Applied Theory of Functional Differential Equations*. Kluwer, Dordrecht, 1992.

292. V. B. Kolmanovskii and A. Myshkis. *Introduction to the Theory and Applications of Functional Differential Equations*. Kluwer, Dordrecht, 1999.

293. V. B. Kolmanovskii and V. R. Nosov. *Stability and Periodic Regimes of Controlled Systems with Delay*. Nauka, Moscow, 1981. Russian.

294. V. B. Kolmanovskii and L. E. Shaikhet. Construction of Lyapunov functionals for stochastic hereditary systems: a survey of some recent results. *Mathematical and Computer Modelling*, 36(6):691–716, 2002.

295. A. N. Kolmogorov and S. V. Fomin. *Introductory Real Analysis*. Dover Publications, New York, 1975.

296. I. G. E. Kordonis and C. G. Dhilosch. Oscillation and nonoscillation in delay or advanced differential equations and in integrodifferential equations. *Georgian Mathematical Journal*, 6(3):263–284, 1999.

297. M. A. Krasnoselskii, V. S. Burd, and Yu. S. Kolesov. *Nonlinear Almost Periodic Oscillations*. John Wiley & Sons Inc., New York, 1973.

298. M. A. Krasnoselskii and P. P. Zabreiko. *Geometrical Methods of Nonlinear Analysis*. Springer-Verlag, Berlin, 1984.

299. N. N. Krasovskii. *Stability of Motion*. Stanford University Press, Stanford, 1963. in Russian, 1959.

300. T. Kulik and C. C. Tisdell. Volterra integral equations on time scales: Basic qualitative and quantitative results with applications to initial value problems on unbounded domains. *International Journal of Difference Equations*, 3(1):103–133, 2008.

301. V. G. Kurbatov. *Functional Differential Operators and Equations*. Kluwer, Dordrecht, 1999.

302. M. Kurihara and T. Suzuki. Galerkin's procedure for quasiperiodic differential difference equations. *Reports of the Faculty of Engineering, Yamanashi University*, 36:77–83, 1985.

303. J. Kurzweil. On the converse of first Liapunov theorem of asymptotic stability. *Czechoslovak Mathematical Journal*, 5:382–398, 1955. Russian.

304. M. Kwapisz. An extension of Bielecki's method of proving global existence and uniqueness results for functional equations. *Annales Universitatis 'Mariae Curie' Sklodowska*, 38(6):59–68, 1984.

305. M. Kwapisz. Bielecki's method, existence and uniqueness results for Volterra integral equations in L^p spaces. *Mathematical Analysis and Applications*, 154(2):403–416, 1991.

306. M. Kwapisz and J. Turo. Some integral-functional equations. *Funkcialaj Ekvacioj*, 18:107–162, 1975.

307. V. Lakshmikantham. Current state of the Lyapunov method for delay differential equations. In P. Borne and V. Matrosov, editors, *The Lyapunov Functions Method and Applications*, volume 10, pages 165–170. I. C. Baltger, Amsterdam, 1990.

308. V. Lakshmikantham. Recent advances in Lyapunov method for delay differential equations. *Differential Equations, Lecture Notes in Pure and Applied Mathematics*, 127:333–343, 1990.

309. V. Lakshmikantham and S. Leela. *Differential and Integral Inequalities: Theory and Applications. Vol. I: Ordinary Differential Equations*. Academic Press, New York, 1969.

310. V. Lakshmikantham and S. Leela. On perturbing Lyapunov functions. *Theory of Computing Systems*, 10(1):85–90, 1976.

311. V. Lakshmikantham, S. Leela, Z. Drici, and F. A. McRae. *Theory of Causal Differential Equations*. World Scientific, Atlantis Press, Amsterdam and Paris, 2009.

312. V. Lakshmikantham, S. Leela, and A. A. Martynyuk. *Stability of Motion by the Comparison Method*. Naukova Dumka, Kiev, 1991.

313. V. Lakshmikantham and X. Z. Liu. *Stability Analysis in Terms of Two Measures*. World Scientific, Singapore, 1993.

314. V. Lakshmikantham, W. Lizhi, and B. G. Zhang. *Theory of Differential Equations with Unbounded Delay*. Kluwer Academic, Dordrecht, 1994.

315. V. Lakshmikantham, V. M. Matrosov, and S. Sivasundaram. *Vector Lyapunov Functions and Stability Analysis of Nonlinear Systems*. Kluwer Academic, Boston, 1991.

316. V. Lakshmikantham and M. Rama Mohana Rao. *Theory of Integro-Differential Equations*. Gordon & Breach Science Publishers, Amsterdam, 1995.

317. V. Lakshmikantham and D. Trigiante. *Theory of Difference Equations: Numerical Methods and Applications*. Academic Press, San Diego, 1988.

318. V. Lakshmikantham and A. S. Vatsala. The present state of uniform asymptotic stability for Volterra and delay equations. In C. Corduneanu and I. W. Sandberg, editors, *Volterra Equations and Applications*, volume 10, pages 83–93. Gordon and Breach, Amsterdam, 2000.

319. T. Lalesco. *Introduction á la Théorie des Équations Intégrales*. Hermann & Fils, Paris, 1912.

320. S. Lang. *Analysis II*. Addison-Wesley, Reading, 1969.

321. B. Lani-Wayda. Erratic solutions of simple delay equations. *Transactions of the American Mathematical Society*, 351(3):901–946, 1999.

322. J. P. LaSalle. *The Stability and Control of Discrete Processes*, volume 62 of *Applied Mathematical Sciences*. Springer, New York, 1994.

323. S. Lefschetz. *Differential Equations: Geometric Theory*. Interscience Publishers, New York, 1962.

324. S. Lei. Boundedness and periodicity of neutral functional differential equations with infinite delay. *Mathematical Analysis and Applications*, 186:338–356, 1994.

325. R. I. Leine. The historical development of classical stability concepts: Lagrange, Poisson, Lyapunov. *Nonlinear Dynamics*, 59:173–182, 2010.

326. B. M. Levitan. *Almost Periodic Functions*. Gostekhizdat, Moscow, 1953. Russian.

327. B. M. Levitan and V. V. Zhikov. *Almost Periodic Functions and Differential Equations*. Cambridge University Press, Cambridge, 1982.

328. Yizeng Li. Existence and integral representation of the second kind initial value problem for functional differential equations with abstract Volterra operators. *Nonlinear Studies*, 3:35–48, 1996.

329. Yizeng Li. Stability problems for functional differential equations with abstract Volterra operators. *Journal of Integral Equations and Applications*, 8:47–63, 1998.

330. Z. Li, Q. Zhao, and D. Liang. Chaotic behavior in a class of delay difference equations. *Advances in Difference Equations*, 2013:1–10, 2013. Article number 99.

331. Jin Liang and Tijun Xiao. Functional differential equations with infinite delay in Banach spaces. *International Journal of Mathematics and Mathematical Sciences*, 14(3):497–508, 1991.

332. A. M. Liapunov. *Stability of Motion*, volume 30. Academic Press, New York, 1966. Mathematics in Science and Engineering (Translated from the Russian).

333. S.-D. Lin and C.-H. Lu. Laplace transform for solving some families of fractional differential equations and its applications. *Advances in Difference Equations*, 2013:137, 2013.

334. J. Lindenstrauss and L. Tzafriri. *Classical Banach Spaces*. Springer-Verlag, Berlin, 1979.

335. E. Litsyn. Elements of the theory of linear Volterra operators in Banach spaces. *Integral Equations and Operator Theory*, 58(2):239–253, 2007.

336. E. Litsyn. Nonlinear Volterra IDE, infinite systems and normal forms of ODE. *Nonlinear Analysis: Theory, Methods & Applications*, 336:1073–1089, 2007.

337. Junwei Liu and Chuanyi Zhang. Composition of piecewise pseudo almost periodic functions and applications to abstract impulsive differential equations. *Advances in Difference Equations*, 2013:1–21, 2013. Article number 11.

338. C. Lizama and J. G. Mesquita. Almost automorphic solutions of non-autonomous difference equations. *Journal of Mathematical Analysis and Applications*, 407(2):339–349, 2013.

339. H. Logemann and E. P. Ryan. Volterra functional differential equations: Existence, uniqueness, and continuation of solution. *American Mathematical Monthly*, 111:490–511, 2010.

340. W. V. Lovitt. *Linear Integral Equations*. Dover Publications, New York, 1950.

341. V. Lupulescu. Causal functional differential equations in Banach spaces. *Nonlinear Analysis*, 69:4787–4795, 2008.

342. V. Lupulescu. On a class of functional differential equations in Banach spaces. *Electronic Journal of Qualitative Theory of Differential Equations*, (64):1–17, 2010.

343. L. A. Lusternik and V. J. Sobolev. *Elements of Functional Analysis*. John Wiley & Sons Inc., New York, 1974.

344. W. Maak. *Fastperiodische Funktionen*. Springer-Verlag, Berlin, 1950.

345. R. C. MacCamy. Stability theorems for a class of functional differential equations. *SIAM Journal of Applied Mathematics*, 30:557–576, 1976.

346. M. Mahdavi. Nonlinear boundary value problems with abstract Volterra operators. *Libertas Mathematica*, 13:17–26, 1993.

347. M. Mahdavi. Linear functional differential equations with abstract Volterra operators. *Differential and Integral Equations*, 8(6):1517–1523, 1995.

348. M. Mahdavi. Neutral equations with causal operators. In *Integral Methods in Science and Engineering*, pages 161–166. Birkhäuser, Boston, 2002.

349. M. Mahdavi. Asymptotic behavior in some classes of functional differential equations. *Nonlinear Dynamics and Systems Theory*, 4(1):51–57, 2004.

350. M. Mahdavi. A note on a second order functional differential equation. *Libertas Mathematica*, 28:37–44, 2008.

351. M. Mahdavi. AP_r-almost periodic solutions to the equation $\dot{x}(t) = ax(t) + (k \star x)(t) + f(t)$. *African Diaspora Journal of Mathematics*, 12(2):43–47, 2011.

352. M. Mahdavi and Yizeng Li. Linear and quasilinear equations with abstract Volterra operators. In C. Corduneanu and I. W. Sandberg, editors, *Volterra Equations and Applications*, volume 10, pages 325–330. Gordon and Breach, Amsterdam, 2000.

353. P. M. Mäkilä, J. R. Partington, and T. Norlander. Bounded power signal spaces for robust control and modeling. *SIAM Journal on Control and Optimization*, 37:92–117, 1998.

354. V. P. Maksimov and A. N. Rumiantsev. Numerical experiment in the investigation of the generalized control of functional differential systems. *Functional Differential Equations*, pages 189–195, 2002. Russian: Special issue of Vestnik Perm Technological University.

355. I. G. Malkin. *Some Problems in the Theory of Nonlinear Oscillations*. Nauka, Moscow, 1956. Russian.

356. I. G. Malkin. *Theory of Stability of Motion*, volume 3352. U. S. Atomic Energy Commission, Berlin and Munich, 1959. AEC Translation:German Edition.
357. V. V. Malygina. On stability of solutions to certain linear differential equations with aftereffect. *Soviet Mathematics: Izv VUZ*, 27:63–74, 1993.
358. C. Marcelli and A. Salvadori. Equivalence of different hereditary structures in ordinary differential equations. *Journal of Differential Equations*, 149:52–68, 1998.
359. R. H. Martin, Jr. *Nonlinear Operators and Differential Equations in Banach Space*. John Wiley & Sons, Inc., New York, 1976.
360. A. Martynyuk, L. Chernetskaya, and V. Martynyuk, editors. *Weakly Connected Nonlinear Systems: Boundedness and Stability of Motion*. CRC Press (Taylor & Francis), Boca Raton, 2013.
361. A. A. Martynyuk. Scalar comparison equations in the theory of the stability of motion. *International Applied Mechanics*, 21(12):1131–1148, 1985.
362. A. A. Martynyuk. A new direction in the method of matrix Lyapunov functions. *Doklady Akademii Nauk SSSR*, 319(3):554–557, 1991.
363. A. A. Martynyuk. Matrix method of comparison in the theory of the stability of motion. *International Applied Mechanics*, 29(10):861–867, 1993.
364. A. A. Martynyuk. Stability and Liapunov's matrix functions method in dynamical systems. *International Applied Mechanics*, 34(10):1057–1065, 1998.
365. A. A. Martynyuk, editor. *Stability by Liapunov Matrix Function Method with Applications*. Marcel Dekker, Inc., New York, 1998.
366. A. A. Martynyuk, editor. *Advances in Stability Theory at the End of the 20th Century*. Taylor & Francis, London, 2003.
367. A. A. Martynyuk. Matrix Liapunov functions and stability analysis of dynamical systems. In *Advances in Stability Theory at the End of the 20th Century*, pages 135–151. Taylor & Francis, London and New York, 2003.
368. A. A. Martynyuk. Comparison principle for a set of equations with a robust causal operator. *Doklady Mathematics MAIK Nauka/Interperiodica*, 80(1):602–605, 2009.
369. A. A. Martynyuk and R. Gutowski, editors. *Integral Inequalities and Stability of Motion*. Naukova Dumka, Kiev, 1979. Russian.
370. A. A. Martynyuk, V. Lakshmikantham, and S. Leela, editors. *Stability of Motion: Method of Integral Inequalities*. Naukova Dumka, Kiev, 1989. Russian.
371. A. A. Martynyuk, V. Lakshmikantham, and S. Leela, editors. *Stability of Motion: The Role of Multicomponent Liapunov's Functions*. Cambridge Scientific Publishers, Cambridge, 2007.
372. Y. A. Martynyuk-Chernienko, editor. *Uncertain Dynamical Systems: Stability and Motion Control*. Feniks, Kyiv, 2009.
373. J. L. Massera. Contributions to stability theory. *The Annals of Mathematics*, 64(1):182–206, 1956.

374. J. L. Massera and J. J. Schäffer. *Linear Differential Equations and Function Spaces*. Academic Press, New York, 1966.

375. V. M. Matrosov. On the theory of stability. *Prikladnaya Matematika i Mekhanika*, 26(5):885–895, 1960. Russian.

376. V. M. Matrosov. On the stability of motion. *Journal of Applied Mathematics and Mechanics*, 26(5):1337–1353, 1962.

377. V. M. Matrosov. On the theory of stability of motion. *Journal of Applied Mathematics and Mechanics*, 26(6):1506–1522, 1962.

378. V. M. Matrosov. Development of the Lyapunov functions method in stability theory. In *Tr. II Vses. S̋ezda Po Teor. i Prikl. Mekh (Proc. II All-Union Congress: Theoretical and Applied Mechanics)*, volume 1, pages 112–125. Nauka, Moscow, 1965.

379. V. M. Matrosov. Principle of comparison with the Liapunov vector function, i. *Differentsialnye Uravneniia*, 4(8):1374–1386, 1968. Russian.

380. V. M. Matrosov. Principle of comparison with the Liapunov vector function, ii. *Differentsialnye Uravneniia*, 6(10):1739–1752, 1968. Russian.

381. V. M. Matrosov. Principle of comparison with the Liapunov vector function, iii. *Differentsialnye Uravneniia*, 5(7):1171–1185, 1969. Russian.

382. V. M. Matrosov. Principle of comparison with the Liapunov vector function, iv. *Differentsialnye Uravneniia*, 5(10):2129–2143, 1969. Russian.

383. V. M. Matrosov. Comparison method in system's dynamics. In P. Janssens, J. Mawhin, and N. Ronche, editors, *Equations Differentielles et Fonctionnelles Non-Lineaires*, pages 407–445. Herman, Paris, 1973.

384. V. M. Matrosov and L. Anapolskii, editors. *Liapunov Vector Functions and Their Construction*. Izdateĺstvo Nauka, Novosibirsk, 1980. Russian.

385. V. M. Matrosov, L. Anapolskii, and S. N. Vassyliev, editors. *Comparison Method in the Theory of Mathematical Systems*. Nauka, Novosibirsk, 1980. Russian.

386. V. M. Matrosov and S. N. Vassyliev, editors. *Liapunov's Function Method and its Applications*. Nauka, Moscow, 1984. Russian.

387. V. M. Matrosov and A. A. Voronov, editors. *The Vector Lyapunov Function Method in Stability Theory*. Nauka, Moscow, 1987. Russian.

388. V. M. Matrosov and V. D. Yrtegov, editors. *Stability of Motion*. Nauka, Siberian Division, 1985. Russian.

389. J. Mawhin. *Topological Degree Methods in Nonlinear Boundary Value Problems*, volume 40. American Mathematical Society, Providence, 1979.

390. R. Medina. Asymptotic behavior of nonlinear difference systems. *Journal of Mathematical Analysis and Applications*, 219(2):294–311, 1998.

391. D. Mentrup and M. Luban. Almost periodic wave packets and wave packets of invariant shape. *American Journal of Physics*, 71:580–584, 2003.

392. Y. Meyer. Quasicrystals, almost periodic patterns, mean-periodic functions and irregular sampling. *African Diaspora Journal of Mathematics*, 13:1–45, 2012.

393. A. N. Michel and R. K. Miller. *Qualitative Analysis of Large Scale Dynamical Systems*. Academic Press, New York, 1977.
394. A. R. Mickelson and D. L. Jaggard. Electromagnetic wave propagation in almost periodic media. *IEEE Transactions on Antennas and Propagation*, 27(1):34–40, 1979.
395. R. E. Mickens. *Difference Equations: Theory and Applications*. CRC Press, Boca Raton, second edition, 1990.
396. G. Micula and P. Pavel. *Differential and Integral Equations through Practical Problems and Exercises*. Kluwer Academic, Dordrecht, 1992.
397. K. S. Miller and B. Ross. *An Introduction to the Fractional Calculus and Fractional Differential Equations*. John Wiley & Sons, Inc., New York, 1993.
398. R. K. Miller and G. R. Sell. *Volterra Integral Equations and Topological Dynamics*, volume 102. American Mathematical Society, Providence, 1970.
399. M. Milman. Stability results for integral operators, i. *Revue Roumaine de Mathematiques Pures et Appliquees*, 22:325–333, 1977.
400. Y. A. Mitropolskii. Etude des equations differentielles non-lineaires 'a Argument Dévié'. In P. Janssens, J. Mawhin, and N. Ronche, editors, *Equations Differentielles et Fonctionnelles Non-Lineaires*, pages 135–152. Herman, Paris, 1973.
401. Y. A. Mitropolskii, A. M. Samoilenko, and V. L. Kulik. *Investigation of Dichotomy of Linear Systems of Differential Equations Using Lyapunov Functions*. Naukova Dumka, Kiev, 1990.
402. S. E. Mohammad. *Stochastic Functional Differential Equations*, volume 99. Pitman, Boston, 1984.
403. G. Morosanu. *Nonlinear Evolution Equations and Applications*. D. Reidel, Dordrecht, 1988.
404. G. Morosanu. Existence results for second-order monotone differential inclusions on the positive half-line. *Journal of Mathematical Analysis and Applications*, 419:94–113, 2014.
405. D. Motreanu and N. H. Pavel. *Tangency, Flow Invariance for Differential Equations, and Optimization Problems*. Marcel Dekker, New York, 1999.
406. S. Murakami. Perturbation theorems for functional differential equations with infinite delay via limiting equations. *Journal of Differential Equations*, 59(3):314–335, 1985.
407. S. Murakami. Exponential stability for fundamental solutions of some linear functional differential equations. In T. Yoshizawa and J. Kato, editors, *Proceedings of the International Symposium: Functional Differential Equations*, pages 259–263. World Scientific, Singapore, 1990.
408. A. S. Muresan. Tonelli's lemma and applications. *Carpathian Journal of Mathematics*, 28:103–110, 2012.
409. V. Muresan. On the solutions of a second order functional-differential equation. *Carpathian Journal of Mathematics*, 28:111–116, 2012.

410. O. G. Mustafa. On the existence of solutions with prescribed asymptotic behavior for perturbed nonlinear differential equations of second order. *Glasgow Mathematical Journal*, 47:177–185, 2005.

411. A. D. Myshkis. *Linear Differential Equations with Retarded Argument*. Nauka, Moscow, 1972. Russian, German Edition: Deutscher Verlag Der Wissenschaften, Berlin, 1955.

412. A. D. Myshkis and L. E. Eĺsgoĺtz. The state and problems of the theory of differential equations with perturbed arguments. *Uspekhi Matematicheskikh Nauk*, 22(134):21–57, 1967. Russian.

413. M. Nagumo. Über die lage der integralkurven gewöhnlicher differentialgleichungen. In *Proceedings of the Physico-Mathematical Society of Japan*, volume 24, pages 551–559, 1942.

414. T. Naito and V. M. Nguyen. Evolution semigroups and spectral criteria for almost periodic solutions of periodic evolution equations. *Journal of Differential Equations*, 152:358–376, 1999.

415. T. Naito, J. S. Shin, and S. Murakami. The generator of the solution semigroup for the general linear fractional differential equations. *Bulletin of the University of Electro-Communications*, pages 29–38, 1998.

416. V. V. Nemytskii and V. F. Stepanov. *Qualitative Theory of Differential Equations*. Princeton University Press, Princeton, 1960.

417. G. M. N'Guerekata. *Almost Automorphic and Almost Periodic Functions in Abstract Spaces*. Kluwer Academic, New York, 2001.

418. G. M. N'Guerekata and A. Pankov. Integral operators in spaces of bounded almost periodic and almost automorphic functions. *Differential and Integral Equations*, 21:1155–1176, 2008.

419. T. H. Nguyen and V. D. Trinh. Integral manifolds for partial functional differential equations in admissible spaces on a half-line. *Journal of Mathematical Analysis and Applications*, 411(2):816–828, 2014.

420. Z. C. Okonkwo. Admissibility and optimal control for difference equations. *Dynamic Systems and Applications*, 5:623–634, 1996.

421. Z. C. Okonkwo, E. Rowe, and A. Adewale. Stability of neutral stochastic functional differential equations with causal operators. *International Journal of Differential Equations and Applications*, 4(1):367–380, 2000.

422. Z. C. Okonkwo and J. C. Turner Jr. On neutral stochastic functional differential equations with causal operators. *International Journal of Applied Mathematics*, 2:131–140, 2000.

423. R. E. O'Malley, Jr. *Singular Perturbation Methods for Ordinary Differential Equations*. Springer-Verlag, New York, 1991.

424. Z. Opluštil. Solvability of a nonlocal boundary value problem for linear functional differential equations. *Advances in Difference Equations*, 2013:244, 2013.

425. D. O'Regan. Abstract operator inclusions. *Functional Differential Equations*, 4:143–154, 1997.

426. D. O'Regan. Abstract Volterra equations. *Pan American Mathematical Journal*, 7:19–28, 1997.

427. D. O'Regan. *Existence Theory for Nonlinear Ordinary Differential Equations*. Kluwer Academic, Dordrecht, 1997.

428. D. O'Regan. Operator equations in Banach spaces relative to the weak topology. *Archiv der Mathematik*, 71:123–136, 1998.

429. D. O'Regan. A note on the topological structure of the solution set of abstract Volterra equations. *Mathematical Proceedings of the Royal Irish Academy*, 99A(1):67–74, 1999.

430. D. O'Regan and R. Precup. *Theorems of Leray-Schauder Type and Applications*, volume 3. Gordon and Breach, Amsterdam, 2001. Series in Mathematical Analysis and Applications.

431. J. A. Ortega. *Teoria General de las Ecuaciones Differenciales Funcionales*. Instituto Politecnico Nacional (Serie en Ciencias e Inggenieria), Mexico City, 1978.

432. V. F. Osipov. *Bohr-Fresnel Almost Periodic Functions*. Peterburg Gos. University, St. Petersburg, 1992.

433. D. Otrocol. Abstract Volterra operators. *Carpathian Journal of Mathematics*, 24(3):370–377, 2008.

434. D. Otrocol. Ulam stabilities of differential equations with abstract Volterra operator in a Banach space. *Nonlinear Functional Analysis and Applications*, 15:613–619, 2010.

435. B. G. Pachpatte. *Inequalities for Differential and Integral Equations*. Academic Press, San Diego, 1998.

436. B. G. Pachpatte. On Volterra integro-differential equations of the Corduneanu-Popov type. manuscript, 7 pages, 2009.

437. L. Pandolfi. Canonical realizations of systems with delays. *SIAM Journal on Control and Optimization*, 21:598–613, 1983.

438. A. A. Pankov. *Bounded and Almost Periodic Solutions of Nonlinear Operator Differential Equations*. Kluwer, Dordrecht, 1990.

439. M. E. Parrott. Representation and approximation of generalized solutions of a nonlinear functional differential equation. *Nonlinear Analysis: Theory, Methods & Applications*, 6(4):307–318, 1982.

440. N. H. Pavel. Approximate solutions of Cauchy problems for some differential equations in Banach spaces. *Funkcialaj Ekvacioj*, 17:85–94, 1974.

441. N. H. Pavel. *Nonlinear Evolution Operators and Semigroups: Applications to Partial Differential Equations*. Springer-Verlag, Berlin, 1987. Lecture Notes in Mathematics 1260.

442. A. Pazy. *Semigroups of Linear Operators and Applications to Partial Differential Equations*. Springer-Verlag, New York, 1983.

443. K. Peiffer and N. Rouche. Liapunov's second method applied to partial stability. *Journal de Mécanique*, 8:323–334, 1969.

444. D. Pennequin. Existence of almost periodic solutions of discrete time equations. *Discrete and Continuous Dynamical Systems*, 7:51–60, 2001.

445. F. Perri and L. Pandolfi. Input identification to a class of nonlinear input-output causal systems. *Computers & Mathematics with Applications*, 51(12):1773–1788, 2006.

446. W. Perruquetti, J. P. Richard, and P. Borne. Vector Lyapunov functions: Recent developments for stability, robustness, partial stability and constrained control. *Nonlinear Times and Digest*, 2:227–258, 1995.

447. K. P. Persidskii. *Selected Works*, volume 1. Izbranye Trudy, Alma-Ata, 1976. Russian.

448. D. Petrovanu. Equations Hammerstein integrales et discretes. *Annali di Matematica Pura ed Applicata*, 70:227–254, 1966.

449. I. G. Petrovskii. *Lectures on the Theory of Ordinary Differential Equations*. Foreign Technology DIV Wright-Patterson AFB, Ohio, 1973.

450. A. Petrusel, I. A. Rus, and M. A. Serban. Fixed points for operators on generalized metric spaces. *CUBO A Mathematical Journal*, 10(4):45–66, 2008.

451. E. Pinney. *Ordinary Difference-Differential Equations*. University of California Press, Berkeley and Los Angeles, 1958.

452. I. Plaskina. About solvability of quasilinear singular functional differential equations. *Functional Differential Equations*, 19:335–339, 2012.

453. I. Podlubny. *Fractional Differential Equations*. Academic Press, San Diego, 1999.

454. H. Poincaré. *Les méthodes nouvelles de la mécanique céleste: Méthodes de MM. Newcomb, Glydén, Lindstedt et Bohlin*, volume 2. Gauthier-Villars et fils, Paris, 1893.

455. A. D. Polyanin and A. V. Manzhirov. *Handbook of Integral Equations*. CRC Press, New York, 1998.

456. V. M. Popov. Absolute stability of nonlinear systems of automatic control. *Automation and Remote Control*, 22(8):857–875, 1962.

457. V. M. Popov. Monotone-gradient systems. *Journal of Differential Equations*, 41:245–261, 1981.

458. C. Popovici. Sur une équation fonctionelle. *Comptes Rendus de l'Academic des Sciences Paris*, 158:1867–1868, 1914.

459. C. Popovici. Nouvelles solutions des équations integro-fonctionelle. *Rendiconti Accademic Nazionale dei Lincei*, 11:1080–1085, 1930.

460. C. Popovici. Intégration des systemés d'équations fonctionelles. *Comptes Rendus de l'Academic des Sciences Paris*, 197:12–14, 1933.

461. H. Poppe and A. Stirk. A compactness criterion for the space of almost periodic functions, ii. *Mathematische Nachrichten*, 108:153–157, 1982.

462. R. Precup. *Methods in Nonlinear Integral Equations*. Kluwer Academic Publishers, Dordrecht, 2002.

463. A. J. Prichard and Yuncheng You. Causal feedback optimal control for Volterra integral equations. *SIAM Journal of Control and Optimization*, 34:1874–1890, 1996.

464. J. Prüss. *Evolutionary Integral Equations and Applications*. Birkhäuser, Basel, 1993.

465. J. Prüss and M. Wilke. *Gewöhnliche Differentialgleichungen und Dynamische Systeme*. Springer, Berlin, 2010.

466. M. L. Puri and D. A. Ralescu. Differentials of fuzzy functions. *Journal of Mathematical Analysis and Applications*, 91(2):552–558, 1983.

467. J. Radon. On linear functional transformations and functional equations. *Uspekhi Matematicheskikh Nauk*, (1):200–227, 1936.

468. T. M. Rassias. On the stability of the linear mapping in Banach spaces. *Proceedings of the American Mathematical Society*, 7(2):297–300, 1978.

469. B. S. Razumikhin. *Stability of Hereditary Systems*. Nauka, Moscow, 1988. Russian.

470. F. Riesz. On linear functional equations. *Uspekhi Matematicheskikh Nauk*, (1):175–199, 1936.

471. A. Rontó and M. Rontó. Successive approximation techniques in non-linear boundary value problems for ordinary differential equations. In F. Battelli and M. Fečkan, editors, *Handbook of Differential Equations: Ordinary Differential Equations*, volume 4, pages 441–592. Elsevier, New York, 2008.

472. M. Roseau. *Vibrations Nonlinéaires et Théorie de la Stabilité*. Springer-Verlag, New York, 1966.

473. N. Rouche, P. Habets, and M. Laloy. *Stability Theory by Liapunov's Direct Method*. Springer-Verlag, New York, 1977.

474. N. Rouche, P. Habets, and K. Pfeiffer. *Stability Theory by Liapunov's Second Method*. Springer-Verlag, New York, 1977.

475. N. Rouche and J. Mawhin. *Ordinary Differential Equations: Stability and Periodic Solutions*. Pitman Advanced Publishing Program, Boston, 1980.

476. W. M. Ruess and W. H. Summers. Weak almost periodicity and the strong ergodic limit theorem for contraction semigroups. *Israel Jounal of Mathematics*, 64:139–157, 1988.

477. V. V. Rumiantsev. On the stability of motion with respect to part of the variables. *Vestnik Moscow University, Mathematics-Mechanics Series, number 4*, pages 9–16, 1957. Russian.

478. V. V. Rumiantsev and A. S. Oziraner. *Stability and Stabilization of Motion with Respect to Part of the Variables*. Nauka, Moscow, 1987. Russian.

479. I. A. Rus. *Principles and Applications of the Fixed Point Theory*. Ed. Dacia, Cluj-Napoca, 1979. Romanian.

480. I. A. Rus. *Differential Equations, Integral Equations and Dynamical Systems*. Ed. Dacia, Cluj-Napoca, 1996. Romanian.

481. I. A. Rus. Ulam stability of ordinary differential equations. *Studia Universitatis "Babes Bolyai" Mathematica*, 54:125–133, 2009.

482. Th. L. Saaty. *Modern Nonlinear Equations*. Dover Publications, New York, 1981.

483. S. Saitoh. *Integral Transforms, Reproducing Kernels and Their Applications*. CRC Press, New York, 1997.

484. L. Salvadori. Famiglie ad un parametro di funzioni di Liapunov nello studio della stabilita. In *Symposia Mathematica, VI*, pages 309–330. 1971. Academic Press, London, 1971.

485. I. W. Sandberg. On causality and linear maps. *IEEE Transactions on Circuits and Systems*, 32:392–393, 1985.

486. I. W. Sandberg and G. J. J. Van Zyl. Evaluation of the response of nonlinear systems to asymptotically almost periodic inputs. In *IEEE International Symposium on Circuits and Systems*, volume 3, pages 77–80, 2001.

487. I. W. Sandberg and G. J. J. Van Zyl. Harmonic balance and almost periodic inputs. *IEEE Transactions on Circuits and Systems*, 49(4):459–464, 2002.

488. G. Sansone. *Orthogonal Functions*, volume 9. Dover Publications, New York, 1991.

489. G. Sansone and R. Conti. *Nonlinear Differential Equations*. Macmillan, New York, 1964.

490. B. Satco. A Cauchy problem on time scale with applications. *Annals of the Alexandru Ioan Cuza University-Mathematics*, LVII:221–234, 2011.

491. N. O. Sedova. Stability in systems with unbounded aftereffect. *Automation and Remote Control*, 70:1553–1564, 2009.

492. S. Seikkala. On the fuzzy initial value problem. *Fuzzy Sets and Systems*, 24(3):319–330, 1987.

493. L. Shaikhet. About Lyapunov functionals construction for difference equations with continuous time. *Applied Mathematics Letters*, 17(8):985–991, 2004.

494. A. A. Shestakov. *Generalized Liapunov's Direct Method for Systems with Distributed Parameters*. Nauka, Moscow, 1990. Russian.

495. A. I. Shtern. Almost periodic functions and representations in locally convex spaces. *Russian Mathematical Surveys*, 60(3):489–557, 2007.

496. M. A. Shubin. Differential and pseudodifferential operators in spaces of almost periodic functions. *Matematicheskii Sbornik*, 137(4):560–587, 1974.

497. M. A. Shubin. Almost periodic functions and partial differential operators. *Uspekni Matematicheskikh Nauk*, 33(2):3–47, 1978.

498. K. S. Sibirskii and A. S. Shuba. *Semi-Dynamical Systems*. Stiinta Press, Chisinau, 1987. Russian.

499. K. S. Sibirsky. *Introduction to Topological Dynamics*. Noordhoff, the Netherlands, 1975.

500. A. Sikorska-Nowak. The set of solutions of Volterra and Urysohn integral equations in Banach spaces. *Rocky Mountain Journal of Mathematics*, 40(4):1313–1331, 2010.

501. L. Simon. Asymptotics for a class of nonlinear evolution equations, with applications to geometric problems. *Annals of Mathematics*, 118(3):525–571, 1983.

502. G. M. Sklyar. On the maximal asymptotics for linear differential equations in Banach spaces. *Taiwanese Journal of Mathematics*, 14:2203–2217, 2010.

503. M. Sofonea, C. Avramescu, and A. Matei. A fixed point result with applications in the study of viscoplastic frictionless contact problems. *Communications on Pure and Applied Analysis*, 7(3):645–658, 2008.

504. A. A. Soliman. On perturbing Liapunov functional. *Applied Mathematics and Computation*, 133(2):319–325, 2002.

505. O. J. Staffans. A direct Lyapunov approach to Volterra integro-differential equations. *SIAM Journal on Mathematical Analysis*, 19:879–901, 1988.

506. G. Stepan. *Retarded Dynamical Systems: Stability and Characteristic Functions*. Longman Scientific & Technical, New York, 1989.

507. V. H. Stettner. Ljapunovsche funktionen und eindeutige auflosbarkeit von funktional-differentialgleichungen. *Bericht, number 64*, pages 86–98, 1976. Forschungzentrun, Graz.

508. M. H. Stone. *Linear Transformations in Hilbert Space and Their Applications to Analysis*, volume 15. American Mathematical Society, Providence, 1932.

509. R. Suarez. Functional differential equations of advanced type. *Boletin Sociedad Matematica Mexicana*, 22:25–34, 1977.

510. N. Sultana and A.F.M.K. Khan. Power series solutions of fuzzy differential equations using Picard method. *Journal of Computer Science and Mathematics*, 8(4):72–77, 2010.

511. E. A. Sumenkov. Conversion of comparison theorems for some dynamical properties. In V. M. Matrosov and A. J. Malikov, editors, *Liapunov Functions and Applications*, pages 93–106. Nauka, Novosibirsk, 1986. Russian.

512. V. I. Sumin. On functional Volterra equations. *Russian Mathematics: Izv VUZ*, 39(9):67–77, 1995.

513. C. Swartz. *Elementary Functional Analysis*. World Scientific, Singapore, 2009.

514. G. P. Szegö and G. Treccani. *Semigruppi di Trasformazioni Multivoche*. Springer-Verlag, Heidelberg, 1969. Lecture Notes in Mathematics 101.

515. J. Tabor. Differential equations in metric spaces. *Mathematica Bohemica*, 127(2):353–360, 2002.

516. H. L. Tidke. Approximate solutions to fractional semilinear evolution equation with nonlocal condition. *International Journal of Mathematical Archive*, 2(11):2403–2412, 2011.

517. H. L. Tidke. Some theorems on fractional semilinear evolution equations. *Journal of Applied Analysis*, 18(2):209–224, 2012.

518. L. Tonelli. Sulle equazioni funzionali del tipo di Volterra. *Bulletin Calcutta Mathematical Society*, 20:31–48, 1928.

519. V. Trénoguine. *Analyse Fonctionnelle*. Ed Mir, Moscow, 1985.

520. F. Tricomi. *Integral Equations*. Interscience, New York, 1957.

521. Z. B. Tsaliuk. Volterra integral equations. *ITOGI NAUKI: Matem. Analiz, (Scientific Inst. Information of the Academy of Science USSR)*, 15:131–198, 1977. Russian.

522. M. Turinici. Iterated operator inequalities on ordered linear spaces. *African Diaspora Journal of Mathematics*, 12:80–88, 2011.

523. S. M. Ulam. *A Collection of Mathematical Problems*. Interscience Publishers, New York, 1960.

524. C. Ursescu. Carathéodory solutions of ordinary differential equations on locally closed sets in finite dimensional spaces. *Mathematica Japonica*, 31(3):483–491, 1986.

525. K. G. Valeev and G. S. Finin. *Construction of Lyapunov Functions*. Naukova Dumka, Kiev, 1981.

526. J. Vasundhara Devi. Comparison theorems and existence results for set causal operators with memory. *Nonlinear Studies*, 18(4):603–610, 2011.

527. M. Vath. *Volterra and Integral Equations of Vector Functions*. Marcel Dekker, New York, 2000.

528. V. Volterra. *Leçons sur les Équations Intégrales et les Équations Intégro-Différentielles*. Gauthier-Villars, Paris, 1913.

529. A. A. Voronov. Present state and problems of stability theory. *Automation and Remote Control*, 43:573–592, 1982.

530. V. I. Vorotnikov. *Stability of Dynamical Systems with Part of the Variables*. Nauka, Moscow, 1991. Russian.

531. V. I. Vorotnikov. *Partial Stability and Control*. Birkhäuser, Boston, 1998.

532. I. Vrkoč. Integral stability. *Czechoslovak Mathematical Journal*, 9:71–129, 1959. Russian.

533. T. Ważewski. Systémes des équations et des inégalités différentielles ordinaires aux deuxiémes membres monotones et leurs applications. *Annales de la Societe Polonaise Mathematique*, 23:112–166, 1950.

534. J. Wiener. *Generalized Solutions of Functional Differential Equations*. World Scientific, Singapore, 1993.

535. N. Wiener. Generalized harmonic analysis. *Acta Mathematica*, 55:117–256, 1930.

536. Wikipedia. Lyapunov stability. *Online*, pages 1–7, 2014.

537. Yingfei Yi. On almost automorphic oscillations. *Fields Institute Communications*, 42:75–99, 2004.

538. J. A. Yorke. Invariance for ordinary differential equations. *Theory of Computing Systems*, 4(1):353–372, 1967.

539. T. Yoshizawa. *Stability Theory by Liapunov's Second Method*. The Mathematical Society of Japan, Tokyo, 1966.

540. T. Yoshizawa. *Stability Theory and The Existence of Periodic Solutions and Almost Periodic Solutions*. Springer-Verlag, New York, 1975.

541. K. Yosida. *Functional Analysis*. Springer-Verlag, Berlin, fifth edition, 1978.

542. H. You and R. Yuan. Random attractor for stochastic partial functional differential equations with finite delay. *Taiwanese Journal of Mathematics*, 18:77–92, 2014.

543. R. M. Young. *An Introduction to Non-harmonic Fourier Series*. Academic Press, San Diego, revised first edition edition, 2001.

544. Rong Yuan. Existence of almost periodic solutions to functional differential equations of neutral type. *Mathematical Analysis and Applications*, 165:524–538, 1992.

545. Rong Yuan. Existence of almost periodic solutions of neutral functional differential equations via Liapunov-Razumikhin function. *Zeitschrift fur Angewandte Mathematic und Physik*, 49:113–136, 1998.

546. S. Zaidman. *Abstract Differential Equations*. Pitman, London, 1979.

547. S. Zaidman. *Almost-Periodic Functions in Abstract Spaces*. Pitman, London, 1985.

548. S. Zaidman. Topics in abstract differential equations. Rapport de recherche, University of Montréal, 1991.

549. M. Zając. Quasicrystals and almost periodic functions. *Annales Polonici Mathematici*, 72:251–259, 1999.

550. E. Zeidler. *Nonlinear Functional Analysis and its Applications (Fixed-Point Theorems)*, volume 1. Springer, Berlin, 1986.

551. E. Zeidler. *Nonlinear Functional Analysis and its Applications (Linear Monotone Operators)*, volume 2. Springer, Berlin, 1990.

552. Chuanyi Zhang. *Pseudo Almost Periodic Functions and Their Applications*. PhD thesis, University of Western Ontario, 1992.

553. Chuanyi Zhang. *Almost Periodic Type Functions and Ergodicity*. Kluwer Academic, Dordrecht, 2003.

554. Chuanyi Zhang. Strong limit power functions. *Journal of Fourier Analysis and Applications*, 12:291–307, 2006.

555. Chuanyi Zhang. A Bohr-like compactification and summability of Fourier series. *Forum of Mathematics*, 21:355–373, 2009.

556. Chuanyi Zhang. Generalized Kronecker's theorem and strong limit power functions. In V. Barbu and O. Cârja, editors, *American Institute of Physics Proceedings: Alexandru Myller Mathematical Centennial Conference*, volume 1329, pages 281–299, Iasi, Romania 2011.

557. Chuanyi Zhang and Weiguo Liu. Uniform limit power-type function spaces. *International Journal of Mathematics and Mathematical Sciences*, ID 17042;Doi 10.1155:1–14, 2006.

558. Chuanyi Zhang and Chenhui Meng. Two new spaces of vector valued limit power functions. *Studia Scientiarum Mathematicarum Hungarica*, 44:423–443, 2007.

559. E. S. Zhukovskii. Continuous dependence on parameters of solutions to Volterra's equations. *Sbornik Mathematics*, 197(10):33–56, 2006.

560. E. S. Zhukovskii and M. J. Alves. Abstract Volterra operators. *Russian Mathematics:Izv VUZ*, 52(3):1–14, 2008.

561. V. J. Zubov. *Methods of AM Liapunov and Their Applications*. P. Noordhoff Ltd, Groningen, 1964.

562. A. Zygmund. *Trigonometric Series*. Cambridge University Press, Cambridge, second edition, 1959.

INDEX

Functional Differential Equations: Advances and Applications, First Edition.
Constantin Corduneanu, Yizeng Li and Mehran Mahdavi

PURE AND APPLIED MATHEMATICS

A Wiley Series of Texts, Monographs, and Tracts

ADÁMEK, HERRLICH, and STRECKER—Abstract and Concreto Catetories
ADAMOWICZ and ZBIERSKI—Logic of Mathematics
AINSWORTH and ODEN—A Posteriori Error Estimation in Finite Element Analysis
AKIVIS and GOLDBERG—Conformal Differential Geometry and Its Generalizations
ALLEN and ISAACSON—Numerical Analysis for Applied Science
*ARTIN—Geometric Algebra
ATKINSON, HAN, and STEWART—Numerical Solution of Ordinary Differential Equations
AUBIN-Applied Functional Analysis, Second Edition
AZIZOV and IOKHVIDOV—Linear Operators in Spaces with an Indefinite Metric
BASENER—Topology and Its Applications
BERG—The Fourier-Analytic Proof of Quadratic Reciprocity
BERKOVITZ—Convexity and Optimization in $\mathbb{R}^n$
BERMAN, NEUMANN, and STERN—Nonnegative Matrices in Dynamic Systems
BOYARINTSEV—Methods of Solving Singular Systems of Ordinary Differential Equations
BRIDGER—Real Analysis: A Constructive Approach
BURK—Lebesgue Measure and Integration: An Introduction
*CARTER—Finite Groups of Lie Type
CASTILLO, COBO, JUBETE, and PRUNEDA—Orthogonal Sets and Polar Methods in Linear Algebra: Applications to Matrix Calculations, Systems of Equations, Inequalities, and Linear Programming
CASTILLO, CONEJO, PEDREGAL, GARCIÁ, and ALGUACIL—Building and Solving Mathematical Programming Models in Engineering and Science
CHATELIN—Eigenvalues of Matrices
CLARK—Mathematical Bioeconomics: The Mathematics of Conservation, Third Edition
CORDUNEANU, LI, and MAHDAVI—Functional Differential Equations: Advances and Applications
COX—Galois Theory, Second Edition
COX—Primes of the Form $x^2 + ny^2$: Fermat, Class Field Theory, and Complex Multiplication, Second Edition
*CURTIS and REINER—Representation Theory of Finite Groups and Associative Algebras
*CURTIS and REINER—Methods of Representation Theory: With Applications to Finite Groups and Orders, Volume I
CURTIS and REINER—Methods of Representation Theory: With Applications to Finite Groups and Orders, Volume II
DINCULEANU—Vector Integration and Stochastic Integration in Banach Spaces
*DUNFORD and SCHWARTZ—Linear Operators
  Part 1—General Theory
  Part 2—Spectral Theory, Self Adjoint Operators in Hilbert Space
  Part 3—Spectral Operators
FARINA and RINALDI—Positive Linear Systems: Theory and Applications
FATICONI—The Mathematics of Infinity: A Guide to Great Ideas, Second Edition
FOLLAND—Real Analysis: Modern Techniques and Their Applications
FRÖLICHER and KRIEGL—Linear Spaces and Differentiation Theory
GARDINER—Teichmüller Theory and Quadratic Differentials
GILBERT and NICHOLSON—Modern Algebra with Applications, Second Edition

*GRIFFITHS and HARRIS—Principles of Algebraic Geometry
GRILLET—Algebra
GROVE—Groups and Characters
GUSTAFSSON, KREISS and OLIGER—Time Dependent Problems and Difference Methods, Second Edition
HANNA and ROWLAND—Fourier Series, Transforms, and Boundary Value Problems, Second Edition
*HENRICI—Applied and Computational Complex Analysis
Volume 1 Power Series–Integration–Conformal Mapping–Location of Zeros
Volume 2 Special Functions–Integral Transforms–Asymptotics–Continued Fractions
Volume 3 Discrete Fourier Analysis, Cauchy Integrals, Construction of Conformal Maps, Univalent Functions
*HILTON and WU—A Course in Modem Algebra
*HOCHSTADT—Integral Equations
JOST—Two-Dimensional Geometric Variational Procedures
KHAMSI and KIRK—An Introduction to Metric Spaces and Fixed Point Theory
*KOBAYASHI and NOMIZU—Foundations of Differential Geometry, Volume I
*KOBAYASHI and NOMIZU—Foundations of Differential Geometry, Volume 11
KOSHY—Fibonacci and Lucas Numbers with Applications
LAX—Functional Analysis
LAX—Linear Algebra and Its Applications, Second Edition
LOGAN—An Introduction to Nonlinear Partial Differential Equations, Second Edition
LOGAN and WOLESENSKY—Mathematical Methods in Biology
MARKLEY–Principles of Differential Equations
MELNIK–Mathematical and Computational Modeling: With Applications in Natural and Social Sciences, Engineering, and The Arts
MORRISON—Functional Analysis: An Introduction to Banach Space Theory
NAYFEH and MOOK—Nonlinear Oscillations
O'LEARY—Revolutions of Geometry
O'NEIL—Beginning Partial Differential Equations, Third Edition
O'NEIL—Solutions Manual to Accompany Beginning Partial Differential Equations, Third Edition
PANDEY—The Hilbert Transform of Schwartz Distributions and Applications
PETKOV—Geometry of Reflecting Rays and Inverse Spectral Problems
*PRENTER—Splines and Variational Methods
PROMISLOW—A First Course in Functional Analysis
RAO—Measure Theory and Integration
RASSIAS and SIMSA—Finite Sums Decompositions in Mathematical Analysis
RENELT—Elliptic Systems and Quasiconformal Mappings
RIVLIN—Chebyshev Polynomials: From Approximation Theory to Algebra and Number Theory, Second Edition
ROCKAFELLAR—Network Flows and Monotropic Optimization
ROITMAN—Introduction to Modem Set Theory
ROSSI—Theorems, Corollaries, Lemmas, and Methods of Proof
*RUDIN—Fourier Analysis on Groups
SENDOV—The Averaged Moduli of Smoothness: Applications in Numerical Methods and Approximations
SENDOV and POPOV—The Averaged Moduli of Smoothness
SEWELL—The Numerical Solution of Ordinary and Partial Differential Equations, Second Edition
SEWELG—Computational Methods of Linear Algebra, Second Edition
SHICK—Topology: Point-Set and Geometric
SHISKOWSKI and FRINKLE—Principles of Linear Algebra with *Maple*™
SHISKOWSKI and FRINKLE—Principles of Linear Algebra with *Mathematica*®

*SIEGEG—Topics in Complex Function Theory
Volume 1—Elliptic Functions and Uniformization Theory
Volume 2—Automorphic Functions and Abelian Integrals
Volume 3—Abelian Functions and Modular Functions of Several Variables
SMITH and ROMANOWSKA—Post-Modern Algebra
ŠOLÍN—Partial Differential Equations and the Finite Element Method
STADE—Fourier Analysis
STAHL and STENSON—Introduction to Topology and Geometry, Second Edition
STAHL—Real Analysis, Second Edition
STAKGOLD—Green's Functions and Boundary Value Problems, Third Edition
STANOYEVITCH—Introduction to Numerical Ordinary and Partial Differential Equations Using MATLAB®
*STOKER—Differential Geometry
*STOKER—Nonlinear Vibrations in Mechanical and Electrical Systems
*STOKER—Water Waves: The Mathematical Theory with Applications
WATKINS—Fundamentals of Matrix Computations, Third Edition
WESSELING—An Introduction to Multigrid Methods
†WHITHAM—Linear and Nonlinear Waves
ZAUDERER—Partial Differential Equations of Applied Mathematics, Third Edition

*Now available in a lower priced paperback edition in the Wiley Classics Library.
†Now available in paperback.